The Contractor's Legal Kit

The Complete, User-Friendly Legal Guide for Home Builders and Remodelers

Gary Ransone, J.D.

Member of California State Bar
Licensed General Building Contractor in the State of California
American Arbitration Association Construction Panel Arbitrator

 A JOURNAL OF LIGHT CONSTRUCTION BOOK

First printing: January 1996

International Standard Book Number: 0-9632268-3-5
Library of Congress Catalog Card Number: 95-83198
Printed in the United States of America

Book design: Deborah Fillion

A *Journal of Light Construction* Book
The Journal of Light Construction is a tradename of
Builderburg Group, Inc.

Builderburg Group, Inc.
P.O. Box 435
Richmond, VT 05477

ACKNOWLEDGMENTS

Special thanks is given to Steven Bliss, Josie Masterson-Glen,
and Bill Brockway at *The Journal of Light Construction* for their
patient and skillful editing. I also appreciate the efforts made by
everyone at *The Journal* to bring practical trade and business
information to the average builder and remodeler.

Thanks is also given to my wife, Janiece, and two children,
Alex and Katy, for their patience and understanding while I worked
on this book. I'm also grateful to my brother, Doug Ransone,
for the learning experiences we've shared over years of
working together in this business.

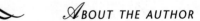

Gary Ransone, J.D., is a licensed general contractor, construction attorney, and construction arbitrator in Soquel, California. His legal practice is devoted almost exclusively to the field of construction law. In 1994 he started the *Construction Law Help Line* to provide contract preparation services and easy access to business and legal information to contractors throughout California. For legal advice he can be reached by phone at 800/98-BUILD (within California only) or 408/476-8784, or by mail at 2805 Porter Street, Soquel, CA 95073. The author welcomes written comments or suggestions regarding this book.

$\mathscr{P}$REFACE

It has been said that perspective is everything. For better or worse, the substance of this book has evolved from my experiences and perspective as both a building contractor and a construction attorney.

My perspective on the construction business has been shaped not just by textbook learning, but also by representing numerous contractors and owners in real-life legal matters. It has been equally shaped by my having worked in the field and later in the office as a construction laborer, carpenter, project manager, and contractor. Over many hundreds of projects, I've had plenty of opportunities to experience firsthand most of the problems and challenges addressed in this book. In short, the construction School of Hard Knocks has tempered my legal and business training with a strong dose of job-site reality.

Despite the legal, and sometimes technical, nature of the subject matter in this book, it is not a legal textbook filled with detailed case law, footnotes, and statistics. Instead, this book is meant to be a practical guide to the most common and important agreements and business practices that builders can use to "level the playing field" before a single nail is driven. It is intended to be used as a hands-on tool under the guidance and final review of your local attorney.

Every effort has been made to arrive at fair and reasonable contract language that protects the interests of the builder, while at the same time being fair to the owner. I've seen plenty of agreements that are grossly unfair to either the contractor or the owner, and I don't think this type of agreement serves anyone's long-term interests.

Don't underestimate the value of being able to furnish your construction agreement to the owner. Both the owner and the builder take considerable risks when they enter into a construction agreement. Being able to furnish the agreement allows the builder to address, assign, and buffer many of his risks so that "Murphy's Law" doesn't end up killing his profits, his relationship with the owner, and his future referrals.

In summary, it's my hope that you can glean some useful lessons from this book, rather than learn everything the hard way as most builders invariably do. If you apply this information to your contracts and basic business practices, you should be able to reduce your liability, maintain clearer communication with the owner, temper unrealistic expectations, save valuable time, and maintain, or even increase, job profits. If so, the effort that went into creating this book will have been well spent.

— *Gary Ransone*

CONTENTS

INTRODUCTION

*A*lthough billions of dollars are spent every year in the United States on residential building and remodeling, many residential contractors face a common problem: finding contracts and other legal documents well suited to their businesses.

Most residential builders who use contracts at all use inadequate "boilerplate" agreements. This is understandable, since it's hard for smaller builders to find the time and money to have an attorney draft good, comprehensive construction agreements, and it's equally hard to find a construction attorney familiar with residential construction.

But don't wait until you are faced with a major lawsuit or collection problem to pay attention to your legal documents. With litigation and arbitration costs on the rise, a medium-sized legal dispute on a $100,000 remodel can easily cost you $10,000 to $20,000 in legal fees, expert witness fees, and related costs. You might also end up having to pay a judgment to the owner as well as part of the owner's legal fees. After paying all those costs, most smaller contractors can kiss their businesses goodbye and start checking out the Help Wanted ads.

Spending a little time educating yourself about how to better control the construction process can't help but pay big dividends regardless of what size business you currently operate. This manual is an attempt to get you off to a good start in that direction. It seeks to bridge the gap between the fully developed sets of construction agreements that are readily available for use on larger commercial projects (such as AIA or AGC documents) and the basic boilerplate agreements typically used by builders and remodelers.

Most importantly, this manual offers specific contract language that takes into account many of the unique risks and challenges that residential builders face. Much of this you will not find in any other "standard" agreements. It is distilled from my experiences — both good and bad — throughout many years as both a residential building contractor and practicing construction attorney. By understanding these agreements and adapting them for use by your own business, or incorporating relevant sections into your existing agreements, you'll be taking an important step toward taking charge of your business relationships.

An Expectations Game

The complex relationship between contractor and client is fertile ground for many kinds of misunderstandings and disputes. Typical problems include disputed change orders, what is and is not part of the scope of work, what are reasonable and unreasonable completion dates, when payments are due, amounts of payments, disputed punch list items, and whether final payment (minus an offset for punch list work) is due upon "substantial completion" or "final completion."

These kinds of problems can cost the builder money and time, and result in loss of reputation and future business. Perhaps more importantly, these and other types of largely avoidable problems weaken whatever trust there is between the owner and builder and can lead to future disputes, lost sleep, and lawsuits.

One major reason why job tensions and disagreements develop during a project is because the owner and the builder often start out with very different *expectations* about how specific aspects of the project will be handled. Often, this stems from a vague written communication system (i.e., a poor contract) furnished by the builder. These differing expectations show up in the form of numerous minor problems and disagreements, and often escalate from there.

A related problem is that most residential owners

understandably have little or no idea of what to expect throughout the many phases of the construction process, both the building phases and the paperwork phases that control the disbursement of the owner's money.

Furthermore, residential owners are ordinarily investing more money in the building of their home than they will ever spend on any other single item in their lifetime. For this reason alone, they have a right to be nervous.

Unfortunately, many builders fuel the fires of potential disaster by not communicating very well with the owner, before, during, or after the project. But builders can't afford to take this approach without suffering the inevitable consequences. The good news is that there is an alternative to this gloomy scenario.

The Solution: The Construction Agreement

The solution to these problems is to establish a system for clarifying and documenting everyone's responsibilities and duties, and thereby avoid many disputes and the loss of time, money, and peace of mind. The contract between the owner and builder, referred to as the *construction agreement*, is the single most valuable tool the builder possesses for this purpose. That is why the focus of this book is on agreements.

Because about 90% of the legal disputes I see could have been avoided if the builder had furnished the owner with a good construction agreement (and then followed the agreement), it's not hard to understand why I think it is critically important for contractors to improve their understanding of basic construction agreements.

If you are like most builders I know, it's tough to find the time and money, and a construction attorney who knows enough about residential construction to draft good, comprehensive agreements for your business. However, now is when you need to find the time, the money, and the attorney — *before* you get into a dispute. Once you have a serious dispute with the owner, it's too late to *re*write the contract.

After you begin work on a job, it's too late to go back and add exclusions that you *should* have written into your agreement at the start. Often, your only

choice is to pay for the items *you* knew were excluded, but the owner assumed were included. Once you forget to include an attorney's-fees clause in your agreement and the owner is refusing to pay you, it's too late to go back and rewrite your agreement. (And your attorney's fees may just equal the amount of money you are owed by the owner!) The only time to include these things in your agreements is each and every time you give a construction agreement to an owner.

Establish the Rules

When you were a young kid and allowed to make up your own rules to a game like Monopoly or Kick the Can, who usually ended up enjoying the game the most and more often than not succeeding? You!

Being in the position to furnish the construction agreement is similar in many ways: it gives you bargaining power. When you furnish the agreement, you are establishing many of the rules that will legally govern the business relationship with the owner. You can draw upon your years of experience in this business and communicate to the owner how you want the business aspects of the project to be handled.

It's my experience over many years in the construction business that if the contractor takes this opportunity to establish reasonable rules, the owner will accept them. It actually puts the owner at ease to know that there is a structure for how the job will proceed. You are giving the owner an education about the construction process, which he needs. However, if you don't establish the rules, the owner will take charge — often to your detriment. I routinely find contractors giving away 5% to 10% of the job profit by not having a comprehensive agreement.

If you fail to put your expectations in writing prior to starting the job, don't assume that the owner will be understanding later on and agree with whatever you want (no matter how reasonable your request may be). Many a competitive low bidder has discovered that once his agreement is signed, the owner is suddenly no longer willing to throw in concessions he would have happily agreed to *prior* to signing the agreement. This is human nature. Once the agreement is signed, both parties lose considerable bargaining power and often find themselves being much more close-minded.

This book is not an attempt to turn contractors into lawyers (there were enough of those around at last count). Nor is this book an attempt to teach contractors how to handle their own construction lawsuits and arbitrations (you'll need a lawyer for that). To the contrary, this book is an attempt to help contractors do a better job of putting together agreements so that they can more successfully do what they enjoy doing and do best — building.

Who Should Use *The Contractor's Legal Kit?*
This manual will be most helpful to the following types of contractors:

• All building contractors who have recently started their own residential construction businesses.

• Residential remodeling contractors who do not have a fully developed set of agreements, subcontracts, and change order forms, and a commonsense procedure for using these forms.

• Residential contractors who build new homes and who do not have a fully developed set of agreements, subcontracts, and change order forms, and a commonsense procedure for using these forms.

• General contractors who do light-commercial construction work where standard AIA-type construction agreements are not used by the owner, and the contractor has the opportunity to furnish the construction agreement to the owner.

• Any building contractor who wants to increase his ability to avoid disputes, litigation, and loss of profits through a review of his own construction agreements and procedures for using them. Any building contractor who wants to improve his ability to identify and shift the inherent risks associated with all building projects more equitably toward the owner.

This manual does *not* attempt to cover the considerably greater liability attached to builders who build and sell "spec" homes. Nor does it attempt to cover the differing liability of subcontractors or general contractors who occasionally work as subcontractors.

How To Use *The Contractor's Legal Kit*
Most of the contracts and other legal documents in this book appear first as blank forms, followed by the same form filled out with sample language (see example on the next page). The filled-out forms are annotated with brief descriptions of each clause, explaining its meaning and usage. By comparing the annotated agreements to the blank forms, you will find it easier to envision how to customize and fill out your own agreements. Some sections of the book, however, contain very short or simple forms which need no annotations.

After reviewing the various sections in this manual, you can draft your own construction documents based on sections of the manual that seem appropriate to your business. A computer disk containing nearly all the forms and contracts in this book is included to help you get started in this process. Once you develop a basic set of agreements, you should have a construction attorney in your area review them to make sure they are suitable for your specific business and in compliance with your state laws.

After developing a set of basic form documents that you are comfortable using, you will become more proficient at analyzing each job as it comes up and deciding which length agreement to use and which specific clauses to include. You'll also learn how to customize an agreement for a specific job.

Don't be afraid to invest a few minutes of time with a construction attorney in your area if you are uncertain about how to proceed with a particular job. Better to spend a few dollars up front than to discover later that certain costly mistakes could have been avoided.

Remember, the sample agreements included in this manual are *a good starting point only* and will ordinarily require some customizing for your business. You also need to include any additional language required for residential construction or home improvement contracts in your state.

Finally, while using the forms or information in this manual, you will be acting as your own attorney. When in doubt about the content or best use of this information, consult a local attorney who is familiar with construction law.

1.2 ■ MEDIUM-FORM FIXED PRICE AGREEMENT

ANNOTATED

Number at top of form corresponds to blank forms and forms on disk.

■ This annotated sample of the medium-form agreement refers to a commercial tenant improvement job. Because the plans did not include a finish schedule, I've included a very thorough description of what is included and not included in the Scope of Work section.

Type in gray boxes indicates annotations added to explain form.

Charlie Contractor Construction, Inc.
123 Hammer Lane
Anywhere, USA 33333
Phone: (123) 456-7890
Fax: (123) 456-7899
Lic#: 11111

Typewriter-style lettering indicates text added by contractor.

DATE: **May 22, 2001**

OWNER'S NAME: **Joe Tenant**
ADDRESS: **689 Commercial Plaza**
   Anywhere, USA 33333

PROJECT ADDRESS: **same**

Type in white areas indicates part of form.

I. PARTIES

This contract (hereinafter referred to as "Agreement") is made and entered into on this **22nd** day of **May**, 20 **01**, by and between **Joe Tenant**, (hereinafter referred to as "Owner"); and **Charlie Contractor Construction, Inc.**, (hereinafter referred to as "Contractor"). In consideration of the mutual promises contained herein, Contractor agrees to perform the following work:

■ The phrase "...in consideration of the mutual promises contained herein, agrees to perform the following work:" is good boilerplate language that should remain in your Agreement. It means there is a "bargained-for exchange," which is necessary for a contract to be valid. The Owner will pay the Contractor the stated sum, and in exchange the Contractor will complete the work described in the Agreement. In the space immediately above, refer to any available plans and specifications on which you have based your bid.

Contractor and owner should initial this stamp on every page of contract.

INITIAL
CC
97

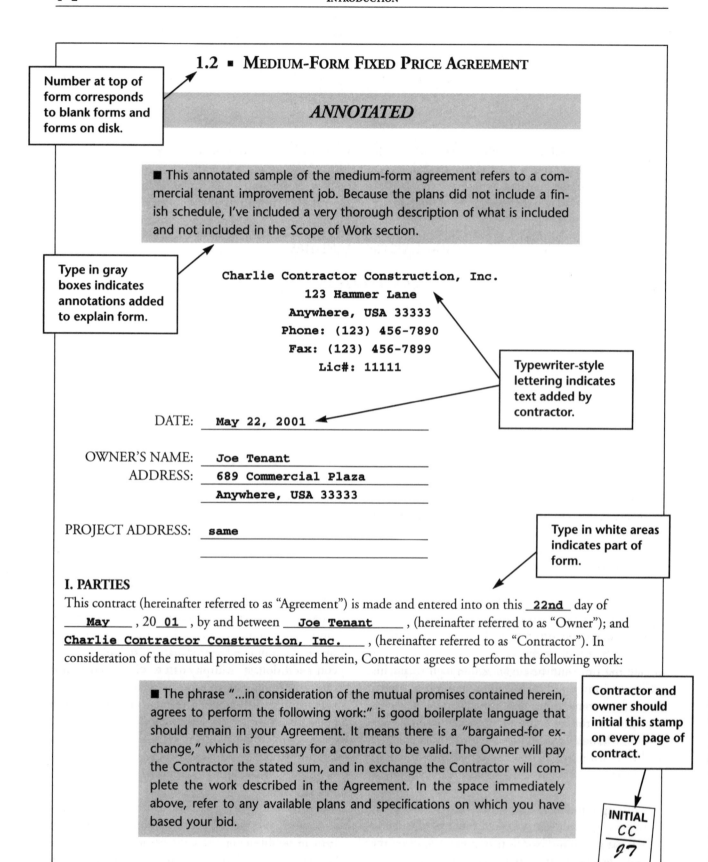

CHAPTER 1

FIXED PRICE AGREEMENTS

1.1 ▪ SHORT-FORM FIXED PRICE AGREEMENT

1.2 ▪ MEDIUM-FORM FIXED PRICE AGREEMENT

1.3 ▪ LONG-FORM FIXED PRICE AGREEMENT

1.4 ▪ ADDENDUM FOR MATCHING EXISTING FINISHES

1.5 ▪ RIGHTS OF RESCISSION AND CANCELLATION

A hidden opportunity often worth thousands of dollars a year awaits residential builders and remodelers. While commercial builders are typically required to sign the owner's construction agreement (usually an AIA agreement, or one prepared by the owner's attorney), residential contractors often have the opportunity to furnish the construction agreement. In doing so, the contractor can establish many of the rules that will legally govern the business relationship with the owner.

This allows you to take advantage of your years of experience in this business and communicate to the owner how you want the business/legal aspects of the project to be handled. But, there's a catch — you have to take the time and be organized enough to put the rules in writing *prior* to beginning the project. Now is the time to think ahead about what typical problems may arise on this type of project and how you would prefer to handle them. Then you can address these problems in the construction agreement you give the owner.

Identifying Hidden Risks

Identifying potential problems on a larger building project is like searching for the fifteen hidden animals in those grade school jungle pictures. It's easy to find three or four of the camouflaged animals — or potential con-

struction problems — at first glance, but finding the other eleven or twelve takes time, patience, and practice. Your job in putting together an agreement is to pull out and identify as many of these hard-to-find hidden risks as possible, and to spell out how they will be handled if and when they show up on your job. Again, the trick is to accomplish this *prior* to signing an agreement with the owner.

By assigning these risks via the agreement (rather than haggling about them later in the project), you'll find it easier to maintain the mutual trust and respect that most jobs start out with. Then, good work on the site and good efforts at regular communication with the owners can build on that foundation of trust.

In addition, if "Murphy's Law" visits your project, as it seems to on most of mine, you will have an agreed upon, contractually binding approach for handling the problems as they arise. Contractors all too often end up absorbing many hidden expenses regardless of what is fair and reasonable.

Avoiding Disputes

Construction disputes are almost always factually complicated and confusing, and expensive to resolve. They also weaken the level of trust between the owner and the contractor and can lead to a loss of reputation and business. Occasional disagreements and disputes are an inevitable part of doing business — any business. However, 90% of the construction disputes I've seen all have one thing in common: they probably could have been avoided if the contractor had taken two simple steps. He should have:

1) furnished the owner with a well-drafted construction agreement, and

2) referred to the agreement during the course of the project to find answers to the questions that arose over payments, extra work, and other obligations.

Many disagreements on a job are not the result of defective or nonconforming work, but rather stem from such things as:

• Ineffective agreements that leave too many foreseeable problems to be "worked out in the field."
• Lack of attention to detail by the contractor during bidding and contract drafting.
• Failure to include enough clauses in the agreement to provide a specific "road map" describing how to proceed if certain problems arise.
• Failure to communicate in the agreement about important aspects of the work and the business/legal relationship so that the owner develops realistic expectations about all phases of the project.
• Inadequate supervision of employees who are performing substandard work or are not working efficiently.

Through the use of a well-drafted construction agreement, the contractor can better control and minimize the friction, disputes, and small disagreements that naturally develop on any construction project. A good agreement benefits the contractor by better communicating mutual expectations between the owner and himself. This will save the contractor valuable time and money because he will spend less time arguing over (and perhaps absorbing or splitting) unforeseen additional expenses.

Some disputes are unavoidable. Just ask around among builders who have been in business for awhile and you'll find without exception that sometimes *who* you build for can be just as important as *how* you build.

There are a few litigious owners who are difficult, if not impossible, to please. And, if those owners aren't pleased, you will have no choice but to fall back on your written construction agreement and change orders in order to be paid. If you ever run into this unfortunate situation, a small problem can cost you plenty — a large problem can cost you your entire business.

If you encounter a litigious owner like this, a good agreement won't necessarily keep you from being sued (or having to sue to be paid), but it will better your chances of having an outcome you can live with.

Good work: a safeguard. It's a given that the work *must* be of good workmanlike quality and conform to plans and contract documents. If the source of the problem on a job is that the work is seriously defective or that it doesn't conform to the plans or contract documents, then that's a problem that only the contractor can fix by reworking the job. The lesson is that most shortcuts will come back to haunt you, sooner or later.

The Benefits of a Good Agreement

Effective agreements supplied by the contractor will also benefit the owner because owners feel more at ease when they are well informed and feel the project is under control from start to finish — with as few surprises as possible. The natural results for the contractor of implementing such an agreement will be:

• Increased retention of job profits.
• Fewer daily problems and hour-long nightly phone conversations with the owner, attempting to justify actions or resolve minor disputes.
• Reduced exposure to costly and time-consuming litigation/arbitration proceedings and attorney's fees.
• Better relationships with all parties involved in the project, including the owner, your employees, and subcontractors.
• A better reputation as a contractor who professionally handles not only the construction, but also the paperwork aspect of the business.
• Better customer referrals.

Know Your State Laws

All applicable state and federal requirements must be placed in your construction agreements. Because laws differ from state to state, be sure to contact your state's contractors' license board (if there isn't one, call your state's information number and ask what agency governs construction and residential home improvement contracts) and get a list of minimal requirements for construction agreements, especially home improvement contracts. You should also be able to get this information from a construction attorney in your area. All applicable state and federal requirements must be placed in your construction agreements or attached to them.

If you provide any type of medium- or long-term financing for your work, you are subject to yet another batch of state and federal requirements. This manual does not address these requirements because the owner typically obtains his own financing independent of the contractor.

After checking with your state, it is likely you'll find you are required to furnish the owner with one or more of the following forms: "Notice to Owner," "Notice of Cancellation," or "Right of Rescission." See Section 1.5 for the federal Notice of Cancellation and Right of Rescission forms. Some states also require that a few specific clauses be placed in every residential home improvement contract as a form of consumer protection. Many contractors ignore these requirements and later regret it.

Failing to include state or federal required contract language can be grounds for disciplinary action by your state contractors' license board and may result in civil penalties and fines, or a costly disciplinary bond when the time comes to renew your contractor's license bond. And failing to provide the federally required notices may technically be grounds for the owner to not pay you.

In addition, the contractor's ability to take appropriate legal action toward an owner who has breached an agreement (e.g., failed to pay retention or progress payments) may be greatly impaired because the contractor doesn't want to "make waves" for fear of having a complaint filed with the state license board. Don't risk putting yourself in this position. It's easy to include state and federal required language in your agreements.

Signing the Contract

This is the procedure I have found to be most practical in getting a signed contract from the owner. First, if at all possible, I allow myself enough time to make sure that my contract is *complete*. In particular, I focus on a detailed scope of work description, a thorough list of exclusions, and a payment schedule that is fair, but gives me enough money to finance operations roughly within the 30-day "float" period I get from suppliers.

While it's common for contractors to rush putting their bids together (sometimes literally up to the very last minute), this can cost you plenty on complicated, competitively priced jobs if you miss many items or omit many exclusions. So the first and most important part of putting together the contract is making sure your bidding has been done carefully, and that your numbers are accurate and complete. Never price the job too low hoping to make it up with extra work.

If, shortly after turning in your bid, you discover you have made an innocent clerical error in your math or have forgotten to transfer a number from your bid sheet into your final calculations, immediately notify the owner of this clerical error by phone and in writing, and request an adjustment to your bid amount or a withdrawal of your bid. If the mistake is truly innocent and clerical in nature and your notification is timely, you may have legal grounds for the withdrawal of your bid, even if you were low bidder. However, the longer you wait to notify the owner, the less legal support you'll have for bid adjustment or withdrawal.

Next, I attach to the filled-out contract any subtrade bids, sketches, or other written clarifications that I want incorporated by reference into the contract. Somewhere in the agreement or the addendum, you must state that "this proposal/sketch/bid is hereby incorporated by reference into this agreement," or similar language.

For example, on a small job, you may have a sketch you have drawn that shows some basic details of the work you will perform. You can make this sketch part of the contract by referencing it in your agreement, attaching it to the agreement, initialing it, and having it initialed by the owner. Or, if the specifications or plans in a given area are vague, you may want to simply copy the relevant subtrade bid(s) and state, for instance, "Plumbing work per attached bid from John's Plumbing: $2,350." Then attach John's bid to the end of your contract and be certain that both you and the owner initial that bid. Use this approach only when the plans do not clearly describe the scope of a subcontractor's work.

I also attach to the contract the "Notice to Owner" and "Right of Cancellation" forms required by law in my state (California).

Then I sign and date the last page of the contract and initial the bottom of each page of the contract and any "attached" pages. I have the owner initial all the pages as well. (This will ensure there is no later confusion over whether or not the owner received and agreed to all pages of the contract.) For a few dollars, you can have an initial stamp like this made up:

Finally, I make two extra copies of the original contract (printed from my computer) and its attachments. I keep one copy for my files and give the owner the original and the other copy, both of which I have initialed on every page and "wet signed." The owner should keep the extra copy for his records and return the original copy to me after signing it and initialing each page. To simplify this procedure, I stamp the original contract at the top of page one with a red stamp that looks like this:

AFTER REVIEWING
PLEASE SIGN THIS
COPY AND RETURN
TO CONTRACTOR

Although not essential, it is better if you have a contract in your files that has been "wet signed" by the owner, rather than just a photocopy of the agreement signed by the owner. Some contractors are more comfortable sending out an unsigned contract for the owners to sign first and return. This is perfectly fine and

gives a little leeway should you wish to make last-minute revisions and issue a revised contract.

On larger jobs, I have often found it helpful to schedule a *preconstruction conference* with the owners (both husband and wife) to go over every page of the agreement prior to signing it. I use this meeting to clarify any questions the owners may have about the scope of work, the exclusions, the schedule, inconveniences and interruptions (for remodels), and any contract language they may have questions about. (See Chapter 7, Section 7.3 for a full discussion of the preconstruction conference.)

This meeting has proven to be a helpful marketing tool by demonstrating to the owners my organizational style and my attention to detail during the "preconstruction" phase of the job. It also helps us avoid misunderstandings that can cause problems later on when the job is well under way.

Be sure to maintain a flexible attitude toward any concerns the owners may express at this meeting. If the owner is uncomfortable with certain language in the agreement, after consideration (and a quick call to your attorney, if you think it necessary), you may decide to modify portions of the contract language to address the concerns of the owner. However, based on my experience, the owner rarely requests more than one or two minor changes in the contract language.

Be sure you have a basic understanding of the meaning of the contract language so that if asked to explain or negotiate a certain clause, you are able to do so in a reasonable way. The annotated versions of the agreements in this book will help you become more proficient at this.

Typically, when you are sitting around the table discussing the contract, the scope of work and the exclusions are of more interest to the owner than the contract language itself. The owner often asks for clarifications of or revisions to the scope of work based on the newly discovered lump sum price. The contractor may be asked to reduce the price by excluding certain items the owner now knows he can't afford. In many cases, cutting out a few expensive finish details is the key to getting the job.

If the changes to the contract requested by the owner are minor, you may write the changes in ink on the agreement. Both you and the owner must initial each and every change. If the scope of the requested change is considerable, you should prepare a new agreement that incorporates the agreed upon changes. With word processors, making even significant changes is fast and easy.

On most larger jobs, the contract is not signed at this first meeting. There may be other bids coming in, or the owners may simply need time to review and discuss the proposed job. But whatever happens, remember one iron-clad rule: Never start a job without having a signed contract! Misunderstandings are typically much harder to clear up after work has commenced than they are before the job starts. Don't be in such a hurry to start the job that you overlook the importance of first getting a signed contract.

Finally, keep your signed contract in a safe place. The most customized, refined, airtight agreement won't do you a bit of good later on if you can't produce a copy of it that has been signed by the owner.

On most medium- and large-sized jobs, I also use a *Preconstruction Conference Form* (Form 7.3). I typically present this *at the final meeting where the contract is to be signed* and have both signed together. The use of this form is entirely optional. It is an informational form, not really an agreement or contract. This form basically indicates that the owner and the contractor have carefully gone over all pages of the agreement. It also puts the owner on notice about important expectations of the contractor that may not appear in the contract, and lets the contractor know how the owner wants him to handle matters such as access to the residence, location of tool storage and dumpsters or debris piles, and dust protection.

What's in This Chapter

Sections 1.1, 1.2, and 1.3 cover short-, medium-, and long-form lump sum (or fixed price) agreements. Section 1.4 discusses the addendum for matching existing finishes, and Section 1.5 explains the rights of rescission and cancellation.

Most residential owners are more comfortable with a lump sum agreement than with a time-and-materials or cost-plus contract. The difference between the three lump sum agreements is the number of clauses that govern the business relationship between the owner and the contractor. Generally, the bigger the job, the longer the agreement should be. However, there are exceptions to this general rule.

When doing work for customers with whom you have a well-developed and trusting relationship, you may opt to use a shorter agreement than you would otherwise. You need to strike a balance here. You want to provide an adequate level of contract language to govern the business relationship, but you don't want to overwhelm people with excessively long agreements for smaller jobs.

However, factor into your decision-making process that about 25% of the residential construction disputes I encounter involve a contractor and owner who were either relatives or "good friends" when the job started. Just because an owner is a friend or relative doesn't mean you will be immune to misunderstandings on the job. Take the time to put your agreement in writing and you'll help to ensure that you're still on speaking terms when the job is completed.

Finally, good construction agreements don't have to be overly complex. A well-drafted simple agreement followed up with written field change orders can be infinitely better than no agreement or a bad agreement that is followed up by verbal change orders.

Short-Form Fixed Price Agreement

The short-form fixed price (lump sum) agreement can give the builder good, basic legal protection on smaller sized jobs such as simple residential and light-commercial remodels. It may also be suitable on somewhat larger jobs when you have worked with the owners before and have established a very good working relationship.

There are two major benefits to using this short-form agreement rather than the commonly used "dime-store" type of proposal/agreement. First, this short-form agreement contains many basic legal protections not found in many short-form boilerplate agreements. Second, the short-form agreement also contains a fairly detailed list of exclusions covering many areas of work that the builder has carefully considered.

Contractually agreeing to exclude detailed areas of work will give the builder a clear legal basis from which to charge for legitimate change orders if these excluded areas of work arise during the job. Many builders rush through their bid preparation on small jobs, overlook these exclusions, and end up personally absorbing what would otherwise have been legitimate change orders. The detailed exclusions section in this agreement will help you to avoid this common trap.

You can line out (and initial) portions of the "Standard Exclusions" section that do not apply to your job and type in any additional exclusions in the section called "Project Specific Exclusions."

The short-form agreement contains less detail than the medium- and long-form agreements, so be sure to review the longer agreements to see if one of those would be more appropriate, or if some of the language in the longer agreements should be incorporated into your short-form agreement.

Keep in mind that there is no absolute rule to follow when trying to determine which length agreement to use. Generally, the more costly and complicated the job, the longer the agreement.

1.1 ■ Short-Form Fixed Price Agreement

CONTRACTOR'S NAME: _____

ADDRESS: _____

_____

PHONE: _____

FAX: _____

LIC #: _____

DATE: _____

OWNER'S NAME: _____

ADDRESS: _____

_____

PROJECT ADDRESS: _____

_____

I. PARTIES

This contract (hereinafter referred to as "Agreement") is made and entered into on this _____ day of
_____ , 19_____ , by and between _____ ,
(hereinafter referred to as "Owner"); and _____ ,
(hereinafter referred to as "Contractor"). In consideration of the mutual promises contained herein,
Contractor agrees to perform the following work:

II. GENERAL SCOPE OF WORK DESCRIPTION

_____

_____

_____

_____

_____

_____

_____

(Additional Scope of Work page(s) attached: _____ Yes _____ No)

LUMP SUM PRICE FOR ALL WORK ABOVE: $_____

III. GENERAL CONDITIONS FOR THE AGREEMENT ABOVE

A. EXCLUSIONS
This Agreement does *not* include *labor or materials* for the following work:

1. PROJECT SPECIFIC EXCLUSIONS:

_____

_____

_____

_____

_____

2. STANDARD EXCLUSIONS: Unless specifically included in the "General Scope of Work" section above, this Agreement does *not* include *labor or materials* for the following work: Plans, engineering fees, or governmental permits and fees of any kind. Testing, removal and disposal of any materials containing asbestos (or any other hazardous material as defined by the EPA). Custom milling of any wood for use in project. Moving Owner's property around the site. Labor or materials required to repair or replace any Owner-supplied materials. Repair of concealed underground utilities not located on prints or physically staked out by Owner which are damaged during construction. Surveying that may be required to establish accurate property boundaries for setback purposes (fences and old stakes may not be located on actual property lines). Final construction cleaning (Contractor will leave site in "broom swept" condition). Landscaping and irrigation work of any kind. Temporary sanitation, power, or fencing. Removal of soils under house in order to obtain 18 inches (or code-required height) of clear space between bottom of joists and soil. Removal of filled ground or rock or any other materials not removable by ordinary hand tools (unless heavy equipment is specified in Scope of Work section above), correction of existing out-of-plumb or out-of-level conditions in existing structure. Correction of concealed substandard framing. Rerouting/removal of vents, pipes, ducts, structural members, wiring or conduits, steel mesh which may be discovered in the removal of walls or the cutting of openings in walls. Removal and replacement of existing rot or insect infestation. Failure of surrounding part of existing structure, despite Contractor's good faith efforts to minimize damage, such as plaster or drywall cracking and popped nails in adjacent rooms or blockage of pipes or plumbing fixtures caused by loosened rust within pipes. Construction of a continuously level foundation around structure (if lot is sloped more than 6 inches from front to back or side to side, Contractor will step the foundation in accordance with the slope of the lot). Exact matching of existing finishes. Public or private utility connection fees. Repair of damage to roadways, driveways, or sidewalks that could occur when construction equipment and vehicles are being used in the normal course of construction.

B. DATE OF WORK COMMENCEMENT AND SUBSTANTIAL COMPLETION
Commence work:_____ . Construction time through substantial completion: Approximately _____ to _____ weeks/months, *not* including delays and adjustments for delays caused by: inclement weather, additional time required for Change Order work, and other delays unavoidable or beyond the control of the Contractor.

C. CHANGE ORDERS: CONCEALED CONDITIONS AND ADDITIONAL WORK

1. CONCEALED CONDITIONS: This Agreement is based solely on the observations Contractor was able to make with the structure in its current condition at the time this Agreement was bid. If additional concealed conditions are discovered once work has commenced which were *not* visible at the time this proposal was bid, Contractor will stop work and point out these unforeseen concealed conditions to Owner so that Owner and Contractor can execute a Change Order for any Additional Work.

2. DEVIATION FROM SCOPE OF WORK: Any alteration or deviation from the Scope of Work referred to in this Agreement involving extra costs of materials or labor (including any overage on **ALLOWANCE** work and any changes in the Scope of Work required by governmental plan checkers or field building inspectors) will be executed upon a written Change Order issued by Contractor and should be signed by Contractor and Owner prior to the commencement of Additional Work by the Contractor.

Contractor to supervise, coordinate, and charge _____% profit and overhead on Owner's separate Subcontractors who are working on site at same time as Contractor. Contractor's profit and overhead, and any supervisory labor will not be credited back to Owner with any deductive Change Orders (work deleted from Agreement by Owner).

D. PAYMENT SCHEDULE AND PAYMENT TERMS

1. PAYMENT SCHEDULE:
* First Payment: $1,000 or 10% of contract amount (whichever is less) due when Agreement
is signed and returned to Contractor: $_____

* Second Payment (Materials Deposits): _____

_____ $_____

* Third Payment: _____

_____ $_____

* Fourth Payment: _____

_____ $_____

* Final Payment: Balance of contract amount due upon Substantial Completion of all
work under contract: $_____

2. PAYMENT OF CHANGE ORDERS: Payment for each Change Order is due upon completion of Change Order work and submittal of invoice by Contractor.

3. ADDITIONAL PAYMENTS FOR ALLOWANCE WORK AND RELATED CREDITS: Payment for work designated in the Agreement as **ALLOWANCE** work has been initially factored into the Lump Sum Price and Payment Schedule set forth in this Agreement. If the actual cost of the **ALLOWANCE** work exceeds the line item **ALLOWANCE** amount in the Agreement, the difference between the cost and the line item **ALLOWANCE** amount stated in the Agreement will be written up by Contractor as a Change Order subject to Contractor's profit and overhead at the rate of _____%.

If the cost of the **ALLOWANCE** work is less than the **ALLOWANCE** line item amount listed in the Agreement, a credit will be issued to Owner after all billings related to this particular line item **ALLOWANCE** work have been received by Contractor. This credit will be applied toward the final payment owing under the Agreement. Contractor profit and overhead and any supervisory labor will not be credited back to Owner for **ALLOWANCE** work.

E. WARRANTY

Contractor provides a limited warranty on all Contractor- and Subcontractor-supplied labor and materials used in this project for a period of one year following substantial completion of all work.

No warranty is provided by Contractor on any materials furnished by the Owner for installation. No warranty is provided on any existing materials that are moved and/or reinstalled by the Contractor within the dwelling (including any warranty that existing/used materials will not be damaged during the removal and reinstallation process). One year after substantial completion of the project, the Owner's sole remedy (for materials and labor) on all materials that are covered by a manufacturer's warranty is strictly with the manufacturer, not with the Contractor.

Repair of the following items is specifically excluded from Contractor's warranty: Damages resulting from lack of Owner maintenance; damages resulting from Owner abuse or ordinary wear and tear; deviations that arise such as the minor cracking of concrete, stucco and plaster; minor stress fractures in drywall due to the curing of lumber; warping and deflection of wood; shrinking/cracking of grouts and caulking; fading of paints and finishes exposed to sunlight.

THE EXPRESS WARRANTIES CONTAINED HEREIN ARE IN LIEU OF ALL OTHER WARRANTIES, EXPRESS OR IMPLIED, INCLUDING ANY WARRANTIES OF MERCHANTABILITY, HABITABILITY, OR FITNESS FOR A PARTICULAR USE OR PURPOSE. THIS LIMITED WARRANTY EXCLUDES CONSEQUENTIAL AND INCIDENTAL DAMAGES AND LIMITS THE DURATION OF IMPLIED WARRANTIES TO THE FULLEST EXTENT PERMISSIBLE UNDER STATE AND FEDERAL LAW.

F. WORK STOPPAGE, TERMINATION OF CONTRACT FOR DEFAULT, AND INTEREST

Contractor shall have the right to stop all work on the project and keep the job idle if payments are not made to Contractor in accordance with the Payment Schedule in this Agreement, or if Owner repeatedly fails or refuses to furnish Contractor with access to the job site and /or product selections or information necessary for the advancement of Contractor's work. Simultaneous with stopping work on the project, the Contractor must give Owner written notice of the nature of Owner's default and must also give the Owner a 14-day period in which to cure this default.

If work is stopped due to any of the above reasons (or for any other material breach of contract by Owner) for a period of 14 days, and the Owner has failed to take significant steps to cure his default, then Contractor may, without prejudicing any other remedies Contractor may have, give written notice of termination of the

Agreement to Owner and demand payment for all completed work and materials ordered through the date of work stoppage, and any other loss sustained by Contractor, including Contractor's Profit and Overhead at the rate of ____% on the balance of the incomplete work under the Agreement. Thereafter, Contractor is relieved from all other contractual duties, including all Punch List and warranty work.

G. DISPUTE RESOLUTION AND ATTORNEY'S FEES

Any controversy or claim arising out of or related to this Agreement involving an amount of less than $5,000 (or the maximum limit of the court) must be heard in the Small Claims Division of the Municipal Court in the county where the Contractor's office is located. Any controversy or claim arising out of or related to this Agreement which is over the dollar limit of the Small Claims Court must be settled by binding arbitration administered by the American Arbitration Association in accordance with the Construction Industry Arbitration Rules. Judgment upon the award may be entered in any Court having jurisdiction thereof.

The prevailing party in any legal proceeding related to this Agreement shall be entitled to payment of reasonable attorney's fees, costs, and expenses.

H. EXPIRATION OF THIS AGREEMENT

This Agreement will expire 30 days after the date at the top of page one of this Agreement if not first accepted in writing by Owner.

I. ENTIRE AGREEMENT

This Agreement represents and contains the entire agreement between the parties. Prior discussions or verbal representations by the parties that are not contained in this Agreement are not a part of this Agreement.

J. ADDITIONAL LEGAL NOTICES REQUIRED BY STATE OR FEDERAL LAW

See page(s) attached: ____Yes ____No

K. ADDITIONAL TERMS AND CONDITIONS

See page(s) attached: ____Yes ____No

I have read and understood, and I agree to, all the terms and conditions contained in the Agreement above.

Date: _____ _____
CONTRACTOR'S SIGNATURE

Date: _____ _____
OWNER'S SIGNATURE

1.1 ▪ Short-Form Fixed Price Agreement

ANNOTATED

Charlie Contractor Construction, Inc.
123 Hammer Lane
Anywhere, USA 33333
Phone: (123) 456-7890
Fax: (123) 456-7899
Lic#: 11111

DATE: **May 22, 2001**

OWNER'S NAME: **Mr. & Mrs. Harry Homeowner**
ADDRESS: **333 Swift St.**
Anywhere, USA 33333

PROJECT ADDRESS: **same**

I. PARTIES

This contract (hereinafter referred to as "Agreement") is made and entered into on this **22nd** day of **May**, 20**01**, by and between **Harry and Helen Homeowner**, (hereinafter referred to as "Owner"); and **Charlie Contractor Construction, Inc.**, (hereinafter referred to as "Contractor"). In consideration of the mutual promises contained herein, Contractor agrees to perform the following work:

> ■ The phrase "...in consideration of the mutual promises contained herein, agrees to perform the following work:" is good boilerplate language that should remain in your Agreement. It means there is a "bargained-for exchange," which is necessary for a contract to be valid. The Owner will pay the Contractor the stated sum, and in exchange the Contractor will complete the work described in the Agreement. In the space immediately above, refer to any available plans and specifications on which you have based your bid.

INITIAL
CC
HH

II. GENERAL SCOPE OF WORK DESCRIPTION

■ Provide general details of the Scope of Work here. Give a very complete description of what is included for the Lump Sum Price stated below. Many disputes occur because the Contractor does not specifically describe the Scope of Work. Spelling this out can also help you remember all the items that should be included in your job costs.

Many Owners feel more comfortable with a detailed line item description of what the Contractor is including in the Scope of Work. Sometimes when two Contractors give very close bids for the same job, the Contractor who gave the more complete Scope of Work description will end up with the job.

You may also want to state that the costs of certain line items are being furnished on an ALLOWANCE basis. For instance, many Owners have not selected specific plumbing fixtures or light fixtures at the time the Agreement is entered into, so the Contractor may want to list these fixtures as ALLOWANCE ITEMS. For instance:

1. Plumbing fixtures for two bathrooms (ALLOWANCE): $ 750
2. Electrical fixtures for entire residence (ALLOWANCE): $2,000

When you include ALLOWANCES, be sure to include contract language that describes how any overage or underage regarding the ALLOWANCE item will be handled. Be sure to state that Contractor Profit and Overhead will be added to any amount that exceeds the ALLOWANCE line item amount stated in the Agreement. The clauses in this manual relating to AL-LOWANCES make this clear, however, you may need to verbally explain how this works to the Owner.

(Additional Scope of Work page(s) attached: _____ Yes __x__ No)

1. Remove existing built-up flat roof on garage. Haul all debris to landfill. Pay dump fees. Clean up job site.

2. Provide temporary sanitation.

3. Furnish materials and labor to frame in new pitched roof on garage per plans. Furnish and install new 1/2-inch CDX plywood to newly framed roof rafters.

4. Furnish and install four clerestory 3'6"x1'6" single-pane, mill-finish, fixed-pane windows in garage. Cut in and provide framing for all of these windows. Trim out all windows on exterior only to match existing trim as closely as possible.

INITIAL
CC
HH

5. Furnish and install materials to extend furnace, hot water heater, and plumbing flues to accommodate height of newly framed roof.

6. Furnish and install gutter, downspouts, flashings, and new roof jacks on new garage roof.

7. Furnish and install felt underlayment and 20-year composition roofing shingles to match color of roof on house as closely as possible.

8. Install roof venting per plans.

9. Furnish and install building paper and new smooth face T1-11 siding on new framing of garage only.

10. Furnish and install 5/8-inch drywall on interior of garage per plans. Tape, texture, and finish drywall with skip trowel texture.

11. Remove and replace rotten framing materials in existing walls on garage (ALLOWANCE): $ 425

* LUMP SUM PRICE FOR ALL WORK ABOVE: $ 13,880

> ■ The Contractor may or may not want to show his profit and overhead and any supervisory labor as a separate line item cost above the Lump Sum Price. In some cases he may want to show line item pricing for all the work, other times he may not want to break the bid out so specifically. Insurance jobs almost always *require* line item pricing.

III. GENERAL CONDITIONS FOR THE AGREEMENT ABOVE

A. EXCLUSIONS

This Agreement does *not* include *labor or materials* for the following work:

> ■ The Exclusions section of your Agreement is just as important as the Scope of Work section. Unless you specifically exclude areas of work that you know you are not going to perform, the Owner might assume they are included.
>
> Contractors lose a great deal of money by failing to exclude work. The Contractor later "eats" items that fall into "gray" areas and weren't specifically excluded in order to get his final check from the Owner. If the Exclusions section is complete, the Contractor will "eat" far fewer of these items. In my experience, paying attention to Exclusions can save a medium-sized Contractor several thousand dollars each year. If you use a computer,

INITIAL
CC
HH

> simply delete the items that clearly don't apply to your particular project. Or you can line them out (and initial them) if you use this form as is. Add any Exclusions to your Agreement that relate specifically to your business or to the job you will be performing. The Exclusions list provided in the form is a good starting point, but it may not be 100% complete for your purposes.

1. PROJECT SPECIFIC EXCLUSIONS:
Paint work or caulking of any kind. Electrical work or fixtures of any kind.

> ■ I list the most obvious Exclusions that specifically relate to the job being contracted as "Project Specific Exclusions." The less obvious, more remote, and "boilerplate" Exclusions are listed below in "Standard Exclusions." If working with a computer, remove any boilerplate Exclusions that don't apply to the job so you don't scare the Owner away with unnecessary Exclusions.

2. STANDARD EXCLUSIONS: Unless specifically included in the "General Scope of Work" section above, this Agreement does *not* include *labor or materials* for the following work: Plans, engineering fees, or governmental permits and fees of any kind. Testing, removal and disposal of any materials containing asbestos (or any other hazardous material as defined by the EPA). Custom milling of any wood for use in project. Moving Owner's property around the site. Labor or materials required to repair or replace any Owner-supplied materials. Repair of concealed underground utilities not located on prints or physically staked out by Owner which are damaged during construction. Surveying that may be required to establish accurate property boundaries for setback purposes (fences and old stakes may not be located on actual property lines). Final construction cleaning (Contractor will leave site in "broom swept" condition). Landscaping and irrigation work of any kind. Temporary sanitation, power, or fencing. Removal of soils under house in order to obtain 18 inches (or code-required height) of clear space between bottom of joists and soil. Removal of filled ground or rock or any other materials not removable by ordinary hand tools (unless heavy equipment is specified in Scope of Work section above), correction of existing out-of-plumb or out-of-level conditions in existing structure. Correction of concealed substandard framing. Rerouting/removal of vents, pipes, ducts, structural members, wiring or conduits, steel mesh which may be discovered in the removal of walls or the cutting of openings in walls. Removal and replacement of existing rot or insect infestation. Failure of surrounding part of existing structure, despite Contractor's good faith efforts to minimize damage, such as plaster or drywall cracking and popped nails in adjacent rooms or blockage of pipes or plumbing fixtures caused by loosened rust within pipes. Construction of a continuously level foundation around structure (if lot is sloped more than 6 inches from front to back or side to side, Contractor will step the foundation in accordance with the slope of the lot). Exact matching of existing finishes. Public or private utility connection fees. Repair of damage to existing roads, sidewalks, and driveways that could occur when construction equipment or vehicles are being used in the normal course of construction.

INITIAL
CC
H H

■ Agreeing to "match existing finishes" is a can of worms requiring such special treatment that an explanation of one approach is given in Form 1.4, Addendum for Matching Existing Finishes.

B. DATE OF WORK COMMENCEMENT AND SUBSTANTIAL COMPLETION

Commence work: **June 1, 2001** . Construction time through substantial completion: Approximately **3 to 4 weeks** , *not* including delays and adjustments for delays caused by: inclement weather, additional time required for Change Order work, and other delays unavoidable or beyond the control of the Contractor.

■ Some states require that the approximate start and completion dates of the project be specified in the Agreement. Every Owner also wants to know this information. However, be sure to give a *range* of days, weeks, or months to complete and not exact dates. If you think the project will take 3 months, state "3 to 4 months" whenever possible so you allow a cushion for delays. Numerous events can increase the time needed to complete the Work. Be sure to add extra contract days to all of your Change Orders in the space provided.

C. CHANGE ORDERS: CONCEALED CONDITIONS AND ADDITIONAL WORK

1. CONCEALED CONDITIONS: This Agreement is based solely on the observations Contractor was able to make with the structure in its current condition at the time this Agreement was bid. If additional concealed conditions are discovered once work has commenced which were *not* visible at the time this proposal was bid, Contractor will stop work and point out these unforeseen concealed conditions to Owner so that Owner and Contractor can execute a Change Order for any Additional Work.

■ Every Agreement should have a Concealed Conditions clause similar to the one above — especially on remodeling jobs. Occasionally Murphy's Law will create some very unforeseeable extra work on projects. If this work is not specifically excluded in the "Exclusions" section of your Agreement, you may need to rely on the "Concealed Conditions" clause in order to write it up as a Change Order.

Examples of Concealed Conditions are subterranean concrete that interferes with your foundation work, spring water that prevents you from pouring concrete, pipes and conduits that interfere with the installation of new doors and windows, asbestos or other toxic compounds discovered during demolition, omitted framing members, etc.

2. DEVIATION FROM SCOPE OF WORK: Any alteration or deviation from the Scope of Work referred to in this Agreement involving extra costs of materials or labor (including any overage on **ALLOWANCE** work and any changes in the Scope of Work required by

INITIAL
CC
HH

governmental plan checkers or field building inspectors) will be executed upon a written Change Order issued by Contractor and should be signed by Contractor and Owner prior to the commencement of Additional Work by the Contractor.

Contractor to supervise, coordinate, and charge __20__% profit and overhead on Owner's separate Subcontractors who are working on site at same time as Contractor. Contractor's profit and overhead, and any supervisory labor will not be credited back to Owner with any deductive Change Orders (work deleted from Agreement by Owner).

■ The first paragraph of this clause is important because it clearly states that any change to the Scope of Work involving extra labor or materials will constitute grounds for a Change Order. The Contractor has the discretion to not charge for a minor change. However, the standard for when he can execute a Change Order is clearly set forth by this section — any time a deviation from the Scope of Work in the Agreement requires additional labor or materials. The Contractor must be careful to always put his Change Orders in writing, pursuant to this section of the Agreement. *Do not* rely on verbal Change Orders!

The second paragraph is also very important. It states that the Contractor will charge 20% profit and overhead on any Subcontractors the Owner wants to have working on the job at the same time the General Contractor is working on the job. You may or may not want to include this clause on all your jobs. Ordinarily, a Contractor will spend just as much time coordinating an Owner's separate Subcontractors as he will his own Subcontractors, so I want the Owner to know that we charge for this. If the Owner wants to avoid this charge, he always has the option of bringing in his own Subcontractors either before we commence work on the project, or after we have completed all of our work on the project.

The sentence that states, "Contractor's profit and overhead, and any supervisory labor will not be credited back to Owner with any deductive Change Orders" is critical because it allows the Contractor to keep the profit and overhead on work that is deleted by the Owner after the Agreement has been signed. Remember, you have already spent considerable time estimating and planning this work.

This can be worth a great deal of money if the Owner (after relying on your contract to obtain financing), decides to delete from your Scope of Work such items as: furnishing windows and doors, cabinets, floor coverings, tile work, paint work, etc., expecting you to credit back all the profit and overhead on this work and any related supervisory labor as well. A $30,000 deletion in work can cost the Contractor between $4,500 and $6,000 in lost profit and overhead.

INITIAL
CC
HH

D. PAYMENT SCHEDULE AND PAYMENT TERMS

1. PAYMENT SCHEDULE:

* First Payment: $1,000 or 10% of contract amount (whichever is less) due when
Agreement is signed and returned to Contractor: $ 1,000

* Second Payment (Materials Deposits): **40% of contract amount due upon
completion of demolition, delivery of materials to job
site, and commencement of framing work:** $ 5,552

* Third Payment: **25% of contract amount due upon completion of framing
work, installation of windows, and sheathing of roof:** $ 3,470

* Fourth Payment: **15% of contract amount due upon completion of
gutters and roofing work:** $ 2,082

* Final Payment: Balance of contract amount due upon Substantial Completion of all
work under contract: $ 1,776

■ The Payment Schedule is one of the most critical parts of any Agreement. It is also regulated by state laws in certain states. The way you draft your Payment Schedule says a lot about how you operate your business. For this reason, I like to keep my payment requests just about equivalent to the amount of work that's been performed under the Agreement. However, a 10% deposit or $1,000 (whichever is less) is always my first contract deposit payment.

Next, I may require a materials deposit for specialized subtrade work such as granite or woodstoves or other special-order items that require a deposit from us. (Beyond that, I don't require materials deposits.) Next, I break down the job into several smaller payments rather than having just a couple of large payments. The final payment should be a fairly small percentage of the contract amount. Leaving a large amount of money owing for the final payment can invite trouble! By the same token, taking up front money for work not completed can weaken the trust the Owner has in the Contractor and can also invite its own set of problems for the Contractor.

I make the Final Payment due upon "Substantial Completion" of the work. "Substantial Completion" is defined as being the point at which the Building/Work of Improvement is suitable for its intended use, or the issuance of an Occupancy Consent, or final building department approval from the city or county building department, whichever occurs first. On smaller jobs, I usually wait until the job is 100% complete to invoice the final payment. However, on larger jobs with a several thousand dollar final draw, if there is a minor Punch List owing, I prefer to invoice the final payment upon "substantial completion." Then the Owner can hold back 150% of the value of the Punch List items from the final contract payment.

INITIAL
CC
HH

2. PAYMENT OF CHANGE ORDERS: Payment for each Change Order is due upon completion of Change Order work and submittal of invoice by Contractor.

3. ADDITIONAL PAYMENTS FOR ALLOWANCE WORK AND RELATED CREDITS: Payment for work designated in the Agreement as **ALLOWANCE** work has been initially factored into the Lump Sum Price and Payment Schedule set forth in this Agreement. If the actual cost of the **ALLOWANCE** work exceeds the line item **ALLOWANCE** amount in the Agreement, the difference between the cost and the line item **ALLOWANCE** amount stated in the Agreement will be written up by Contractor as a Change Order subject to Contractor's profit and overhead at the rate of __20__ %.

If the cost of the **ALLOWANCE** work is less than the **ALLOWANCE** line item amount listed in the Agreement, a credit will be issued to Owner after all billings related to this particular line item **ALLOWANCE** work have been received by Contractor. This credit will be applied toward the final payment owing under the Agreement. Contractor profit and overhead and any supervisory labor will not be credited back to Owner for **ALLOWANCE** work.

> ■ It's very important to have a clear written explanation of how an ALLOWANCE item works. The statement above explains that any overage on an ALLOWANCE item will be subject to profit and overhead. But, the Owner will not receive a credit of profit and overhead for any underage on ALLOWANCE items.
>
> I don't credit back this profit and overhead for the same reason I don't credit back profit and overhead on deleted work — the Contractor has often already done a great deal of work bidding and arranging to have the work done, and has already earned a good part of this profit and overhead. It also discourages the Owner from deciding to be an Owner/Builder in the middle of his project after the Contractor has already done most of the work.

E. WARRANTY

Contractor provides a limited warranty on all Contractor- and Subcontractor-supplied labor and materials used in this project for a period of one year following substantial completion of all work.

No warranty is provided by Contractor on any materials furnished by the Owner for installation. No warranty is provided on any existing materials that are moved and/or reinstalled by the Contractor within the dwelling (including any warranty that existing/used materials will not be damaged during the removal and reinstallation process). One year after substantial completion of the project, the Owner's sole remedy (for materials and labor) on all materials that are covered by a manufacturer's warranty is strictly with the manufacturer, not with the Contractor.

Repair of the following items is specifically excluded from Contractor's warranty: Damages resulting from lack of Owner maintenance; damages resulting from Owner abuse or ordinary wear and tear; deviations that arise such as the minor cracking of concrete, stucco and plaster; minor stress fractures in drywall due to the curing of lumber; warping and deflection of wood; shrinking/cracking of grouts and caulking; fading of paints and finishes exposed to sunlight.

INITIAL
CC
HH

THE EXPRESS WARRANTIES CONTAINED HEREIN ARE IN LIEU OF ALL OTHER WARRANTIES, EXPRESS OR IMPLIED, INCLUDING ANY WARRANTIES OF MERCHANTABILITY, HABITABILITY, OR FITNESS FOR A PARTICULAR USE OR PURPOSE. THIS LIMITED WARRANTY EXCLUDES CONSEQUENTIAL AND INCIDENTAL DAMAGES AND LIMITS THE DURATION OF IMPLIED WARRANTIES TO THE FULLEST EXTENT PERMISSIBLE UNDER STATE AND FEDERAL LAW.

> ■ It's a good idea to state the scope of your company's warranty on work performed. State any limitations to your warranty — such as excluding warranty on Owner-supplied materials.

F. WORK STOPPAGE, TERMINATION OF CONTRACT FOR DEFAULT, AND INTEREST

Contractor shall have the right to stop all work on the project and keep the job idle if payments are not made to Contractor in accordance with the Payment Schedule in this Agreement, or if Owner repeatedly fails or refuses to furnish Contractor with access to the job site and /or product selections or information necessary for the advancement of Contractor's work. Simultaneous with stopping work on the project, the Contractor must give Owner written notice of the nature of Owner's default and must also give the Owner a 14-day period in which to cure this default.

If work is stopped due to any of the above reasons (or for any other material breach of contract by Owner) for a period of 14 days, and the Owner has failed to take significant steps to cure his default, then Contractor may, without prejudicing any other remedies Contractor may have, give written notice of termination of the Agreement to Owner and demand payment for all completed work and materials ordered through the date of work stoppage, and any other loss sustained by Contractor, including Contractor's Profit and Overhead at the rate of __20__ % on the balance of the incomplete work under the Agreement. Thereafter, Contractor is relieved from all other contractual duties, including all Punch List and warranty work.

> ■ The Owner's primary duty under a Construction Agreement is to pay for properly performed work in accordance with the payment schedule in the Agreement. What if the Owner fails to do this and yet still expects the Contractor to keep working? Whether or not the failure to pay money amounts to a major breach of contract — one that will justify your stopping work and considering the Agreement terminated by the Owner's failure to pay — will depend upon the particular facts of your situation.
>
> However, by having a termination clause like the one above, the Contractor will have a much better chance of being able to stop work, consider the Agreement terminated, and have a contractual basis for collecting interest on unpaid amounts owed to him along with his lost profit and overhead on the amount of work left to complete at the time of the Owner's breach of contract.

INITIAL
CC
HH

G. DISPUTE RESOLUTION AND ATTORNEY'S FEES

Any controversy or claim arising out of or related to this Agreement involving an amount of *less* than $5,000 (or the maximum limit of the court) must be heard in the Small Claims Division of the Municipal Court in the county where the Contractor's office is located. Any controversy or claim arising out of or related to this Agreement which is over the dollar limit of the Small Claims Court must be settled by binding arbitration administered by the American Arbitration Association in accordance with the Construction Industry Arbitration Rules. Judgment upon the award may be entered in any Court having jurisdiction thereof.

The prevailing party in any legal proceeding related to this Agreement shall be entitled to payment of reasonable attorney's fees, costs, and expenses.

▪ Whether a dispute will be resolved in the normal court system or through binding arbitration will be determined by your Agreement. For small disputes, I prefer a Small Claims Court where you can ordinarily get a court date within a few months, where the maximum dollar limit is between $2,500 and $5,000, and where you can't take a lawyer. It's fast (usually less than a 20-minute hearing in my area), easy, and cheap.

Very limited appeals, if any, are available (depending upon whether you are the plaintiff or defendant). You can't take a lawyer into a Small Claims Court proceeding in many jurisdictions, but prior to the hearing you can consult a lawyer familiar with construction law to help you identify the legal issues and prepare an oral outline to follow. He can also help you come up with a written statement of your position so you can present your case well and thereby increase your odds of receiving a judgment in your favor.

For disputes over the jurisdictional limits of the local Small Claims Court, I prefer binding arbitration through either privately selected arbitrators or through the American Arbitration Association (AAA). AAA arbitration is relatively fast (usually several months from application to completion of hearing — which is ordinarily much faster than the court system), perhaps a bit cheaper than the court system (much cheaper if the court system involves appeals), and allows you the option of bringing an attorney to the arbitration hearing.

In addition, the rules of evidence are much more relaxed in arbitration. The hearing is less formal than the court system hearing and a big advantage is that you may end up with an arbitrator who is familiar with the construction business.

One *disadvantage* to arbitration is that some arbitrators have a tendency to "split the difference" if they don't feel the case is clearly in favor of one side or the other. This makes it important to get legal representation (or at least advice) *prior* to going into an arbitration hearing. The failure to get some legal advice has made many a contractor unhappy when they open

INITIAL
CC
HH

up the letter that contains the decision against them! Once again, an ounce of prevention...

Attorney's fees are ordinarily awarded only if they have been agreed to in the Construction Agreement, although the AAA arbitrator has the authority to award attorney's fees and costs. An attorney's fees clause is a very good idea if you do good work and don't think you will be in the wrong if a dispute arises. Being able to tell the other side, "You'll not only have to pay what you owe me, but also my attorney's fees," can be a helpful negotiating lever when a dispute arises.

H. EXPIRATION OF THIS AGREEMENT

This Agreement will expire 30 days after the date at the top of page one of this Agreement if not first accepted in writing by Owner.

■ You might furnish a signed Agreement to an Owner and then not hear back from that Owner for several months. Perhaps during that time, a low-priced Subcontractor you relied on in your bid has left the area and the price of lumber has gone up 12%, so you may not want to be bound to the old price. The clause above gives you the option to consider the Agreement or offer to contract expired if not signed (accepted) by the Owner within 30 days of his receipt of that Agreement.

I. ENTIRE AGREEMENT

This Agreement represents and contains the entire agreement between the parties. Prior discussions or verbal representations by the parties that are not contained in this Agreement are not a part of this Agreement.

■ The clause above is necessary because it states that the agreement of the parties is limited to what is actually in the written contract. Pre-contract signing or verbal representations from the Contractor to the Owner or from the Owner to the Contractor that are not included in the Agreement are not legal and binding parts of the Agreement.

This clause reduces the possibility of the Owner coming back to you in the paint phase, for instance, and saying, "You told me that if I gave you the job, you'd consider changing all the paint-grade trim, casings, and baseboards to stain-grade oak; now do it!" In this type of situation the Contractor can point to this clause and say that the contract contained their entire Agreement and did not include the extra work the Owner is now demanding.

J. ADDITIONAL LEGAL NOTICES REQUIRED BY STATE OR FEDERAL LAW

See page(s) attached: __X__ Yes _____ No

INITIAL
CC
HH

■ Be sure to include all notices required by your Contractor's state license board or other state or federal agency which governs home improvement contracts entered into by a Prime Contractor and a Homeowner! Failure to include these notices in some states subjects the Contractor to a civil penalty fine and a much more expensive disciplinary bond at the time of bond renewal. About 90% of the contracts I see don't include these notices, but should!

If you contract directly with a Homeowner who uses the house as his personal residence, consider yourself a Prime Contractor who needs to include any and all required notices.

The purpose of these notices is primarily to inform the Homeowner of his rights and legal exposure when contracting for residential home improvements.

These notices can be changed by state legislatures on a regular basis. Also, these required notices vary from state to state. Consult your construction attorney and your state agency that governs Contractors to find out what requirements apply in your state for your type of contracting business.

For purposes of this Agreement, the "attached page" is not included here, but you should be sure to check "Yes," above, and attach the appropriate legal notices.

K. ADDITIONAL TERMS AND CONDITIONS

See page(s) attached: _____ Yes __X__ No

■ The clause above has been placed in the Agreement to remind you that you are acting as your own attorney in using these agreements and to remind you that the forms as presented are a starting point and not necessarily entirely suited in their present form to your purpose without some additional contract language or modifications.

This clause specifically indicates that you may need to add additional clauses to the Agreement based on the unique needs of your business, the disposition of the Owner, or the particular job that is the subject of the contract. The additional contract language you may need to add is beyond the scope of any form agreement in this book.

You may also decide to delete certain clauses from the Agreement depending upon the same factors mentioned above. If you work with a word processor, modifications to the Agreement will be fast and simple to make. Simply add (or delete) any clauses necessary to your Agreement and then delete the last clause, "ADDITIONAL TERMS AND CONDITIONS." Remember to consult an attorney familiar with construction before making significant changes to any agreement.

INITIAL
CC
HH

I have read and understood, and I agree to, all the terms and conditions contained in the Agreement above.

> ■ You need a statement which indicates that the parties have read and agree to the terms and conditions of the Agreement. Be sure to have the Agreement signed by the Owner *prior* to the time you commence work. Make sure you keep a signed copy of the Agreement in your records — an unsigned copy won't do you any good later on if you have a dispute. Finally, have the Owner initial every page of your Agreement, including any supplemental attachments, such as materials or Subcontractor bids.

Date: _5/22/01_   *CHARLIE CONTRACTOR, PRESIDENT*
                          CHARLIE CONTRACTOR, PRESIDENT
                          CHARLIE CONTRACTOR CONSTRUCTION, INC.

> ■ If your business is a corporation, be sure to sign the Agreement using your corporate title and place the word, "Inc." after your company's name. If you fail to do this, you may have personal liability under the Agreement.

Date: _5/22/01_   *Harry Homeowner*
                          HARRY HOMEOWNER

Date: _5/22/01_   *Helen Homeowner*
                          HELEN HOMEOWNER

> ■ Be sure you have an Agreement in your files which is signed by the Owner and contains the Owner's initials on each page.

The medium-form fixed price (lump sum) agreement is good to use on medium-sized jobs and on larger jobs where you have established very good working relationships with the owners. The medium-form agreement contains all the beneficial language found in the short-form agreement, lengthens some, and adds several new clauses more appropriate to larger jobs.

A clause has been added (III.D.2) to clarify how changes in the work will be handled and charged to the customer. This educates the owners about the financial impact of their changing their minds in the middle of a project.

Another clause about changes (III.D.4) contains expanded language allowing you to charge for changes required by code officials, which are not uncommon on larger projects. This is particularly important if the approved set of plans requires additional work that was not shown in your bid set of plans.

The Conflict of Documents clause (III. G) clarifies which document governs if there is an intentional or unintentional conflict. For example, if your scope of work states that the windows are metal, but the plans show wood windows, you are only bound to provide metal, which is what you bid on.

An abbreviated clause on the matching of existing finishes (III.H) has been added to specify a reasonable procedure for handling this subjective and often contentious issue.

The medium-form fixed price agreement works well on both residential and light-commercial remodeling jobs. With some customization, it can also be used for new construction with owners with whom you have a very good working relationship and high level of trust.

Keep in mind that there is no absolute rule to help you decide which length agreement to use. Review the long-form agreement to see if it is more appropriate or if any of the language should be incorporated into your medium-form agreement. Generally, the more costly and complicated the job, the longer the agreement.

1.2 ∎ Medium-Form Fixed Price Agreement

CONTRACTOR'S NAME: _____

ADDRESS: _____

_____

PHONE: _____

FAX: _____

LIC #: _____

DATE: _____

OWNER'S NAME: _____

ADDRESS: _____

_____

PROJECT ADDRESS: _____

_____

I. PARTIES

This contract (hereinafter referred to as "Agreement") is made and entered into on this _____ day of _____ , 19_____ , by and between _____ , (hereinafter referred to as "Owner"); and _____ , (hereinafter referred to as "Contractor"). In consideration of the mutual promises contained herein, Contractor agrees to perform the following work:

II. GENERAL SCOPE OF WORK DESCRIPTION

_____

_____

_____

_____

_____

_____

_____

_____

(Additional Scope of Work page(s) attached: _____ Yes _____ No)

LUMP SUM PRICE FOR ALL WORK ABOVE: $_____

III. GENERAL CONDITIONS FOR THE AGREEMENT ABOVE

A. EXCLUSIONS
This Agreement does *not* include *labor or materials* for the following work (unless Owner selects one of these items as an Additional Alternate):

1. PROJECT SPECIFIC EXCLUSIONS:

_____

_____

_____

_____

_____

_____

2. STANDARD EXCLUSIONS: Unless specifically included in the "General Scope of Work" section above, this Agreement does *not* include *labor or materials* for the following work: Plans, engineering fees, or governmental permits and fees of any kind. Testing, removal and disposal of any materials containing asbestos (or any other hazardous material as defined by the EPA). Custom milling of any wood for use in project. Moving Owner's property around the site. Labor or materials required to repair or replace any Owner-supplied materials. Repair of concealed underground utilities not located on prints or physically staked out by Owner which are damaged during construction. Surveying that may be required to establish accurate property boundaries for setback purposes (fences and old stakes may not be located on actual property lines). Final construction cleaning (Contractor will leave site in "broom swept" condition). Landscaping and irrigation work of any kind. Temporary sanitation, power, or fencing. Removal of soils under house in order to obtain 18 inches (or code-required height) of clear space between bottom of joists and soil. Removal of filled ground or rock or any other materials not removable by ordinary hand tools (unless heavy equipment is specified in Scope of Work section above), correction of existing out-of-plumb or out-of-level conditions in existing structure. Correction of concealed substandard framing. Rerouting/removal of vents, pipes, ducts, structural members, wiring or conduits, steel mesh which may be discovered in the removal of walls or the cutting of openings in walls. Removal and replacement of existing rot or insect infestation. Failure of surrounding part of existing structure, despite Contractor's good faith efforts to minimize damage, such as plaster or drywall cracking and popped nails in adjacent rooms, or blockage of pipes or plumbing fixtures caused by loosened rust within pipes. Construction of continuously level foundation around structure (if lot is sloped more than 6 inches from front to back or side to side, Contractor will step the foundation in accordance with the slope of the lot). Exact matching of existing finishes. Public or private utility connection fees. Repair of damage to roadways, driveways, or sidewalks that could occur when construction equipment and vehicles are being used in the normal course of construction.

B. DATE OF WORK COMMENCEMENT AND SUBSTANTIAL COMPLETION
Commence work:_____ Construction time through substantial completion: Approximately _____ to _____ weeks/months, *not* including delays and adjustments for delays caused by: inclement weather, accidents, additional time required for performance of Change Order work (as specified in each Change Order), delays caused by Owner, and other delays unavoidable or beyond the control of the Contractor.

C. EXPIRATION OF THIS AGREEMENT

This Agreement will expire 30 days after the date at the top of page one of this Agreement if not accepted in writing by Owner and returned to Contractor within that time.

D. CHANGE ORDERS: CONCEALED CONDITIONS, ADDITIONAL WORK, AND CHANGES IN THE WORK

1. CONCEALED CONDITIONS: This Agreement is based solely on the observations Contractor was able to make with the structure in its current condition at the time this Agreement was bid. If additional concealed conditions are discovered once work has commenced which were *not* visible at the time this proposal was bid, Contractor will stop work and point out these unforeseen concealed conditions to Owner so that Owner and Contractor can execute a Change Order for any Additional Work.

2. CHANGES IN THE WORK: During the course of the project, Owner may order changes in the work (both additions and deletions). The cost of these changes will be determined by the Contractor and the cost of this Additional Work will be added to Contractor's profit and overhead at the rate of _____% in order to arrive at the net amount of any Additional Change Order Work.

Contractor to supervise, coordinate, and charge _____% profit and overhead on Owner's separate Subcontractors who are working on site at same time as Contractor. Contractor's profit and overhead, and any supervisory labor will not be credited back to Owner with any deductive Change Orders (work deleted from Agreement by Owner).

3. DEVIATION FROM SCOPE OF WORK: Any alteration or deviation from the Scope of Work referred to in the Contract Documents involving extra costs of materials or labor (including any overage on **ALLOWANCE** work) will be executed upon a written Change Order issued by Contractor and should be signed by Contractor and Owner prior to the commencement of Additional Work by the Contractor. This Change Order will become an extra charge over and above the Lump Sum Price referred to at the beginning of this Agreement.

4. CHANGES REQUIRED BY PLAN CHECKERS OR FIELD INSPECTORS: Any increase in the Scope of Work set forth in these Contract Documents which is required by plan checkers or field inspectors with city or county building/planning departments will be treated as Additional Work to this Agreement for which the Contractor will issue a Change Order.

E. PAYMENT SCHEDULE AND PAYMENT TERMS

1. PAYMENT SCHEDULE:

* First Payment: $1,000 or 10% of contract amount (whichever is less) due when Agreement is signed and returned to Contractor: $ _____

* Materials Deposits: Within 2 days of submittal of invoice by Contractor, the following materials deposits are due: _____ $ _____

* Second Payment: _____

_____ $ _____

* Third Payment: _____

_____ $ _____

* Fourth Payment: _____

_____ $ _____

* Fifth Payment: _____

_____ $ _____

* Sixth Payment: _____

_____ $ _____

* Final Payment: Balance of contract amount due upon Substantial Completion of all work
under contract: $ _____

2. PAYMENT OF CHANGE ORDERS: Payment for each Change Order is due upon completion of
Change Order work and submittal of invoice by Contractor.

3. ADDITIONAL PAYMENTS FOR ALLOWANCE WORK AND RELATED CREDITS: Payment for
work designated in the Agreement as **ALLOWANCE** work has been initially factored into the Lump Sum
Price and Payment Schedule set forth in this Agreement. If the actual cost of the **ALLOWANCE** work
exceeds the line item **ALLOWANCE** amount in the Agreement, the difference between the cost and the
line item **ALLOWANCE** amount stated in the Agreement will be written up by Contractor as a Change
Order subject to Contractor's profit and overhead at the rate of _____%.

If the cost of the **ALLOWANCE** work is *less* than the **ALLOWANCE** line item amount listed in the
Agreement, a credit will be issued to Owner after all billings related to this particular line item
ALLOWANCE work have been received by Contractor. This credit will be applied toward the final
payment owing under the Agreement. Contractor profit and overhead and any supervisory labor will *not*
be credited back to Owner for **ALLOWANCE** work.

F. WARRANTY

Contractor provides a limited warranty on all Contractor- and Subcontractor-supplied labor and materials
used in this project for a period of one year following substantial completion of all work.

No warranty is provided by Contractor on any materials furnished by the Owner for installation. No warranty
is provided on any existing materials that are moved and/or reinstalled by the Contractor within the dwelling
(including any warranty that existing/used materials will not be damaged during the removal and reinstallation
process). One year after substantial completion of the project, the Owner's sole remedy (for materials and
labor) on all materials that are covered by a manufacturer's warranty is strictly with the manufacturer, not with
the Contractor.

Repair of the following items is specifically excluded from Contractor's warranty: Damages resulting from lack of Owner maintenance; damages resulting from Owner abuse or ordinary wear and tear; deviations that arise such as the minor cracking of concrete, stucco and plaster; minor stress fractures in drywall due to the curing of lumber; warping and deflection of wood; shrinking/cracking of grouts and caulking; fading of paints and finishes exposed to sunlight.

THE EXPRESS WARRANTIES CONTAINED HEREIN ARE IN LIEU OF ALL OTHER WARRANTIES, EXPRESS OR IMPLIED, INCLUDING ANY WARRANTIES OF MERCHANTABILITY, HABITABILITY, OR FITNESS FOR A PARTICULAR USE OR PURPOSE. THIS LIMITED WARRANTY EXCLUDES CONSEQUENTIAL AND INCIDENTAL DAMAGES AND LIMITS THE DURATION OF IMPLIED WARRANTIES TO THE FULLEST EXTENT PERMISSIBLE UNDER STATE AND FEDERAL LAW.

G. CONFLICT OF DOCUMENTS

If any conflict should arise between the plans, specifications, addenda to plans, and this Agreement, then the terms and conditions of this Agreement shall be controlling and binding upon the parties to this Agreement.

H. MATCHING EXISTING FINISHES

Where Contractor's work involves the "matching of existing finishes or materials," Contractor will use his best efforts to match existing finishes and materials. However, an exact match is not guaranteed by Contractor due to such factors as discoloration due to the aging process, difference in dye lots, and difficulty of exactly matching certain finishes, colors, and planes.

I. WORK STOPPAGE, TERMINATION OF CONTRACT FOR DEFAULT, AND INTEREST

Contractor shall have the right to stop all work on the project and keep the job idle if payments are not made to Contractor in accordance with the Payment Schedule in this Agreement, or if Owner repeatedly fails or refuses to furnish Contractor with access to the job site and /or product selections or information necessary for the advancement of Contractor's work. Simultaneous with stopping work on the project, the Contractor must give Owner written notice of the nature of Owner's default and must also give the Owner a 14-day period in which to cure this default.

If work is stopped due to any of the above reasons (or for any other material breach of contract by Owner) for a period of 14 days, and the Owner has failed to take significant steps to cure his default, then Contractor may, without prejudicing any other remedies Contractor may have, give written notice of termination of the Agreement to Owner and demand payment for all completed work and materials ordered through the date of work stoppage, and any other loss sustained by Contractor, including Contractor's Profit and Overhead at the rate of _____% on the balance of the incomplete work under the Agreement. Thereafter, Contractor is relieved from all other contractual duties, including all Punch List and warranty work.

J. DISPUTE RESOLUTION AND ATTORNEY'S FEES

Any controversy or claim arising out of or related to this Agreement involving an amount of *less* than $5,000 (or the maximum limit of the court) must be heard in the Small Claims Division of the Municipal Court in the county where the Contractor's office is located. Any controversy or claim arising out of or related to this Agreement which is over the dollar limit of the Small Claims Court must be settled by binding arbitration

administered by the American Arbitration Association in accordance with the Construction Industry Arbitration Rules. Judgment upon the award may be entered in any Court having jurisdiction thereof.

The prevailing party in any legal proceeding related to this Agreement shall be entitled to payment of reasonable attorney's fees, costs, and expenses.

K. ENTIRE AGREEMENT, SEVERABILITY, AND MODIFICATION

This Agreement represents and contains the entire agreement between the parties. Prior discussions or verbal representations by the parties that are not contained in this Agreement are *not* a part of this Agreement. In the event that any provision of this Agreement is at any time held by a Court to be invalid or unenforceable, the parties agree that all other provisions of this Agreement will remain in full force and effect. Any future modification of this Agreement must be executed in writing in order to be valid and binding upon the parties.

L. ADDITIONAL LEGAL NOTICES REQUIRED BY STATE OR FEDERAL LAW

See page(s) attached: _____Yes _____No

M. ADDITIONAL TERMS AND CONDITIONS

See page(s) attached: _____Yes _____No

I have read, and I understand and agree to, all the terms and conditions contained in the Agreement above.

Date: _____   _____
                           CONTRACTOR'S SIGNATURE

Date: _____   _____
                           OWNER'S SIGNATURE

1.2 ■ MEDIUM-FORM FIXED PRICE AGREEMENT

ANNOTATED

■ This annotated sample of the medium-form agreement refers to a commercial tenant improvement job. Because the plans did not include a finish schedule, I've included a very thorough description of what is included and not included in the Scope of Work section.

Charlie Contractor Construction, Inc.
123 Hammer Lane
Anywhere, USA 33333
Phone: (123) 456-7890
Fax: (123) 456-7899
Lic#: 11111

DATE: **May 22, 2001**

OWNER'S NAME: **Joe Tenant**
ADDRESS: **689 Commercial Plaza**
           Anywhere, USA 33333

PROJECT ADDRESS: **same**

I. PARTIES

This contract (hereinafter referred to as "Agreement") is made and entered into on this **22nd** day of **May**, 20**01**, by and between **Joe Tenant**, (hereinafter referred to as "Owner"); and **Charlie Contractor Construction, Inc.**, (hereinafter referred to as "Contractor"). In consideration of the mutual promises contained herein, Contractor agrees to perform the following work:

■ The phrase "...in consideration of the mutual promises contained herein, agrees to perform the following work:" is good boilerplate language that should remain in your Agreement. It means there is a "bargained-for exchange," which is necessary for a contract to be valid. The Owner will pay the Contractor the stated sum, and in exchange the Contractor will complete the work described in the Agreement. In the space immediately above, refer to any available plans and specifications on which you have based your bid.

INITIAL
CC
97

II. GENERAL SCOPE OF WORK DESCRIPTION

■ Provide general details of the Scope of Work here. Refer to any plans and specifications that are available. Give a very complete description of what is included for the Lump Sum Price stated below. Many disputes occur because the Contractor does not specifically describe the Scope of Work. Spelling this out can also help you remember all the items that should be included in your job costs.

Many Owners feel more comfortable with a detailed line item description of what the Contractor is including in the Scope of Work. Sometimes when two Contractors give very close bids for the same job, the Contractor who gave the more complete Scope of Work description will end up with the job.

You can also state that labor or materials are being furnished "per the attached bid from XYZ Subcontractor or Materials Supplier." This is helpful if the plans are vague and you want to make very clear to the Owner exactly what a particular Subcontractor or materials supplier has bid to you, the General Contractor.

You may also want to state that the costs of certain line items are being furnished on an ALLOWANCE basis. For instance, many Owners have not selected specific plumbing fixtures or light fixtures at the time the Agreement is entered into, so the Contractor may want to list these fixtures as ALLOWANCE ITEMS. For instance:

1. Plumbing fixtures for two bathrooms (ALLOWANCE):	$ 750
2. Electrical fixtures for entire residence (ALLOWANCE):	$2,000

When you include ALLOWANCES, be sure to include contract language that describes how any overage or underage regarding the ALLOWANCE item will be handled. Be sure to state that Contractor Profit and Overhead will be added to any amount that exceeds the ALLOWANCE line item amount stated in the Agreement. The clauses in this manual relating to ALLOWANCES make this clear, however, you may need to verbally explain how this works to the Owner.

1. Misc. clean-up labor and dumpsters (does not include final project cleaning): $ 450

2. All paint and paint preparation work — by Owner: not included

3. Protection of carpets. Owner to install 6-mil plastic which is taped at all edges and seams. Owner to monitor and maintain this protective plastic on floors so that carpet is not damaged by construction process: not included

INITIAL
CC
97

* Contractor strongly recommends that Owner remove all existing
floor coverings prior to commencement of any work by Contractor
in order to prevent damage to existing floor coverings. Owner
should contact Landlord to see if he will pay to remove and
reinstall these carpets so Contractor can frame directly on top
of concrete slab, not ruin carpets, save Owner the expense of
protecting the carpets throughout construction, and inspect
condition of concrete slab floor.

4. Demolition of existing walls and removal (not reinstallation)
   of iron interior railing in lobby: $ 75

5. Furnish framing, drywall, tape, texture (Landlord to frame in
   and drywall 21 feet of demising wall at rear of Owner's space): $ 19,002

6. Furnish and install rough electrical (ALLOWANCE): $ 3,500

7. Electrical fixtures; Owner to specify 8 wall sconces, 2 emergency
   lights. (Contractor recommends that wall sconces be fluorescent
   and that a night lighting circuit be wired. Contractor recommends
   existing track lights be installed only in stage/theater room and
   lobby area. Recommend new fixtures in all other locations.
   Refer to utility company rebate information furnished to Owner.)
   The fixture ALLOWANCE does not reflect the cost of these new
   fixtures or the rebate for the new fixtures available from utility
   company. In addition, the figures above do not reflect the
   considerable monthly savings in electricity the Owner would
   experience by using new energy-efficient fixtures. All electrical
   on existing walls and ceilings will be run via surface-mounted
   EMT conduit. Contractor still needs electrical plan from Owner.
   Contractor can design this plan with Owner if Contractor is
   awarded contract for this work. Owner should contact Landlord
   to see if he will split the cost of new energy-efficient light
   fixtures and labor to install them. (ALLOWANCE): $ 1,250

8. Furnish rough and finish plumbing required for installation of
   sink in lounge. (Concrete cutter must core and drill slab. Drain
   must run through basement area and connect to waste line in
   basement area. Includes basic stainless-steel sink w/chrome faucet.

INITIAL
CC
97

Hot and cold water will be taken from water heater closet next to lounge area. Owner to verify with Landlord that it is okay to take water from the water heater closet area.) (ALLOWANCE): $ 1,250

9. Furnish and install labor and materials to run new gas line from Owner's meter to two new furnaces in furnace closet. Remove existing ceiling-mounted furnace and cap off gas line (existing flue to remain): $ 1,250

10. Furnish and install heating system w/exposed ducts (also includes venting break room area exhaust fan): $ 17,223

11. Furnish and install oak base cabinet (lounge) and 6-foot oak upper cabinet: $ 1,158

12. Furnish and install Formica countertop in break area: $ 478

13. Furnish and install 12 doors (Standard wood jamb [fingerjoint], 20-minute fire rated and labeled solid core birch [labeled: 20-minute], paint grade doors [eleven 3'0"x6'8" doors]; and one pair of 3'0"x6'8" double doors — same specification. Includes door casing [paint grade] around all new doors, door sweeps. Architect to verify code conformance of doors above in one-hour corridor. Smoke gasket included. This is a rated door/jamb assembly.): $ 3,886

14. Furnish and install 13 lever handsets for doors above (4 locking and 8 passage, 1 dummy handset, oil-rubbed bronze finish): $ 1,275

15. Furnish and install 13 door closers (Cal Royal, medium duty, bronze finish): $ 1,093

16. Supervisory labor: $ 1,500

17. Furnish and install 6 windows (1/4-inch clear wire safety glass set in metal jamb assembly and metal casing. 3 @ 2x5 feet; 1 @ 4x3 feet; 2 @ 6x5 feet. Top of window at header height. Architect to verify code conformance of windows in one-hour corridor. This assembly is rated at 45 minutes. Approximately 10- to 14-day lead time on these windows.): $ 5,482

18. Scaffold rental: $ 275

19. Oak floor in dance academy. (Glue down 12x12-inch oak parquet flooring, sanded and finished. Glue down over existing slab. Does not include floor preparation required due to irregularities in

INITIAL
CC
97

slab that may appear once existing carpet is removed. Does not

include reducers or moisture sealing slab. Slab moisture

could warp floor.): $ 1,940

20. Subtotal: $ 61,087

21. Contractor Profit and Overhead @ 20%: $ 12,217

(Additional Scope of Work page(s) attached: _____ Yes __x__ No)

LUMP SUM PRICE FOR ALL WORK ABOVE: $ 73,304

> ■ The Contractor may or may not want to show his profit and overhead and any supervisory labor as a separate line item cost above the Lump Sum Price.

ALTERNATES

1. For air conditioning added to heating system above (does not

   include electrical): $ 4,800

2. Furnish and install MDF 3 1/4-inch colonial baseboard throughout

   new area (primed, but not painted, approx. 950 linear feet): $ 1,522

3. Furnish and install 4-inch rubber topset baseboard throughout

   new area: $ 1,750

4. Furnish and install 2 1/2-inch rubber topset baseboard: $ 1,500

5. Furnish and install prefinished 1 7/8-inch walnut baseboard: $ 1,100

6. Insulate walls (for sound) around director's office, computer

   lab, dance academy, accounting office (ALLOWANCE): $ 300

7. Furnish and install R/C channel for sound-insulating purposes

   around dance academy (ALLOWANCE): $ 200

8. Paint all interior walls, doors, and casings with Kelly Moore

   or equal: $ 3,800

> ■ Sometimes an Owner wants to know the cost of alternate work, but doesn't want it placed in the bid. You can show this type of work as "Alternates" in this location without factoring the cost of the alternate items into your lump sum bid.

III. GENERAL CONDITIONS FOR THE AGREEMENT ABOVE

INITIAL
CC
97

A. EXCLUSIONS

This Agreement does *not* include *labor or materials* for the following work (unless Owner selects one of these items as an Additional Alternate):

> ■ The Exclusions section of your Agreement is just as important as the Scope of Work section. Unless you specifically exclude areas of work that you know you are not going to perform, the Owner might assume they are included.
>
> Contractors lose a great deal of money by failing to exclude work. The Contractor later "eats" items that fall into "gray" areas and weren't specifically excluded in order to get his final check from the Owner. If the Exclusions section is complete, the Contractor will "eat" far fewer of these items. In my experience, paying attention to Exclusions can save a medium-sized Contractor several thousand dollars each year. If you use a computer, simply delete the items that clearly don't apply to your particular project. Or you can line them out (and initial them) if you use this form as is. Add any Exclusions to your Agreement that relate specifically to your business or to the job you will performing. The Exclusions list provided in the form is a good starting point, but it may not be 100% complete for your purposes.

1. PROJECT SPECIFIC EXCLUSIONS:

Exterior concrete flatwork of any kind; landscape and irrigation work.

Plans, permits, insulation, doorstops, hot tar roof patching (cold patch of flue vents is included), floor covering repair or work of any kind, carpet cleaning, protection of carpets or existing floor coverings. Replacement of existing door or hardware to HVAC room. Fire sprinkler work of any kind.

Wallpaper removal and repair of walls where wallpaper has been removed.

Replacement of 21 feet of demising wall (framing and drywall) at rear of space adjacent to lounge area. Work of any kind on loft in reception area.

Additional drywall work (including repair of existing defective texture and tape joints) of any kind in areas that are not directly affected by new framing. Computer, security, and phone wiring; paint work of any kind.

Baseboard of any kind; mirrors; grab rails; work in any bathrooms; appliances or installation of appliances; panic bar door hardware; window coverings; handicap ramp work of any kind; any work on exterior of building; signage of any kind (interior or exterior); reinstalling metal railing; construction cleaning, final cleaning, window cleaning; repair of existing defective light fixtures or plumbing

INITIAL
CC
97

fixtures (including replacing burnt out light bulbs). Any work not specifically called out in the Scope of Work section above.

■ I list the most obvious Exclusions that specifically relate to the job being contracted as "Project Specific Exclusions." The less obvious, more remote, and "boilerplate" Exclusions are listed below in "Standard Exclusions." If working with a computer, remove any boilerplate Exclusions that don't apply to the job so that you don't scare the Owner away with unnecessary Exclusions.

2. STANDARD EXCLUSIONS: Unless specifically included in the "General Scope of Work" section above, this Agreement does *not* include *labor or materials* for the following work: Plans, engineering fees, or governmental permits and fees of any kind. Testing, removal and disposal of any materials containing asbestos (or any other hazardous material as defined by the EPA). Custom milling of any wood for use in project. Moving Owner's property around the site. Labor or materials required to repair or replace any Owner-supplied materials. Repair of concealed underground utilities not located on prints or physically staked out by Owner which are damaged during construction. Surveying that may be required to establish accurate property boundaries for setback purposes (fences and old stakes may not be located on actual property lines). Final construction cleaning (Contractor will leave site in "broom swept" condition). Landscaping and irrigation work of any kind. Temporary sanitation, power, or fencing. Removal of soils under house in order to obtain 18 inches (or code-required height) of clear space between bottom of joists and soil. Removal of filled ground or rock or any other materials not removable by ordinary hand tools (unless heavy equipment is specified in Scope of Work section above), correction of existing out-of-plumb or out-of-level conditions in existing structure. Correction of concealed substandard framing. Rerouting/removal of vents, pipes, ducts, structural members, wiring or conduits, steel mesh which may be discovered in the removal of walls or the cutting of openings in walls. Removal and replacement of existing rot or insect infestation. Failure of surrounding part of existing structure, despite Contractor's good faith efforts to minimize damage, such as plaster or drywall cracking and popped nails in adjacent rooms, or blockage of pipes or plumbing fixtures caused by loosened rust within pipes. Construction of continuously level foundation around structure (if lot is sloped more than 6 inches from front to back or side to side, Contractor will step the foundation in accordance with the slope of the lot). Exact matching of existing finishes. Public or private utility connection fees. Repair of damage to roadways, driveways, or sidewalks that could occur when construction equipment and vehicles are being used in the normal course of construction.

■ Agreeing to "match existing finishes" is a can of worms requiring such special treatment that an explanation of one approach is given in Form 1.4, Addendum for Matching Existing Finishes.

B. DATE OF WORK COMMENCEMENT AND SUBSTANTIAL COMPLETION

Commence work: **June 1, 2001.** Construction time through substantial completion: **Approximately 7 to 9 weeks**, *not* including delays and adjustments for delays caused by: inclement weather, accidents, additional time required for performance of Change Order work (as specified in each Change Order), delays caused by Owner, and other delays unavoidable or beyond the control of the Contractor.

INITIAL
CC
97

■ Some states require that the approximate start and completion dates of the project be specified in the Agreement. Every Owner also wants to know this information. However, be sure to give a *range* of days, weeks, or months to complete and not exact dates. If you think the project will take 3 months, state "3 to 4 months" whenever possible so you allow a cushion for delays. Numerous events can increase the time needed to complete the Work. Be sure to add extra contract days to all of your Change Orders in the space provided.

C. EXPIRATION OF THIS AGREEMENT

This Agreement will expire 30 days after the date at the top of page one of this Agreement if not accepted in writing by Owner and returned to Contractor within that time.

■ You may furnish an Owner with a signed Agreement and not hear back from him for several months. Perhaps during that time, a low-priced Subcontractor you relied on in your bid has left the area and the price of lumber has gone up 12%. So you do not want to be bound to the old price. The clause above gives you the option to consider the Agreement expired if not signed (accepted) by the Owner within 30 days of his receipt of that Agreement.

D. CHANGE ORDERS: CONCEALED CONDITIONS, ADDITIONAL WORK, AND CHANGES IN THE WORK

1. CONCEALED CONDITIONS: This Agreement is based solely on the observations Contractor was able to make with the structure in its current condition at the time this Agreement was bid. If additional concealed conditions are discovered once work has commenced which were *not* visible at the time this proposal was bid, Contractor will stop work and point out these unforeseen concealed conditions to Owner so that Owner and Contractor can execute a Change Order for any Additional Work.

■ Every Agreement should have a Concealed Conditions clause similar to the one above — especially on remodeling jobs. Occasionally Murphy's Law will create some very unforeseeable extra work on projects. If this work is not specifically excluded in the "Exclusions" section of your Agreement, you may need to rely on the "Concealed Conditions" clause in order to be able to write up the additional unforeseen work as a Change Order.

Examples of Concealed Conditions could be subterranean concrete that interferes with your foundation work, spring water that prevents you from pouring concrete, pipes and conduits that interfere with the installation of new doors and windows, asbestos or other toxic compounds discovered during demolition, omitted framing members, etc.

INITIAL
CC
97

2. CHANGES IN THE WORK: During the course of the project, Owner may order changes in the work (both additions and deletions). The cost of these changes will be determined by the Contractor and the cost of this Additional Work will be added to Contractor's profit and overhead at the rate of __20__ % in order to arrive at the net amount of any Additional Change Order Work.

Contractor to supervise, coordinate, and charge __20__ % profit and overhead on Owner's separate Subcontractors who are working on site at same time as Contractor. Contractor's profit and overhead, and any supervisory labor will not be credited back to Owner with any deductive Change Orders (work deleted from Agreement by Owner).

> ■ The clause above simply states that the Owner can both add and delete work. For additional work, the cost will be determined by the Contractor and will be subject to Contractor's profit and overhead at the rate of 20%. Even if the profit and overhead rate in the primary Agreement is as low as 10%, charging 20% profit and overhead is ordinarily justified and necessary as it takes extra time to explain the Change Order to the Owner, wait for him to sign it, and then execute the Change Order work.
>
> The statement "Contractor will supervise, coordinate, and charge 20% profit and overhead on Owner's separate Subcontractors who are working on site at same time as Contractor," is very important. You may or may not want to include this clause on all your jobs. Ordinarily, a Contractor will spend just as much time coordinating an Owner's separate Subcontractors as he will his own Subcontractors, so I want the Owner to know that we charge for this. If the Owner wants to avoid this charge, he always has the option of bringing in his own Subcontractors either before we commence work on the project, or after we have completed all of our work on the project.
>
> The sentence that states, "Contractor's profit and overhead, and any supervisory labor will not be credited back to Owner with any deductive Change Orders" is critical because it allows the Contractor to keep the profit and overhead on work that is deleted by the Owner after the Agreement has been signed. Remember, you have already spent considerable time estimating and planning this work.
>
> This can be worth a great deal of money if the Owner (after relying on your contract to obtain financing), decides to delete from your Scope of Work such items as: furnishing windows and doors, cabinets, floor coverings, tile work, paint work, etc., expecting you to credit back all the profit and overhead on this work and any related supervisory labor as well. A $30,000 deletion in work can cost the Contractor between $4,500 and $6,000 in lost profit and overhead.

3. DEVIATION FROM SCOPE OF WORK: Any alteration or deviation from the Scope of Work referred to in the Contract Documents involving extra costs of materials or labor

INITIAL
CC
97

(including any overage on **ALLOWANCE** work) will be executed upon a written Change Order issued by Contractor and should be signed by Contractor and Owner prior to the commencement of Additional Work by the Contractor. This Change Order will become an extra charge over and above the Lump Sum Price referred to at the beginning of this Agreement.

> ■ The clause above is important because it clearly states that any change to the Scope of Work involving extra labor or materials will constitute grounds for a Change Order. The Contractor has the discretion to not charge for a minor change. However, the standard for when he can execute a Change Order is clearly set forth by this section — any time a deviation from the Scope of Work in the Agreement requires additional labor or materials. The Contractor must be careful to always put his Change Orders in writing, pursuant to this section of the Agreement. *Do not* rely on verbal Change Orders!

4. CHANGES REQUIRED BY PLAN CHECKERS OR FIELD INSPECTORS: Any increase in the Scope of Work set forth in these Contract Documents which is required by plan checkers or field inspectors with city or county building/planning departments will be treated as Additional Work to this Agreement for which the Contractor will issue a Change Order.

> ■ This exclusion can be very important to the Contractor. Often you will bid the job before the final approved copy of the job plan is issued. If the approved set of plans requires additional work that was not shown in your bid set of plans, this additional work is excluded from your Agreement and will be written up as a Change Order. In addition, if a field building inspector decides after inspecting the site that additional work not shown on the plans is required, this additional work is also excluded from your agreement and will also be written up as a Change Order.

INITIAL
CC
97

E. PAYMENT SCHEDULE AND PAYMENT TERMS

1. PAYMENT SCHEDULE:

* First Payment: $1,000 or 10% of contract amount (whichever is less) due when Agreement is signed and returned to Contractor: $ **1,000**

* Materials Deposits: Within 2 days of submittal of invoice by Contractor, the following materials deposits are due: _____ **None Due**

* Second Payment: **15% of contract amount due upon completion of demolition work, delivery of framing materials to site, and commencement of framing work:** $ **10,995**

* Third Payment: **20% of contract amount due upon completion**
of rough framing work: **$ 14,660**

* Fourth Payment: **30% of contract amount due upon completion of**
rough plumbing, rough electrical, installation of rough
mechanical work, and delivery of drywall to site: **$ 21,991**

* Fifth Payment: **20% of contract amount due upon completion of**
drywall installation, taping and texturing of
drywall, and installation of all doors: **$ 14,660**

* Sixth Payment: **10% of contract amount due upon installation of**
cabinets, baseboard, and countertops: **$ 7,330**

* Final Payment: Balance of contract amount due upon Substantial Completion of all
work under contract: **$ 2,668**

■ The Payment Schedule is one of the most critical parts of any Agreement. It also is regulated by state laws in certain states. The way you draft your Payment Schedule says a lot about how you operate your business. For this reason, I like to keep my payment requests just about equivalent to the amount of work that's been performed under the Agreement. However, a 10% deposit or $1,000 (whichever is less) is always my first contract deposit payment.

Next, I may require a materials deposit for specialized subtrade work such as granite or woodstoves or other special-order items that require a deposit from us. (Beyond that, I don't require materials deposits.) Next, I break down the job into several smaller payments rather than having just a couple of large payments. The final payment should be a fairly small percentage of the contract amount. Leaving a large amount of money owing for the final payment can invite trouble. By the same token, taking up-front money for work not completed can also weaken the trust the Owner has in the Contractor and can invite its own set of problems for the Contractor.

I make the Final Payment due upon "substantial completion" of the work. "Substantial completion" is defined as being the point at which the Building/Work of Improvement is suitable for its intended use, or the issuance of an Occupancy Consent, or final building department approval from the city or county building department, whichever occurs first. On smaller jobs, I usually wait until the job is 100% complete to invoice the final payment. However, on larger jobs with a several thousand dollar final draw, if there is a minor Punch List owing, I prefer to be able to invoice the final payment upon "substantial completion" and have the Owner hold back 150% of the value of the Punch List items from the final contract payment.

INITIAL
CC
97

2. PAYMENT OF CHANGE ORDERS: Payment for each Change Order is due upon completion of Change Order work and submittal of invoice by Contractor.

3. ADDITIONAL PAYMENTS FOR ALLOWANCE WORK AND RELATED CREDITS: Payment for work designated in the Agreement as **ALLOWANCE** work has been initially factored into the Lump Sum Price and Payment Schedule set forth in this Agreement. If the actual cost of the **ALLOWANCE** work *exceeds* the line item **ALLOWANCE** amount in the Agreement, the difference between the cost and the line item **ALLOWANCE** amount stated in the Agreement will be written up by Contractor as a Change Order subject to Contractor's profit and overhead at the rate of __20__%.

If the cost of the **ALLOWANCE** work is *less* than the **ALLOWANCE** line item amount listed in the Agreement, a credit will be issued to Owner after all billings related to this particular line item **ALLOWANCE** work have been received by Contractor. This credit will be applied toward the final payment owing under the Agreement. Contractor profit and overhead and any supervisory labor will *not* be credited back to Owner for **ALLOWANCE** work.

> ■ It's very important to have a clear written explanation of how an AL-LOWANCE item works. The statement above explains that any overage on an ALLOWANCE item will be subject to profit and overhead. But the Owner will not receive a credit of profit and overhead for any underage on AL-LOWANCE items.
>
> I don't credit back this profit and overhead for the same reason I don't credit back profit and overhead on deleted work — the Contractor has often already done a great deal of work bidding and arranging to have the work done, and has already earned a good part of this profit and overhead. It also discourages the Owner from deciding to be an Owner/Builder in the middle of his project after the Contractor has already done most of the work.

F. WARRANTY

Contractor provides a limited warranty on all Contractor- and Subcontractor-supplied labor and materials used in this project for a period of one year following substantial completion of all work.

No warranty is provided by Contractor on any materials furnished by the Owner for installation. No warranty is provided on any existing materials that are moved and/or reinstalled by the Contractor within the dwelling (including any warranty that existing/used materials will not be damaged during the removal and reinstallation process). One year after substantial completion of the project, the Owner's sole remedy (for materials and labor) on all materials that are covered by a manufacturer's warranty is strictly with the manufacturer, not with the Contractor.

Repair of the following items is specifically excluded from Contractor's warranty: Damages resulting from lack of Owner maintenance; damages resulting from Owner abuse or ordinary wear and tear; deviations that arise such as the minor cracking of concrete, stucco and plaster; minor stress

INITIAL
CC
97

fractures in drywall due to the curing of lumber; warping and deflection of wood; shrinking/cracking of grouts and caulking; fading of paints and finishes exposed to sunlight.

THE EXPRESS WARRANTIES CONTAINED HEREIN ARE IN LIEU OF ALL OTHER WARRANTIES, EXPRESS OR IMPLIED, INCLUDING ANY WARRANTIES OF MERCHANTABILITY, HABITABILITY, OR FITNESS FOR A PARTICULAR USE OR PURPOSE. THIS LIMITED WARRANTY EXCLUDES CONSEQUENTIAL AND INCIDENTAL DAMAGES AND LIMITS THE DURATION OF IMPLIED WARRANTIES TO THE FULLEST EXTENT PERMISSIBLE UNDER STATE AND FEDERAL LAW.

> ■ It's a good idea to state the scope of your company's warranty on work performed. State any limitations to your warranty — such as excluding warranty on Owner-supplied materials.

G. CONFLICT OF DOCUMENTS

If any conflict should arise between the plans, specifications, addenda to plans, and this Agreement, then the terms and conditions of this Agreement shall be controlling and binding upon the parties to this Agreement.

> ■ With most construction projects, the Construction Agreement, plans, any written specifications, any written addenda, and sometimes engineering reports, are all considered "contract documents" to which the Contractor is bound. The problem is that these documents may either intentionally or unintentionally contradict each other regarding specific work the Contractor is to perform.
>
> The question then becomes whether or not the Contractor is legally required to perform the work in question. I try to settle that question with the clause above because it simply states that this Agreement is the one contract document that controls the various other contract documents in the project.
>
> For instance, if the plans call out for wood windows but the Owner has directed the Contractor to bid metal windows and the Contractor specifically states metal windows in his Agreement, then metal windows are what the Contractor is bound to, even though the plans call out for wood windows. Or, if the plans call out for an item such as French drains, and the Contractor specifically excludes it, then the Contractor is not held to furnishing the item simply because it may show up somewhere as a note in one of the other contract documents. However, don't exclude too many items called for in the plans or your bid may be considered nonresponsive and rejected on that basis.

INITIAL
CC
97

H. MATCHING EXISTING FINISHES

Where Contractor's work involves the "matching of existing finishes or materials," Contractor will use his best efforts to match existing finishes and materials. However, an exact match is not guaranteed by Contractor due to such factors as discoloration due to the aging process, difference in dye lots, and difficulty of exactly matching certain finishes, colors, and planes.

> ■ Agreeing to "match existing finishes" is a can of worms requiring such special treatment that an explanation of one approach is given in Form 1.4, Addendum for Matching Existing Finishes.

I. WORK STOPPAGE, TERMINATION OF CONTRACT FOR DEFAULT, AND INTEREST

Contractor shall have the right to stop all work on the project and keep the job idle if payments are not made to Contractor in accordance with the Payment Schedule in this Agreement, or if Owner repeatedly fails or refuses to furnish Contractor with access to the job site and /or product selections or information necessary for the advancement of Contractor's work. Simultaneous with stopping work on the project, the Contractor must give Owner written notice of the nature of Owner's default and must also give the Owner a 14-day period in which to cure this default.

If work is stopped due to any of the above reasons (or for any other material breach of contract by Owner) for a period of 14 days, and the Owner has failed to take significant steps to cure his default, then Contractor may, without prejudicing any other remedies Contractor may have, give written notice of termination of the Agreement to Owner and demand payment for all completed work and materials ordered through the date of work stoppage, and any other loss sustained by Contractor, including Contractor's Profit and Overhead at the rate of __20__ % on the balance of the incomplete work under the Agreement. Thereafter, Contractor is relieved from all other contractual duties, including all Punch List and warranty work.

> ■ The Owner's primary duty under a Construction Agreement is to pay for properly performed work in accordance with the payment schedule in the Agreement. What if the Owner fails to do this and yet still expects the Contractor to keep working? Whether or not the failure to pay money amounts to a major breach of contract — one that will justify your stopping work and considering the Agreement terminated by the Owner's failure to pay — will depend upon the particular facts of your situation.
>
>   However, by having a termination clause like the one above, the Contractor will have much better grounds for stopping work and considering the Agreement terminated. He also will have a contractual basis for collecting interest on unpaid amounts owed to him along with his lost profit and overhead on the amount of work left to complete at the time of the Owner's breach of contract.

J. DISPUTE RESOLUTION AND ATTORNEY'S FEES

Any controversy or claim arising out of or related to this Agreement involving an amount of *less* than $5,000 (or the maximum limit of the court) must be heard in the Small Claims Division of the

INITIAL
CC
97

Municipal Court in the county where the Contractor's office is located. Any controversy or claim arising out of or related to this Agreement which is over the dollar limit of the Small Claims Court must be settled by binding arbitration administered by the American Arbitration Association in accordance with the Construction Industry Arbitration Rules. Judgment upon the award may be entered in any Court having jurisdiction thereof.

The prevailing party in any legal proceeding related to this Agreement shall be entitled to payment of reasonable attorney's fees, costs, and expenses.

■ Whether a dispute will be resolved in the normal court system or through binding arbitration will be determined by your Agreement. For small disputes I prefer a Small Claims Court where you can ordinarily get a court date within a few months, where the maximum dollar limit is $2,500 to $5,000, and where you can't take a lawyer. It's fast (usually less than a 20-minute hearing in my area), easy, and cheap.

Very limited appeals, if any, are available (depending upon whether you are the plaintiff or defendant). You can't take a lawyer into a Small Claims Court proceeding in many jurisdictions, but prior to the hearing, you can consult a lawyer familiar with construction law to help you identify the legal issues and prepare an oral outline to follow. He can also help you come up with a written statement of your position so you can present your case well, and thereby increase your odds of receiving a judgment in your favor.

For disputes over the jurisdictional limits of the local Small Claims Court, I prefer binding arbitration through either privately selected arbitrators or through the American Arbitration Association (AAA). AAA arbitration is relatively fast (usually several months from application to completion of hearing — which is ordinarily much faster than the court system), perhaps a bit cheaper than the court system (much cheaper if the court system involves appeals), and allows you the option of bringing an attorney to the arbitration hearing.

In addition, the rules of evidence are much more relaxed in arbitration. The hearing is less formal than the court system hearing and a big advantage is that you may end up with an arbitrator who is familiar with the construction business.

One *disadvantage* to arbitration is that some arbitrators have a tendency to "split the difference" if they don't feel the case is clearly in favor of one side or the other. This makes it important to get legal representation (or at least advice) *prior* to going into an arbitration hearing. The failure to get some legal advice has made many a contractor unhappy when they open up the letter that contains the decision against them! Once again, an ounce of prevention...

Attorney's fees are ordinarily awarded only if they have been agreed to

INITIAL
CC
97

> in the Construction Agreement, although the AAA arbitrator has the authority to award attorney's fees and costs. An attorney's fees clause is a very good idea if you do good work and don't think you will be in the wrong if a dispute arises. Being able to tell the other side, "You'll not only have to pay what you owe me, but also my attorney's fees," can be a helpful negotiating lever when a dispute arises.

K. ENTIRE AGREEMENT, SEVERABILITY, AND MODIFICATION

This Agreement represents and contains the entire agreement between the parties. Prior discussions or verbal representations by the parties that are not contained in this Agreement are *not* a part of

this Agreement. In the event that any provision of this Agreement is at any time held by a Court to be invalid or unenforceable, the parties agree that all other provisions of this Agreement will remain in full force and effect. Any future modification of this Agreement must be executed in writing in order to be valid and binding upon the parties.

> ▪ The clause above is necessary because it states that the agreement of the parties is limited to what is actually in the written contract. Pre-contract signing and verbal representations from the Contractor to the Owner or from the Owner to the Contractor that are not included in the Agreement are not legal and binding parts of the Agreement.
>
> This clause decreases the possibility of the Owner coming back to you in the paint phase, for instance, and saying, "You told me that if I gave you the job, you'd consider changing all the paint-grade trim, casings, and baseboards to stain-grade oak; now do it!" In this type of situation the Contractor can point to this clause and say that the contract contained their entire agreement and did not include the extra work the Owner is now demanding.
>
> The section above also contains a clause stating that future modifications to the Agreement must be made in writing and signed by both parties. This means the terms of the Agreement can't be verbally modified. This provides obvious protection to both parties — but the Contractor still should not make oral promises to the Owner that he may later have a hard time keeping (e.g., "This concrete sidewalk won't show a single crack for the next 20 years because I have the very best concrete man in the entire area"). Don't say this kind of thing because it may create warranty or other legal obligations that otherwise wouldn't exist.

L. ADDITIONAL LEGAL NOTICES REQUIRED BY STATE OR FEDERAL LAW

See page(s) attached: __**x**__ Yes _____ No

INITIAL
CC
97

■ Be sure to include all notices required by the Contractor's state license board or other state or federal agency which governs home improvement contracts entered into by a prime contractor and a homeowner! Failure to include these notices in some states subjects the Contractor to a civil penalty fine and a much more expensive disciplinary bond at the time of bond renewal. About 90% of the contracts I see don't include these notices, but should!

If you contract directly with a homeowner who uses the house as his personal residence, consider yourself a prime contractor who needs to include any and all required notices.

The purpose of these notices is primarily to inform the homeowner of his rights and legal exposure when contracting for residential home improvements.

These notices can be changed by state legislatures on a regular basis. Also, these required notices vary from state to state. Consult your construction attorney and your state agency that governs contractors to find out what requirements apply in your state for your type of contracting business.

M. ADDITIONAL TERMS AND CONDITIONS

See page(s) attached: _____ Yes __**X**__ No

■ The clause above has been placed in the Agreement to remind you that you are acting as your own attorney in using these agreements and to remind you that the forms as presented are a starting point and not necessarily entirely suited in their present form to your purpose without some additional contract language or modifications.

This clause specifically indicates that you may need to add additional clauses to the Agreement based on the unique needs of your business, the disposition of the Owner, or the particular job that is the subject of the contract. The additional contract language you may need to add is beyond the scope of any form agreement in this book.

You may also decide to delete certain clauses from the Agreement depending upon the same factors mentioned above. If you work with a word processor, modifications to the Agreement will be fast and simple to make. Simply add (or delete) any clauses necessary to your Agreement and then delete the last clause, "ADDITIONAL TERMS AND CONDITIONS." Remember to consult an attorney familiar with construction before making significant changes to any agreement.

INITIAL
CC
97

I have read and understood, and I agree to, all the terms and conditions contained in the Agreement above.

■ You need a statement which indicates that the parties have read and agree to the terms and conditions of the Agreement. Be sure to have the Agreement signed by the Owner *prior* to the time you commence work. Make sure you keep a signed copy of the Agreement in your records — an unsigned copy won't do you any good later on if you have a dispute. Finally, have the Owner initial every page of your Agreement, including any supplemental attachments, such as materials or Subcontractor bids.

Date: 5/22/01

CHARLIE CONTRACTOR, PRESIDENT
CHARLIE CONTRACTOR, PRESIDENT
CHARLIE CONTRACTOR CONSTRUCTION, INC.

■ If your business is a corporation, be sure to sign the Agreement using your corporate title and place the word, "Inc." after your company's name. If you fail to do this, you may have personal liability under the Agreement.

Date: 5/22/01

Joe Tenant
JOE TENANT

■ Be sure you have an Agreement in your files which is signed by the Owner and contains the Owner's initials on each page.

The long-form fixed price (lump sum) agreement is well suited for use on new residential construction and on larger residential and light-commercial remodels. It has extensive language to help identify many of the risks a builder faces on larger projects and clarifies how to handle the many problems that can, and often do, arise on larger jobs.

Many of these risks are hidden from the unwary or inexperienced builder, but many lurk just around the next corner. This agreement assigns many of these risks in favor of the builder so he doesn't end up unfairly absorbing all the costs associated with "job surprises."

The long-form agreement contains all the beneficial language found in the short- and medium-form agreements, and expands and adds many clauses that may be helpful on a larger or complicated job. Noteworthy changes include the following:

Clause III.A.3 reinforces and adds to the Exclusions section of the contract. Among other things, it excludes the cost of utility connections, since they are difficult to bid and can cost you a lot if you guess wrong about what lies under the street.

All three contracts give the contractor the right to supervise the owner's subs and charge overhead and profit on their work. Clause III.D, in this version, adds the right of the contractor to prequalify any subs the owner wants to bring on the job. Another approach is to allow the owner's subs to work directly for the owner, but only before you start or after you complete your work.

A clause has been added that requires the owner to indicate who can authorize change orders (III.E.1) which can be important when both husband and wife are involved in a project.

Clause III.E.6 clarifies that the contractor will treat time-and-materials change orders like other change orders and, therefore, charge for overhead and profit. The same markup is established for any allowance work that exceeds the allowance amount.

Section III.F concerning payment schedules and terms has a number of important changes and additions, including Final Contract Payment (III.F.4), Hold Back From Final Payment for Punch List Work (III.F.5), Payment for Completed Punch List Work (III.F.6), and Interest Charges (III.F.7). The main impact of these additions is to make sure the final draw is not held up by a few minor punch list items. It also limits the retention (or hold back) for punch list work to 150% of the value of the work to be done.

Other changes include the addition of a Lien Release clause (III.I), an Insurance clause (III.K) requiring the owner to obtain Builder's Risk or Course of Construction coverage, and expanded language in the Warranty clause (III.L) and Entire Agreement clause (III.M).

Keep in mind that there is no absolute rule to help you decide which length agreement to use on a given job. Generally, the more costly and complicated the job, the longer the agreement.

1.3 ■ Long-Form Fixed Price Agreement

CONTRACTOR'S NAME: _____

ADDRESS: _____

_____

PHONE: _____

FAX: _____

LIC #: _____

DATE: _____

OWNER'S NAME: _____

ADDRESS: _____

_____

PROJECT ADDRESS: _____

_____

I. PARTIES

This contract (hereinafter referred to as "Agreement") is made and entered into on this _____ day of _____ , 19_____ , by and between _____ , (hereinafter referred to as "Owner"); and _____ , (hereinafter referred to as "Contractor").

In consideration of the mutual promises contained herein, Contractor agrees to perform the following work:

II. GENERAL SCOPE OF WORK DESCRIPTION

_____

_____

_____

_____

_____

(Additional Scope of Work page(s) attached: _____ Yes _____ No)

LUMP SUM PRICE FOR ALL WORK ABOVE: $_____

III. GENERAL CONDITIONS FOR THE AGREEMENT ABOVE

A. EXCLUSIONS

This Agreement does *not* include *labor or materials* for the following work:

1. PROJECT SPECIFIC EXCLUSIONS:

_____

_____

_____

_____

2. STANDARD EXCLUSIONS: Unless specifically included in the "General Scope of Work" section above, this Agreement does *not* include *labor or materials* for the following work (any Exclusions in this paragraph which have been lined out and initialed by the parties do not apply to this Agreement): Removal and disposal of any materials containing asbestos or any other hazardous material as defined by the EPA. Custom milling of any wood for use in project. Moving Owner's property around the site. Labor or materials required to repair or replace any Owner-supplied materials. Repair of concealed underground utilities not located on prints or physically staked out by Owner which are damaged during construction. Surveying that may be required to establish accurate property boundaries for setback purposes (fences and old stakes may not be located on actual property lines). Final construction cleaning (Contractor will leave site in "broom swept" condition). Landscaping and irrigation work of any kind. Temporary sanitation, power, or fencing. Removal of soils under house in order to obtain 18 inches (or code-required height) of clear space between bottom of joists and soil. Removal of filled ground or rock or any other materials not removable by ordinary hand tools (unless heavy equipment is specified in Scope of Work section above), correction of existing out-of-plumb or out-of-level conditions in existing structure. Correction of concealed substandard framing. Removal and replacement of existing rot or insect infestation. Construction of a continuously level foundation around structure (if lot is sloped more than 6 inches from front to back or side to side, Contractor will step the foundation in accordance with the slope of the lot). Exact matching of existing finishes. Repair of damage to existing roads, sidewalks, or driveways that could occur when construction equipment and vehicles are being used in the normal course of construction.

3. PERMITS, PLANS, ENGINEERING & ARCHITECTURAL FEES, UTILITY CONNECTION FEES, AND SPECIAL TESTING FEES: This Agreement does not include the cost of coordinating or submitting for the permits, fees, and services referred to above. If Owner requests Contractor to coordinate any of these services or obtain any of the permits above, Contractor will perform this work on an hourly basis at the hourly rate of: $_____

Owner (not Contractor) is to enter into contracts for all of the above-mentioned services and provide direct payment to the people or agencies contracted with for all of the services and permit fees in the paragraph above.

If Owner requests that Contractor meet with Owner and architect or other design professionals to review the construction plans and specifications prior to completion of the final design documents, Contractor will perform this work on an hourly basis at the hourly rate of: $_____

B. DATE OF WORK COMMENCEMENT AND SUBSTANTIAL COMPLETION

Commence work: _____ . Construction time through substantial completion: Approximately _____ to _____ weeks/months, *not* including delays and adjustments for delays caused by: holidays, inclement

weather, accidents, shortage of labor or materials, additional time required for completion of Change Order work (as specified in each Change Order), delays caused by Owner, and other delays unavoidable or beyond the control of the Contractor.

C. EXPIRATION OF THIS AGREEMENT

This Agreement will expire 30 days after the date at the top of page one of this Agreement if not accepted in writing by Owner and returned to Contractor within that time.

D. WORK PERFORMED BY OWNER OR OWNER'S SEPARATE SUBCONTRACTORS

Any labor or materials provided by the Owner's separate Subcontractors while Contractor is still working on this project *must be* supervised by Contractor. Profit and overhead at the rate of _____% will be charged on all labor and materials provided by Owner's separate Subcontractors while Contractor is still working on the project. Contractor has right to qualify and approve Owner's Subcontractors and require evidence of work experience, proper licensing, and insurance. If Owner wants to avoid paying Contractor's profit and overhead per this section, Owner must then bring in his separate Subcontractors only *before* or *after* Contractor has performed all of his work.

E. CHANGE ORDERS: CONCEALED CONDITIONS, ADDITIONAL WORK, AND CHANGES IN THE WORK

1. PEOPLE AUTHORIZED TO SIGN CHANGE ORDERS: The following people are authorized to sign Change Orders:

_____

_____

(Please fill in line(s) above at time of signing Agreement)

2. CONCEALED CONDITIONS: This Agreement is based solely on the observations Contractor was able to make with the structure in its current condition at the time this Agreement was bid. If additional Concealed Conditions are discovered once work has commenced which were *not* visible at the time this proposal was bid, Contractor will stop work and point out these unforeseen Concealed Conditions to Owner so that Owner and Contractor can execute a Change Order for any Additional Work.

3. CHANGES IN THE WORK: During the course of the project, Owner may order changes in the work (both additions and deletions). The cost of these changes will be determined by the Contractor and the cost of this Additional Work will be added to Contractor's profit and overhead at the rate of _____% in order to arrive at the net amount of any additional Change Order work.

Contractor's profit and overhead, and supervisory labor will not be credited back to Owner with any deductive Change Orders (work deleted from Agreement by Owner).

4. DEVIATION FROM SCOPE OF WORK IN CONTRACT DOCUMENTS: Any alteration or deviation from the Scope of Work referred to in the Contract Documents involving extra costs of materials or labor (including any overage on **ALLOWANCE** work) will be executed upon a written Change Order

issued by Contractor and should be signed by Contractor and Owner prior to the commencement of any Additional Work. This Change Order will become an extra charge over and above the Lump Sum Contract Price referred to at the beginning of this Agreement.

5. CHANGES REQUIRED BY PLAN CHECKERS OR FIELD INSPECTORS: Any increase in the Scope of Work set forth in the Contract Documents which is required by plan checkers or field inspectors with city or county building/planning departments will be treated as Additional Work to this Agreement for which the Contractor will issue a Change Order.

6. RATES CHARGED FOR ALLOWANCE-ONLY AND TIME-AND-MATERIALS WORK:
Journeyman Carpenter: _____ per hour; Apprentice Carpenter: _____ per hour; Laborer: _____ per hour; Contractor: _____ per hour; Subcontractor: Amount charged by Subcontractor. *Note*: Contractor will charge for profit and overhead at the rate of _____ on all work performed on a Time-and-Materials basis (on both materials and labor rates set forth in this paragraph) and on all costs that exceed specifically stated **ALLOWANCE** estimates in the Agreement.

F. PAYMENT SCHEDULE AND PAYMENT TERMS

1. PAYMENT SCHEDULE:
*First Payment: $1,000 or 10% of contract amount (whichever is less) due when Agreement is signed and returned to Contractor.

Contract Deposit Payment: $_____
* Second Payment (Materials Deposits): Any materials deposits — required for such items as woodstoves, cabinetry, carpets and vinyl, granite, tile, and any and all special order items that require the payment of a materials deposit — must be paid within 3 days of submittal of invoice by Contractor. These items will not be ordered until the deposits set forth below are received by Contractor. Materials Deposits required on this project include the following:

_____

_____

* Total Due for All Materials Deposits: $_____

* Third Payment: _____

_____ $_____

* Fourth Payment: _____

_____ $_____

* Fifth Payment: _____

_____ $_____

* Sixth Payment: _____

_____ $_____

* Seventh Payment: _____

_____ $ _____

* Eighth Payment: _____

_____ $ _____

* Final Payment: Due upon Substantial Completion of all work under this Agreement: $ _____

2. PAYMENT OF CHANGE ORDERS: Payment for each Change Order is due upon completion of Change Order work and submittal of invoice by Contractor for this work.

3. ADDITIONAL PAYMENTS FOR ALLOWANCE WORK AND RELATED CREDITS: Payment for work designated in the Agreement as **ALLOWANCE** work has been initially factored into the Lump Sum Price and Payment Schedule set forth in this Agreement. If the actual cost of the **ALLOWANCE** work *exceeds* the line item **ALLOWANCE** amount in the Agreement, the difference between the cost and the line item **ALLOWANCE** amount stated in the Agreement will be written up by Contractor as a Change Order subject to Contractor's profit and overhead at the rate of _____%.

If the cost of the **ALLOWANCE** work is *less* than the **ALLOWANCE** line item amount listed in the Agreement, a credit will be issued to Owner after all billings related to this particular line item **ALLOWANCE** work have been received by Contractor. This credit will be applied toward the final payment owing under the Agreement. Contractor profit and overhead and any supervisory labor will *not* be credited back to Owner for **ALLOWANCE** work.

4. FINAL CONTRACT PAYMENT: The final contract payment is due and payable upon "Substantial Completion" (not Final Completion) of all work under contract. "Substantial Completion" is defined as being the point at which the Building/Work of Improvement is suitable for its intended use, or the issuance of an Occupancy Consent, or final building department approval from the city or county building department, whichever occurs first.

5. HOLD BACK FROM FINAL PAYMENT FOR PUNCH LIST WORK: At time of making the final contract payment, Owner may hold back 150% of the value of all Punch List work. Owner and Contractor will place a fair and reasonable value on each Punch List item at time of Punch List walk-through with Owner. Contractor and Owner will then execute the Punch List form. This 150% hold back for Punch List work assures Owner that all Punch List work will be completed by Contractor in a timely manner.

6. PAYMENT FOR COMPLETED PUNCH LIST WORK: Payment for completed Punch List items is due and payable upon submittal of invoice for those completed items, even though entire punch list is not completed.

7. INTEREST CHARGES: Interest in the amount of _____% per month will be charged on all late payments under this Agreement. "Late Payments" are defined as any payment not received within _____ days of receipt of invoice from Contractor.

G. MISCELLANEOUS CONDITIONS

1. MATCHING EXISTING FINISHES: Contractor will use his best efforts to match existing finishes and materials. However, an exact match is *not* guaranteed by Contractor due to such factors as discoloration from aging, a difference in dye lots, discontinuation of product lines, and the difficulty of exactly matching certain finishes, colors, and planes.

Custom milling of materials has not been included in this Agreement, unless specifically stated in the Scope of Work section above. Unless *custom milling of materials* is specifically called out in the plans, specifications, or Scope of Work description above, any material not readily available at local lumberyards or suppliers is *not* included in this Agreement.

If Owner requires an exact match of materials or textures in a particular area, Owner must inform Contractor of this requirement in writing within seven (7) days of signing this Agreement. Contractor will then provide Owner with either a materials sample or a test patch prior to the commencement of work involving the matching of existing finishes.

Owner must then approve or disapprove of the suitability of the match within 24 hours. After that time, or after Contractor has provided Owner with two or more test patches that have been rejected by Owner, all further test patches, materials submittals, or any removal and replacement of materials already installed in accordance with the terms of this section will be performed strictly as Extra Work on a time-and-materials basis by Contractor.

2. CONFLICT OF DOCUMENTS: If any conflict should arise between the plans, specifications, addenda to plans, and this Agreement, then the terms and conditions of this Agreement shall be controlling and binding upon the parties to this Agreement.

3. INSTALLATION OF OWNER-SUPPLIED FIXTURES AND MATERIALS: Contractor cannot warrant any Owner-Supplied materials or fixtures (whether new or used). If Owner-supplied fixtures or materials fail due to a defect in the materials or fixtures themselves, Contractor will charge for all labor and materials required to repair or replace both the defective materials or fixtures, and any surrounding work that is damaged by these defective materials or fixtures.

4. CONTROL AND DIRECTION OF EMPLOYEES AND SUBCONTRACTORS: Contractor, or his appointed Supervisor, shall be the sole supervisor of Contractor's Employees and Subcontractors. Owner must not order or request Contractor's Employees or Subcontractors to make changes in the work. All changes in the work are to be first discussed with Contractor and then performed according to the Change Order process as set forth in this Agreement.

5. OWNER COORDINATION WITH CONTRACTOR: Owner agrees to promptly furnish Contractor with all details and decisions about unspecified construction finishes, and to consent to or deny changes in the Scope of Work that may arise so as not to delay the progress of the Work. Owner agrees to furnish Contractor with continual access to the job site.

6. CONTRACTOR NOT TO BE RELIED UPON AS ARCHITECT, ENGINEER, OR DESIGNER: The Contractor is *not* an architect, engineer, or designer. Contractor is not being hired to perform any of these services. To the extent that Contractor makes any suggestions in these areas, the Owner acknowledges and agrees that Contractor's suggestions are merely options that the Owner may want to review with the appropriate design professional for consideration. Contractor's suggestions are not a substitute for professional engineering, architectural, or design services, and are not to be relied on as such by Owner.

H. WORK STOPPAGE AND TERMINATION OF AGREEMENT FOR DEFAULT

Contractor shall have the right to stop all work on the project and keep the job idle if payments are not made to Contractor in accordance with the Payment Schedule in this Agreement, or if Owner repeatedly fails or refuses to furnish Contractor with access to the job site and /or product selections or information necessary for the advancement of Contractor's work. Simultaneous with stopping work on the project, the Contractor must give Owner written notice of the nature of Owner's default and must also give the Owner a 14-day period in which to cure this default.

If work is stopped due to any of the above reasons (or for any other material breach of contract by Owner) for a period of 14 days, and the Owner has failed to take significant steps to cure his default, then Contractor may, without prejudicing any other remedies Contractor may have, give written notice of termination of the Agreement to Owner and demand payment for all completed work and materials ordered through the date of work stoppage, and any other loss sustained by Contractor, including Contractor's Profit and Overhead at the rate of _____% on the balance of the incomplete work under the Agreement. Thereafter, Contractor is relieved from all other contractual duties, including all Punch List and warranty work.

I. LIEN RELEASES

Upon request by Owner, Contractor and Subcontractors will issue appropriate lien releases prior to receiving final payment from Owner.

J. DISPUTE RESOLUTION AND ATTORNEY'S FEES

Any controversy or claim arising out of or related to this Agreement involving an amount of *less* than $5,000 (or the maximum limit of the court) must be heard in the Small Claims Division of the Municipal Court in the county where the Contractor's office is located. Any controversy or claim arising out of or related to this Agreement which is over the dollar limit of the Small Claims Court must be settled by binding arbitration administered by the American Arbitration Association in accordance with the Construction Industry Arbitration Rules. Judgment upon the award may be entered in any Court having jurisdiction thereof.

The prevailing party in any legal proceeding related to this Agreement shall be entitled to payment of reasonable attorney's fees, costs, and expenses.

K. INSURANCE

Owner shall pay for and maintain "Course of Construction" or "Builder's Risk" or any other insurance that provides the same type of coverage to the Contractor's work in progress during the course of the project. It is Owner's express responsibility to insure dwelling and all work in progress against all damage caused by fire and Acts of God such as earthquakes, floods, etc.

L. WARRANTY

Contractor provides a limited warranty on all Contractor- and Subcontractor-supplied labor and materials used in this project for a period of one year following substantial completion of all work.

No warranty is provided by Contractor on any materials furnished by the Owner for installation. No warranty is provided on any existing materials that are moved and/or reinstalled by the Contractor within the dwelling (including any warranty that existing/used materials will not be damaged during the removal and reinstallation process). One year after substantial completion of the project, the Owner's sole remedy (for materials and labor) on all materials that are covered by a manufacturer's warranty is strictly with the manufacturer, not with the Contractor.

Repair of the following items is specifically excluded from Contractor's warranty: Damages resulting from lack of Owner maintenance; damages resulting from Owner abuse or ordinary wear and tear; deviations that arise such as the minor cracking of concrete, stucco and plaster; minor stress fractures in drywall due to the curing of lumber; warping and deflection of wood; shrinking/cracking of grouts and caulking; fading of paints and finishes exposed to sunlight.

THE EXPRESS WARRANTIES CONTAINED HEREIN ARE IN LIEU OF ALL OTHER WARRANTIES, EXPRESS OR IMPLIED, INCLUDING ANY WARRANTIES OF MERCHANTABILITY, HABITABILITY, OR FITNESS FOR A PARTICULAR USE OR PURPOSE. THIS LIMITED WARRANTY EXCLUDES CONSEQUENTIAL AND INCIDENTAL DAMAGES AND LIMITS THE DURATION OF IMPLIED WARRANTIES TO THE FULLEST EXTENT PERMISSIBLE UNDER STATE AND FEDERAL LAW.

M. ENTIRE AGREEMENT, SEVERABILITY, AND MODIFICATION

This Agreement represents and contains the entire agreement between the parties. Prior discussions or verbal representations by Contractor or Owner that are not contained in this Agreement are *not* a part of this Agreement. In the event that any provision of this Agreement is at any time held by a Court to be invalid or unenforceable, the parties agree that all other provisions of this Agreement will remain in full force and effect. Any future modification of this Agreement must be made in writing and executed by Owner and Contractor in order to be valid and binding upon the parties.

N. ADDITIONAL LEGAL NOTICES REQUIRED BY STATE OR FEDERAL LAW

See page(s) attached: _____ Yes   _____ No

O. ADDITIONAL TERMS AND CONDITIONS

See page(s) attached: _____Yes _____No

I have read and understood, and I agree to, all of the terms and conditions in the Agreement above.

Date: _____   _____
                     CONTRACTOR'S SIGNATURE

Date: _____   _____
                     OWNER'S SIGNATURE

Date: _____   _____
                     OWNER'S SIGNATURE

1.3 ■ LONG-FORM FIXED PRICE AGREEMENT

ANNOTATED

■ This Long-Form Agreement is useful for larger projects. It helps clarify how you will handle most aspects of the construction process.

This is a sample of an Agreement I used on a recent remodel project that had some of the vaguest plans I've ever seen. They contained fewer details than you can imagine. However, the job still needed to be bid. To make up for the lack of detail and to assure I wasn't just guessing at prices, I designated a number of items as ALLOWANCE items.

In addition, I bid a number of items according to the subtrade proposal or materials supplier bid that was given to me. I simply copied those bids and referred to the corresponding line items as being bid "per the attached enclosure from Joe Supplier or John Subcontractor."

I then attached the copies of the subtrade and materials bids to the Agreement. Due to the lack of detail in the plans, I wasn't able to bid the job precisely. However, I at least clearly stated exactly what I had bid, and contractually obligated myself to furnish no more or less than what was on the subtrade or materials supplier enclosures.

Charlie Contractor Construction, Inc.
123 Hammer Lane
Anywhere, USA 33333
Phone: (123) 456-7890
Fax: (123) 456-7899
Lic#: 11111

DATE: **May 22, 2001**

OWNER'S NAME: **Mr. & Mrs. Harry Homeowner**
ADDRESS: **333 Swift St.**
    Anywhere, USA 33333

PROJECT ADDRESS: **same**

I. PARTIES

This contract (hereinafter referred to as "Agreement") is made and entered into on this **22nd** day of **May** , 20 **01** , by and between **Harry and Helen Homeowner** , (hereinafter referred to as "Owner"); and **Charlie Contractor Construction, Inc.** , (hereinafter referred to as "Contractor"). In consideration of the mutual promises contained herein, Contractor agrees to perform the following work **in accordance with 6 pages of plans by Art Architect dated April 1, 2001** :

INITIAL
CC
HH

■ The phrase "...in consideration of the mutual promises contained herein, agrees to perform the following work:" is good boilerplate language that should remain in your Agreement. It means there is a "bargained-for exchange," which is necessary for a contract to be valid. The Owner will pay the Contractor the stated sum, and in exchange the Contractor will complete the work described in the Agreement. In the space immediately above, refer to any available plans and specifications on which you have based your bid.

II. GENERAL SCOPE OF WORK DESCRIPTION

■ Provide general details of the Scope of Work here. Refer to any plans and specifications that are available. Give a very complete description of what is included for the Lump Sum Price stated below. Many disputes occur because the Contractor does not specifically describe the Scope of Work. Spelling this out can also help you remember all the items that should be included in your job costs.

Many Owners feel more comfortable with a detailed line item description of what the Contractor is including in the Scope of Work. Sometimes when two Contractors give very close bids for the same job, the Contractor who gave the more complete Scope of Work description will end up with the job.

You can also state that labor or materials are being furnished "per the attached bid from XYZ Subcontractor or Materials Supplier." This is helpful if the plans are vague and you want to make very clear to the Owner exactly what a particular Subcontractor or materials supplier has bid to you, the General Contractor.

You may also want to state that the costs of certain line items are being furnished on an ALLOWANCE basis. For instance, many Owners have not selected specific plumbing fixtures or light fixtures at the time the Agreement is entered into, so the Contractor may want to list these fixtures as ALLOWANCE ITEMS. For instance:

1. Plumbing fixtures for two bathrooms (ALLOWANCE): $ 750
2. Electrical fixtures for entire residence (ALLOWANCE): $2,000

When you include ALLOWANCES, be sure to include contract language that describes how any overage or underage regarding the ALLOWANCE item will be handled. Be sure to state that Contractor Profit and Overhead will be added to any amount that exceeds the ALLOWANCE line item amount stated in the Agreement. The clauses in this manual relating to ALLOWANCES make this clear, however, you may need to verbally explain how this works to the Owner.

INITIAL
CC
HH

1. Temporary Sanitation: $ 250

2. Dumpsters, debris removal, clean-up labor: $ 5,051

3. Foundation (materials and labor per plans): $ 1,830

4. Framing materials: $ 9,544

5. Framing labor: $ 13,230

6. Scaffolding rental (ALLOWANCE): $ 250

7. Norco Wood Windows and Sliding Doors; skylight (quantity and style per enclosure from Art's Glass.) Does not include interior arched casing: $ 12,212

8. Furnish and install roofing per enclosure from Ronnie's Roofing: $ 4,945

9. Furnish and install 2 coat stucco on new addition and tie-in to existing stucco on first floor (2 coat stucco; match existing as closely as possible; Contractor recommends 3 coat stucco as an additional alternate). $ 4,500

10. Furnish and install insulation: $ 1,140

11. Furnish and install drywall, tape & strd. texture: $ 4,200

12. Heating system, sheet metal, gutters per enclosure from Mike's Mechanical: $ 4,655

13. Furnish and install rough plumbing per enclosure from Paula's Plumbing. Includes labor to set finish fixtures ($600 included toward gas work): $ 4,245

14. Plumbing fixture (ALLOWANCE): $ 3,000

15. Furnish and install rough electrical (include 125-amp service and panel). Includes labor to set finish fixtures: $ 3,345

16. Furnish finish light fixtures (ALLOWANCE): $ 2,000

17. Interior and exterior painting per enclosure from Pete's Painting (does not include painting interior cabinetry -- see alternate below -- or handrail or decks): $ 3,940

18. Furnish interior doors, door hardware, and wood trim per enclosure from Dan's Door: $ 1,958

19. Furnish and install paint-grade cabinets (all paint-grade birch, Blum European hinges, melamine interiors okay, Revere doors (not painted); kitchen cabinets per plans; 3 lavie cabinets (one w/linen hamper); upper and lower laundry cabinet; and hutch/room divider cabinet: $ 6,500

INITIAL
CC
HH

20. Furnish and install new wood floors, carpet, and pad, refinish and patch hardwood floor in master bedroom (ALLOWANCE):	$ 3,500
21. Furnish and install oak handrail and guardrail (ALLOWANCE):	$ 2,000
22. Furnish and install tile per enclosure (laundry room floor; 3 lavie cabinets; 3 bathroom floors; shower stall including ceiling; backsplash in kitchen):	$ 5,080
23. Furnish and install appliances — electric range, dishwasher, and disposal (ALLOWANCE):	$ 2,150
24. Furnish and install towel bars, mirrors, tissue holders, and shower door (ALLOWANCE):	$ 1,250
25. Finish carpentry labor:	$ 601
26. Labor to install doors and door hardware:	$ 795
27. Supervisory Labor:	$ 2,000
28. Fire sprinklers to include: fire riser, flow switch, drain and valves (riser begins at ground level); plans, pipe, fittings, sprinkler heads and trim rings; Inspector's test assembly (gate valve, pressure gauge, sprinkler head). Does not include water service and related work required from the water supply in the street to the fire riser at the house or required electrical connection (see Alternate Section below):	$ 1,800
29. Exterior second-floor redwood decks w/guardrail posts per plans (no pickets or plexiglass included due to lack of detail on plans):	$ 5,235
30. Furnish and install Corian countertops per enclosure from Smith Brothers:	$ 4,110
31. Labor to install windows:	$ 1,075
32. Subtotal:	$116,391
33. Contractor Overhead @ 10%:	$ 11,639
34. Contractor Profit @ 10%:	$ 11,639

(Additional Scope of Work page(s) attached: _____ Yes __x__ No)

LUMP SUM PRICE FOR ALL WORK ABOVE: $139,669

■ The Contractor may or may not want to show his profit and overhead and any supervisory labor as a separate line item cost above the Lump Sum Price.

INITIAL
CC
HH

ALTERNATES

(Note: All prices below have not been factored into the Lump Sum Price above. Contractor's profit and overhead have been added to the prices below.)

1. Upper birch cabinet over 7-foot peninsula in kitchen:	$ 300
2. 4x4-inch American Olean tile set on mortar bed w/4-inch tile backsplash on base cabinet in laundry room:	$ 550
3. Paint all interior cabinets (no paint on interiors). Cabinets will be primed and painted, doors and drawer fronts sprayed, face frames brushed. Paint will be Benjamin Moore oil-base; bid based on one coat primer and one coat oil enamel (per enclosure from Pete's Painting):	$ 3,890
4. Fire sprinkler connection from water supply at street to fire riser at house per enclosure from Fred's Firesprinkler Co. (ALLOWANCE):	$ 3,050
5. Corian sink (single):	$ 621
6. Corian sink (double):	$ 924

III. GENERAL CONDITIONS FOR THE AGREEMENT ABOVE

A. EXCLUSIONS

This Agreement does *not* include *labor or materials* for the following work:

> ∎ The Exclusions section of your Agreement is just as important as the Scope of Work section. Unless you specifically exclude areas of work that you know you are not going to perform, the Owner might assume they are included.
>
> Contractors lose a great deal of money by failing to exclude work. The Contractor later "eats" items that fall into "gray" areas and weren't specifically excluded in order to get his final check from the Owner. If the Exclusions section is complete, the Contractor will "eat" far fewer of these items. In my experience, paying attention to Exclusions can save a medium-sized Contractor several thousand dollars each year.
>
> I list the most obvious exclusions that specifically relate to the job being contracted as "Project Specific Exclusions." The less obvious, more remote, and "boilerplate" Exclusions are listed below in "Standard Exclusions." If working with a computer, remove any Standard Exclusions that don't apply to the job so that you don't scare the Owner away with unnecessary Exclusions. Add any Standard Exclusions to your Agreement that relate specifically to your business or to the job you will be performing. The Exclusions list provided in the form is a good starting point, but it may not be 100% complete for your purposes.

INITIAL
CC
HH

1. PROJECT SPECIFIC EXCLUSIONS:

Exterior concrete flatwork of any kind; landscape and irrigation work of any kind; patio work; any utility trenching and work outside footprint of house; work of any kind on garage/studio unit. All work required for fire sprinkler system on street side of sidewalk (underground contracting work for fire service supply — unless Owner selects alternate above).

2. STANDARD EXCLUSIONS: Unless specifically included in the "General Scope of Work" section above, this Agreement does *not* include *labor or materials* for the following work (any Exclusions in this paragraph which have been lined out and initialed by the parties do not apply to this Agreement): Removal and disposal of any materials containing asbestos or any other hazardous material as defined by the EPA. Custom milling of any wood for use in project. Moving Owner's property around the site. Labor or materials required to repair or replace any Owner-supplied materials. Repair of concealed underground utilities not located on prints or physically staked out by Owner which are damaged during construction. Surveying that may be required to establish accurate property boundaries for setback purposes (fences and old stakes may not be located on actual property lines). Final construction cleaning (Contractor will leave site in "broom swept" condition). Landscaping and irrigation work of any kind. Temporary sanitation, power, or fencing. Removal of soils under house in order to obtain 18 inches (or code-required height) of clear space between bottom of joists and soil. Removal of filled ground or rock or any other materials not removable by ordinary hand tools (unless heavy equipment is specified in Scope of Work section above), correction of existing out-of-plumb or out-of-level conditions in existing structure. Correction of concealed substandard framing. Removal and replacement of existing rot or insect infestation. Construction of a continuously level foundation around structure (if lot is sloped more than 6 inches from front to back or side to side, Contractor will step the foundation in accordance with the slope of the lot.) Exact matching of existing finishes. Repair of damage to existing roads, sidewalks, or driveways that could occur when construction equipment and vehicles are being used in the normal course of construction.

3. PERMITS, PLANS, ENGINEERING & ARCHITECTURAL FEES, UTILITY CONNECTION FEES, AND SPECIAL TESTING FEES: This Agreement does not include the cost of coordinating or submitting for the permits, fees, and services referred to above. If Owner requests Contractor to coordinate any of these services or obtain any of the permits above, Contractor will perform this work on an hourly basis at the hourly rate of: $ __35.00__

Owner (not Contractor) is to enter into contracts for all of the above-mentioned services and provide direct payment to the people or agencies contracted with for all of the services and permit fees in the paragraph above.

If Owner requests that Contractor meet with Owner and architect or other design professionals to review the construction plans and specifications prior to completion of the final design documents, Contractor will perform this work on an hourly basis at the hourly rate of: $ __35.00__

INITIAL
CC
HH

■ Carefully check the Exclusions in the paragraph above and determine which ones belong in your Agreement. In our area, the Owner typically pulls and pays for the permit, and pays for his own plans and any engineering or special testing fees that are required. Another area that I prefer to either exclude or treat as an ALLOWANCE item is connection and permit fees for public and private utilities, and the cost of labor and materials to connect all these utilities outside the footprint of the structure. This is an area that can be difficult to bid because the connections are often buried somewhere deep in the street. It also can be a very expensive area if you base your bid on inaccurate information.

B. DATE OF WORK COMMENCEMENT AND SUBSTANTIAL COMPLETION

Commence work: __June 1, 2001__ . Construction time through substantial completion: Approximately __5 to 7 months__ , *not* including delays and adjustments for delays caused by: holidays, inclement weather, accidents, shortage of labor or materials, additional time required for completion of Change Order work (as specified in each Change Order), delays caused by Owner, and other delays unavoidable or beyond the control of the Contractor.

■ Some states require that the approximate start and completion dates of the project be specified in the Agreement. Every Owner also wants to know this information. However, be sure to give a *range* of days, weeks, or months to complete and not exact dates. If you think the project will take 3 months, state "3 to 4 months" whenever possible so you allow a cushion for delays. Numerous events can increase the time needed to complete the Work. Be sure to add extra contract days to all of your Change Orders in the space provided.

C. EXPIRATION OF THIS AGREEMENT

This Agreement will expire 30 days after the date at the top of page one of this Agreement if not accepted in writing by Owner and returned to Contractor within that time.

■ You may furnish the Owner with a signed Agreement and not hear back from the Owner for several months. Perhaps during that time, a low-priced Subcontractor you relied on in your bid has left the area and the price of lumber has gone up 12%. So you do not want to be bound to the old price. The clause above gives you the option to consider the Agreement expired if not signed (accepted) by the Owner within 30 days of his receipt of that Agreement.

INITIAL

CC

HH

D. WORK PERFORMED BY OWNER OR OWNER'S SEPARATE SUBCONTRACTORS

Any labor or materials provided by the Owner's separate Subcontractors while Contractor is still working on this project *must be* supervised by Contractor. Profit and overhead at the rate of ___**20**___% will be charged on all labor and materials provided by Owner's separate Subcontractors while Contractor is still working on the project. Contractor has right to qualify and approve Owner's Subcontractors and require evidence of work experience, proper licensing, and insurance. If Owner wants to avoid paying Contractor's profit and overhead per this section, Owner must then bring in his separate Subcontractors only *before* or *after* Contractor has performed all of his work.

> ■ The clause above is very important and you may want to include it on all your jobs. Ordinarily, a Contractor will spend just as much time coordinating an Owner's separate Subcontractors as he will with his own, so I want the Owner to know that we charge for this. The Owner always has the option of bringing in his own separate Subcontractors either before we commence work on the project, or after we have completed all of our work on the project.

E. CHANGE ORDERS: CONCEALED CONDITIONS, ADDITIONAL WORK, AND CHANGES IN THE WORK

1. PEOPLE AUTHORIZED TO SIGN CHANGE ORDERS: The following people are authorized to sign Change Orders:

Harry Homeowner

Helen Homeowner

(Please fill in line(s) above at time of signing Agreement)

> ■ On a job with many Change Orders, it can be helpful to have in writing the people who are authorized to sign Change Orders. On one job we contracted, the wife authorized a number of Change Orders. When the husband's business took a nose dive, the husband refused to pay the Change Orders and later claimed his wife had not been authorized to approve them.

2. CONCEALED CONDITIONS: This Agreement is based solely on the observations Contractor was able to make with the structure in its current condition at the time this Agreement was bid. If additional Concealed Conditions are discovered once work has commenced which were *not* visible at the time this proposal was bid, Contractor will stop work and point out these unforeseen Concealed Conditions to Owner so that Owner and Contractor can execute a Change Order for any Additional Work.

INITIAL
CC
HH

■ Every Agreement should have a Concealed Conditions clause similar to the one above — especially on remodeling jobs. Occasionally Murphy's Law will create some very unforeseeable extra work on projects. If this work is not specifically excluded in the "Exclusions" section of your Agreement, you may need to rely on the "Concealed Conditions" clause in order to be able to write up the additional unforeseen work as a Change Order.

Examples of Concealed Conditions could be subterranean concrete that interferes with your foundation work, spring water that prevents you from pouring concrete, pipes and conduits that interfere with the installation of new doors and windows, asbestos or other toxic compounds discovered during demolition, omitted framing members, etc.

3. CHANGES IN THE WORK: During the course of the project, Owner may order changes in the work (both additions and deletions). The cost of these changes will be determined by the Contractor and the cost of this Additional Work will be added to Contractor's profit and overhead at the rate of __20__ % in order to arrive at the net amount of any additional Change Order work.

Contractor's profit and overhead, and supervisory labor will not be credited back to Owner with any deductive Change Orders (work deleted from Agreement by Owner).

■ The clause above simply states that the Owner can both add and delete work. The cost of all additional work will be determined by the Contractor and will be subject to Contractor's profit and overhead at the rate of 20%. Even if the profit and overhead rate in the primary Agreement is as low as 10%, charging 20% profit and overhead is ordinarily justified and necessary as it takes extra time to explain the Change Order to the Owner, wait for him to sign it, and then execute the Change Order work.

The sentence above that states, "Contractor's profit and overhead, and any supervisory labor will not be credited back to Owner with any deductive Change Orders" is critical because it allows the Contractor to keep the profit and overhead on work that is deleted by the Owner after the Agreement has been signed. Remember, you have already spent considerable time estimating and planning this work.

This can be worth a great deal of money if the Owner (after relying on your contract to obtain financing), decides to delete from your Scope of Work such items as: furnishing windows and doors, cabinets, floor coverings, tile work, paint work, etc., expecting you to credit back all the profit and overhead on this work and any related supervisory labor as well. A $30,000 deletion in work can cost the Contractor between $4,500 and $6,000 in lost profit and overhead.

INITIAL
CC
HH

4. DEVIATION FROM SCOPE OF WORK IN CONTRACT DOCUMENTS: Any alteration or deviation from the Scope of Work referred to in the Contract Documents involving extra costs of materials or labor (including any overage on **ALLOWANCE** work) will be executed upon a written Change Order issued by Contractor and should be signed by Contractor and Owner prior to the commencement of any Additional Work. This Change Order will become an extra charge over and above the Lump Sum Contract Price referred to at the beginning of this Agreement.

> ■ The clause above is important because it clearly states that any change to the Scope of Work involving extra labor or materials will constitute grounds for a Change Order. The Contractor has the discretion to not charge for a minor change. However, the standard for when he can execute a Change Order is clearly set forth by this section — any time a deviation from the Scope of Work in the Agreement requires additional labor or materials. The Contractor must be careful to always put his Change Orders in writing, pursuant to this section of the Agreement. *Do not* rely on verbal Change Orders!

5. CHANGES REQUIRED BY PLAN CHECKERS OR FIELD INSPECTORS: Any increase in the Scope of Work set forth in the Contract Documents which is required by plan checkers or field inspectors with city or county building/planning departments will be treated as Additional Work to this Agreement for which the Contractor will issue a Change Order.

> ■ This Exclusion can be very important to the Contractor. Often you will bid the job before the final approved copy of the job plans is issued. If the approved set of plans requires additional work that was not shown in your bid set of plans, this additional work is excluded from your Agreement and will be written up as a Change Order. In addition, if a field building inspector decides after inspecting the site that additional work not shown on the plans is required, this additional work is also excluded from your Agreement and will also be written up as a Change Order.

6. RATES CHARGED FOR ALLOWANCE-ONLY AND TIME-AND-MATERIALS WORK: Journeyman Carpenter: **$26** per hour; Apprentice Carpenter: **$21** per hour; Laborer: **$17** per hour; Contractor: **$35** per hour; Subcontractor: Amount charged by Subcontractor. *Note:* Contractor will charge for profit and overhead at the rate of **20%** on all work performed on a Time-and-Materials basis (on both materials and labor rates set forth in this paragraph) and on all costs that exceed specifically stated **ALLOWANCE** estimates in the Agreement.

INITIAL
CC
HH

■ In general, I prefer to give a fixed price for Change Orders rather than bill on a Time-and-Materials basis because there are usually fewer problems when the Owner knows the cost of the work and authorizes it before the work is performed. However, where a certain task is extremely difficult to estimate, performing certain Change Orders on a Time-and-Materials basis is best for both the Owner and the Contractor.

When this might be the case, I put the rates for T&M work in the Agreement so that the Owner knows and agrees to these rates at the time the primary Agreement is signed.

Because a T&M Change Order is still a Change Order, I also charge 20% profit and overhead on my in-house labor and on all materials and sub-contractors that are a part of the Change Order.

I also make it clear that the overage on an ALLOWANCE item is essentially extra work and this overage is treated like any other Change Order, i.e., written up with a 20% profit and overhead charge.

F. PAYMENT SCHEDULE AND PAYMENT TERMS

1. PAYMENT SCHEDULE:

* First Payment: $1,000 or 10% of contract amount (whichever is less) due when Agreement is signed and returned to Contractor.

Contract Deposit Payment: **$ 1,000**

* Second Payment (Materials Deposits): Any materials deposits — required for such items as woodstoves, cabinetry, carpets and vinyl, granite, tile, and any and all special order items that require the payment of a materials deposit — must be paid within 3 days of submittal of invoice by Contractor. These items will not be ordered until the deposits set forth below are received by Contractor. Materials Deposits required on this project include the following:

           1. Corian Deposit: **$ 2,000**
* Total Due for All Materials Deposits: **$ 2,000**

* Third Payment: **10% of contract amount due upon completion of demolition, foundation work, and delivery of initial framing materials to site:**   **$13,996**

* Fourth Payment: **25% of contract amount due upon completion of rough framing:**   **$34,917**

* Fifth Payment: **15% of contract amount due upon completion of roofing, installation of gutters and exterior sheet metal, and installation of all exterior doors and windows:**   **$20,950**

INITIAL
CC
HH

* Sixth Payment: **15% of contract amount due upon completion of rough plumbing, rough electrical, rough-in of fire sprinklers inside house, insulation, and hanging of drywall (not taping or texturing):**   **$20,950**

* Seventh Payment: **15% of contract amount due upon installation of lath, stucco scratch coat, stucco finish coat, taping and texturing of drywall, installation of interior doors, closet poles and shelves, and installation of heating system:**   **$20,950**

* Eighth Payment: **15% of contract amount due upon completion of interior paint, exterior paint, installation of cabinets, installation of rough and finish electrical fixtures:**   **$20,950**

* Final Payment: Due upon Substantial Completion of all work under this Agreement: **$ 1,986**

> ■ The Payment Schedule is one of the most critical parts of any Agreement. It is also regulated by state laws in certain states. The way you draft your Payment Schedule says a lot about how you operate your business. For this reason, I like to keep my payment requests just about equivalent to the amount of work that's been performed under the Agreement. However, a 10% deposit or $1,000 (whichever is less) is always my first contract deposit payment.
>
> Next, I may require a materials deposit for specialized subtrade work such as granite or woodstoves or other special-order items that require a deposit from us. (Beyond that, I don't require materials deposits.) Next, I break down the job into several smaller payments rather than having just a couple of large payments. The final payment should be a fairly small percentage of the contract amount. Leaving a large amount of money owing for the final payment can invite trouble! By the same token, taking up front money for work not completed can weaken the trust the Owner has in the Contractor and can also invite its own set of problems for the Contractor.

2. PAYMENT OF CHANGE ORDERS: Payment for each Change Order is due upon completion of Change Order work and submittal of invoice by Contractor for this work.

> ■ I like to have it understood that payment for Change Orders is due and payable upon completion of the Change Order work and submittal of invoice by Contractor.

INITIAL
CC
HH

3. ADDITIONAL PAYMENTS FOR ALLOWANCE WORK AND RELATED CREDITS: Payment for work designated in the Agreement as **ALLOWANCE** work has been initially factored into the Lump Sum Price and Payment Schedule set forth in this Agreement. If the actual cost of the **ALLOWANCE** work *exceeds* the line item **ALLOWANCE** amount in the Agreement, the difference between the cost and the line item **ALLOWANCE** amount stated in the Agreement will be written up by Contractor as a Change Order subject to Contractor's profit and overhead at the rate of __**20**__%.

If the cost of the **ALLOWANCE** work is *less* than the **ALLOWANCE** line item amount listed in the Agreement, a credit will be issued to Owner after all billings related to this particular line item **ALLOWANCE** work have been received by Contractor. This credit will be applied toward the final payment owing under the Agreement. Contractor profit and overhead and any supervisory labor will *not* be credited back to Owner for **ALLOWANCE** work.

> ■ It's very important to have a clear written explanation of how an AL-LOWANCE item works. The statement above explains that any overage on an ALLOWANCE item will be subject to profit and overhead. But the Owner will not receive a credit of profit and overhead for any underage on ALLOWANCE items.
>
> I don't credit back this profit and overhead for the same reason I don't credit back profit and overhead on deleted work — the Contractor has often already done a great deal of work bidding and arranging to have the work done, and has already earned a good part of this profit and overhead. It also discourages the Owner from deciding to be an Owner/Builder in the middle of his project after the Contractor has already done most of the work.

4. FINAL CONTRACT PAYMENT: The final contract payment is due and payable upon "Substantial Completion" (not Final Completion) of all work under contract. "Substantial Completion" is defined as being the point at which the Building/Work of Improvement is suitable for its intended use, or the issuance of an Occupancy Consent, or final building department approval from the city or county building department, whichever occurs first.

> ■ On a larger job I prefer to have the language above which indicates that, for instance, my $5,000 final draw will not be held up by a $250 Punch List. By allowing a hold back of 150% of the fair value of the Punch List, the Owner should not be too reluctant to agree to pay the Contractor upon "substantial completion" rather than "final completion" of the project. Occasionally you can run into a nervous Owner who has a Punch List that seems to never end. Sometimes I wonder if that's because when the Punch List is complete, he knows he'll have to pay up. The language above avoids this problem.

INITIAL
CC
HH

5. HOLD BACK FROM FINAL PAYMENT FOR PUNCH LIST WORK: At time of making the final contract payment, Owner may hold back 150% of the value of all Punch List work. Owner and Contractor will place a fair and reasonable value on each Punch List item at time of Punch List walk-through with Owner. Contractor and Owner will then execute the Punch List form. This 150% hold back for Punch List work assures Owner that all Punch List work will be completed by Contractor in a timely manner.

> ■ This language is to assure the Owner that he has more than enough money to complete the Punch List if the Contractor fails to do so after receiving the bulk of the final payment. If the Owner is hesitant about this phrase during contract review, I offer to make the hold back 200% of the value of the Punch List.

6. PAYMENT FOR COMPLETED PUNCH LIST WORK: Payment for any completed Punch List items is due and payable upon submittal of invoice for those completed items, even though entire Punch List is not completed.

> ■ With a long Punch List, this simply gives the Contractor the right to invoice for completed parts of the Punch List at his discretion.

7. INTEREST CHARGES: Interest in the amount of __1.5__ % per month will be charged on all late payments under this Agreement. "Late Payments" are defined as any payment not received within __5__ days of receipt of invoice from Contractor.

> ■ This clause defines when a payment is "late" and states that the Contractor is entitled to interest on the late payment at the rate of 1.5% per month. As a practical matter, I almost never actually charge the interest. However, it's important to have this clause in your Agreement in case you are ever in a lengthy dispute over the Owner's failure to pay you a large sum of money. The interest on the money you are owed can add up if you have to wait a year or two to get paid.

G. MISCELLANEOUS CONDITIONS

1. MATCHING EXISTING FINISHES: Contractor will use best efforts to match existing finishes and materials. However, an exact match is not guaranteed by Contractor due to such factors as discoloration from aging, a difference in dye lots, and the difficulty of exactly matching certain finishes, colors, and planes.

Unless custom milling of materials is specifically called out in the plans, specifications, or Scope of Work description, any material not readily available at local lumberyards or suppliers is *not* included in this Agreement.

INITIAL
CC
HH

If Owner requires an exact match of materials or textures in a particular area, Owner must inform Contractor of this requirement in writing within seven (7) days of signing this Agreement. Contractor will then provide Owner with either a materials sample or a test patch prior to the commencement of work involving the matching of existing finishes.

Owner must then approve or disapprove of the suitability of the match within 24 hours. After that time, or after Contractor has provided Owner with two or more test patches that have been rejected by Owner, all further test patches, materials submittals, or any removal and replacement of materials already installed in accordance with the terms of this section will be performed strictly as Extra Work on a time-and-materials basis by Contractor.

> ■ Agreeing to match existing finishes is a can of worms that requires special treatment. The clause above is one way to deal with some of the potential problems that can arise if an Owner is difficult to please. Also see Form 1.4, Addendum for Matching Existing Finishes.

2. CONFLICT OF DOCUMENTS: If any conflict should arise between the plans, specifications, addenda to plans, and this Agreement, then the terms and conditions of this Agreement shall be controlling and binding upon the parties to this Agreement.

> ■ With most construction projects, the Construction Agreement, plans, any written specifications, any written addenda, and sometimes engineering reports, are all considered "contract documents" to which the Contractor is bound. The problem is that these documents may either intentionally or unintentionally contradict each other regarding specific work the Contractor is to perform.
>
> The question then becomes whether or not the Contractor is legally required to perform the work in question. I try to settle this question with the clause above because it simply states that this Agreement is the one contract document that controls the various other contract documents in the project.
>
> For instance, if the plans call out for wood windows but the Owner has directed the Contractor to bid metal windows and the Contractor specifically states metal windows in his Agreement, then metal windows are what the Contractor is bound to, even though the plans call out for wood windows. Or, if the plans call out for an item such as French drains, and the Contractor specifically excludes it, then the Contractor is not held to furnishing the item simply because it may show up somewhere as a note in one of the other contract documents.

3. INSTALLATION OF OWNER-SUPPLIED FIXTURES AND MATERIALS: Contractor cannot warrant any Owner-Supplied materials or fixtures (whether new or used). If Owner-supplied fixtures or materials fail due to a defect in the materials or fixtures themselves,

INITIAL
CC
HH

Contractor will charge for all labor and materials required to repair or replace both the defective materials or fixtures, and any surrounding work that is damaged by these defective materials or fixtures.

■ This clause states that when the Owner supplies materials or fixtures, the Contractor does not warrant these items. If you didn't purchase the items, you may have trouble returning them, and the warranty may not extend to you because you were not the original purchaser. In addition, any surrounding parts of the structure that are damaged by Owner-supplied fixtures are excluded from the Contractor's warranty, because if it weren't for the Owner's defective material, the surrounding area would not have been damaged.

4. CONTROL AND DIRECTION OF EMPLOYEES AND SUBCONTRACTORS: Contractor, or his appointed Supervisor, shall be the sole supervisor of Contractor's Employees and Subcontractors. Owner must not order or request Contractor's Employees or Subcontractors to make changes in the work. All changes in the work are to be first discussed with Contractor and then performed according to the Change Order process as set forth in this Agreement.

■ This clause attempts to deal with the classic problem of the Owner ordering your employees and Subcontractors to change work or perform additional work when you are not on the job site. This shouldn't happen for many good reasons, the least of which is that you have a written procedure in the Agreement for how changes in the work are to be made.
   Every Contractor has had a little trouble at one time or another with this situation and this clause reminds the Owner to give all changes in the work to the Contractor and not to his employees and Subcontractors (unless specific arrangements have been made to turn over this responsibility to the Contractor's job foreman or supervisor).

5. OWNER COORDINATION WITH CONTRACTOR: Owner agrees to promptly furnish Contractor with all details and decisions about unspecified construction finishes, and to consent to or deny changes in the Scope of Work that may arise so as not to delay the progress of the Work. Owner agrees to furnish Contractor with continual access to the job site.

■ This clause states the Owner's second duty in the construction process, which is to make sure the Contractor is furnished with proper construction details in a timely manner so that the progress of the work is not unreasonably delayed.

INITIAL
CC
HH

6. CONTRACTOR NOT TO BE RELIED UPON AS ARCHITECT, ENGINEER, OR DESIGNER: The Contractor is *not* an architect, engineer, or designer. Contractor is not being hired to perform any of these services. To the extent that Contractor makes any suggestions in these areas, the Owner acknowledges and agrees that Contractor's suggestions are merely options that the Owner may want to review with the appropriate design professional for consideration. Contractor's suggestions are not a substitute for professional engineering, architectural, or design services, and are not to be relied on as such by Owner.

> ■ The clause above should help prevent the Owner from later coming back to hang you with any suggestions you made about how to handle certain construction details. If your suggestions were really off base, you may be still held liable for suggesting a bad design, or for executing a bad design that you should have known was defective.
>
> However, the clause above tells the Owner that your suggestions are merely ideas which the design professional should consider — and nothing more or less than that. Unless you are a licensed architect, don't pretend to be one. From a legal standpoint, you are much safer being just the Contractor and not the "Design/Builder."
>
> Even where you are not the designer, be sure to notify the Owner if you encounter a design detail that you think won't work — you may later find out you had a duty to do this. If you notice a faulty detail but fail to inform the Owner and Architect of it, and then build the work according to this faulty detail, you may share in the liability.

H. WORK STOPPAGE AND TERMINATION OF AGREEMENT FOR DEFAULT

Contractor shall have the right to stop all work on the project and keep the job idle if payments are not made to Contractor in accordance with the Payment Schedule in this Agreement, or if Owner repeatedly fails or refuses to furnish Contractor with access to the job site and /or product selections or information necessary for the advancement of Contractor's work. Simultaneous with stopping work on the project, the Contractor must give Owner written notice of the nature of Owner's default and must also give the Owner a 14-day period in which to cure this default.

If work is stopped due to any of the above reasons (or for any other material breach of contract by Owner) for a period of 14 days, and the Owner has failed to take significant steps to cure his default, then Contractor may, without prejudicing any other remedies Contractor may have, give written notice of termination of the Agreement to Owner and demand payment for all completed work and materials ordered through the date of work stoppage, and any other loss sustained by Contractor, including Contractor's Profit and Overhead at the rate of __20__ % on the balance of the incomplete work under the Agreement. Thereafter, Contractor is relieved from all other contractual duties, including all Punch List and warranty work.

> ■ The Owner's primary duty under a Construction Agreement is to pay for properly performed work in accordance with the Payment Schedule in the Agreement. What if the Owner fails to do this and yet still expects the Contractor to keep working? Whether or not the failure to pay money amounts to a major breach of contract — one that will justify your stopping work and considering the Agreement terminated by the Owner's failure to pay — will depend upon the particular facts of your situation.
>
> However, by having a termination clause like the one above, the Con-

INITIAL
CC
HH

> tractor will have much better grounds for stopping work and considering the Agreement terminated. He will also have a contractual basis for collecting interest on unpaid amounts owed to him, along with his lost profit and overhead on the amount of work left to complete at the time of the Owner's breach of contract.

I. LIEN RELEASES

Upon request by Owner, Contractor and Subcontractors will issue appropriate lien releases prior to receiving final payment from Owner.

> ■ This clause simply informs the Owner that (assuming the Owner is not in default under the Agreement) the Contractor will be happy to furnish Conditional Lien Releases prior to final payment.

J. DISPUTE RESOLUTION AND ATTORNEY'S FEES

Any controversy or claim arising out of or related to this Agreement involving an amount of *less* than $5,000 (or the maximum limit of the court) must be heard in the Small Claims Division of the Municipal Court in the county where the Contractor's office is located. Any controversy or claim arising out of or related to this Agreement which is over the dollar limit of the Small Claims Court must be settled by binding arbitration administered by the American Arbitration Association in accordance with the Construction Industry Arbitration Rules. Judgment upon the award may be entered in any Court having jurisdiction thereof.

The prevailing party in any legal proceeding related to this Agreement shall be entitled to payment of reasonable attorney's fees, costs, and expenses.

> ■ Whether a dispute will be resolved in the normal court system or through binding arbitration will be determined by your Agreement. For small disputes, I prefer a Small Claims Court where you can ordinarily get a court date within a few months, where the maximum dollar limit is $2,500 to $5,000, and where you can't take a lawyer. It's fast (usually less than a 20-minute hearing in my area), easy, and cheap.
>
> Very limited appeals, if any, are available (depending upon whether you are the plaintiff or defendant). You can't take a lawyer into a Small Claims Court proceeding in many jurisdictions, but prior to the hearing, you can consult a lawyer familiar with construction law to help you identify the legal issues and prepare an oral outline to follow. He can also help you come up with a written statement of your position so you can present your case well, and thereby increase your odds of receiving a judgment in your favor.
>
> For disputes too large for the local Small Claims Court, I prefer binding arbitration through either privately selected arbitrators or through the American Arbitration Association (AAA). AAA arbitration is relatively fast (usually several months from application to completion of hearing — which

INITIAL
CC
HH

is ordinarily much faster than the court system), perhaps a bit cheaper than the court system (much cheaper if the court system involves appeals), and allows you the option of bringing an attorney to the arbitration hearing.

In addition, the rules of evidence are much more relaxed in arbitration. The hearing is less formal than the court system hearing and a big advantage is that you may end up with an arbitrator who is familiar with the construction business.

One disadvantage to arbitration is that some arbitrators have a tendency to "split the difference" if they don't feel the case is clearly in favor of one side or the other. This makes it important to get legal representation (or at least advice) prior to going into an arbitration hearing. The failure to get some legal advice has made many a contractor unhappy when they open up the letter that contains the decision against them! Once again, an ounce of prevention...

Attorney's fees are ordinarily awarded only if they have been agreed to in the Construction Agreement, although the AAA arbitrator has the authority to award attorney's fees and costs. An attorney's fees clause is a very good idea if you do good work and don't think you will be in the wrong if a dispute arises. Being able to tell the other side, "You'll not only have to pay what you owe me, but also my attorney's fees," can be a helpful negotiating lever when a dispute arises.

K. INSURANCE

Owner shall pay for and maintain "Course of Construction" or "Builder's Risk" or any other insurance that provides the same type of coverage to the Contractor's work in progress during the course of the project. It is Owner's express responsibility to insure dwelling and all work in progress against all damage caused by fire and Acts of God such as earthquakes, floods, etc.

■ On a larger project, it is actually helpful to both parties to make sure the Owner has contacted his insurance company and informed them of the work about to take place so that the insurance company can insure the work in progress. By doing so, if the work in progress is destroyed though no fault of the Contractor, the Owner will have the money to rebuild the project and the Contractor will have a much better chance of being paid for any money left owing to him on the partially completed project — and probably the chance to rebuild it.

L. WARRANTY

Contractor provides a limited warranty on all Contractor- and Subcontractor-supplied labor and materials used in this project for a period of one year following substantial completion of all work.

No warranty is provided by Contractor on any materials furnished by the Owner for installation. No warranty is provided on any existing materials that are moved and/or reinstalled by the Contractor

INITIAL
CC
HH

within the dwelling (including any warranty that existing/used materials will not be damaged during the removal and reinstallation process). One year after substantial completion of the project, the Owner's sole remedy (for materials and labor) on all materials that are covered by a manufacturer's warranty is strictly with the manufacturer, not with the Contractor.

Repair of the following items is specifically excluded from Contractor's warranty: Damages resulting from lack of Owner maintenance; damages resulting from Owner abuse or ordinary wear and tear; deviations that arise such as the minor cracking of concrete, stucco and plaster; minor stress fractures in drywall due to the curing of lumber; warping and deflection of wood; shrinking/cracking of grouts and caulking; fading of paints and finishes exposed to sunlight.

THE EXPRESS WARRANTIES CONTAINED HEREIN ARE IN LIEU OF ALL OTHER WARRANTIES, EXPRESS OR IMPLIED, INCLUDING ANY WARRANTIES OF MERCHANTABILITY, HABITABILITY, OR FITNESS FOR A PARTICULAR USE OR PURPOSE. THIS LIMITED WARRANTY EXCLUDES CONSEQUENTIAL AND INCIDENTAL DAMAGES AND LIMITS THE DURATION OF IMPLIED WARRANTIES TO THE FULLEST EXTENT PERMISSIBLE UNDER STATE AND FEDERAL LAW.

> ■ It's a good idea to state the scope of your company's warranty on work performed. State any limitations to your warranty — such as excluding warranty on Owner-supplied materials, inability to guarantee that materials being removed and later reinstalled won't suffer some damage while being removed, and the fact that after the Contractor's warranty period is up, any extended product warranties are to be pursued with the manufacturer of the product, not with the Contractor.

M. ENTIRE AGREEMENT, SEVERABILITY, AND MODIFICATION

This Agreement represents and contains the entire agreement between the parties. Prior discussions or verbal representations by Contractor or Owner that are not contained in this Agreement are *not* a part of this Agreement. In the event that any provision of this Agreement is at any time held by a Court to be invalid or unenforceable, the parties agree that all other provisions of this Agreement will remain in full force and effect. Any future modification of this Agreement must be made in writing and executed by Owner and Contractor in order to be valid and binding upon the parties.

> ■ The clause above is necessary because it states that the agreement of the parties is limited to what is actually in the written contract. Pre-contract signing and verbal representations from the Contractor to the Owner or from the Owner to the Contractor that are not included in the Agreement are not legal and binding parts of the Agreement.
>   This clause reduces the possibility of the Owner coming back to you in the paint phase, for instance, and saying, "You told me if I gave you the job, you'd consider changing all the paint-grade trim, casings and baseboards to stain-grade oak; now do it!" In this type of situation the Con-

INITIAL
CC
HH

tractor can point to this clause and say that the contract contained their entire agreement and did not include the extra work the Owner is now demanding.

The Section above also contains a clause stating that future modifications to the Agreement must be made in writing and signed by both parties. This means the terms of the Agreement can't be verbally modified. This provides obvious protection to both parties — but the Contractor still should not make oral promises to the Owner that he may later have a hard time keeping (e.g., "This concrete sidewalk won't show a single crack for the next 20 years because I have the very best concrete man in the entire area." Don't say these kinds of things because they may create warranty or other legal obligations that otherwise wouldn't exist!

N. ADDITIONAL LEGAL NOTICES REQUIRED BY STATE OR FEDERAL LAW

See page attached: __X__ Yes _____ No

■ Be sure to include all notices required by the Contractor's state license board or other state or federal agency which governs home improvement contracts entered into by a prime contractor and a homeowner! Failure to include these notices in some states subjects the Contractor to a civil penalty fine and a much more expensive disciplinary bond at the time of bond renewal. About 90% of the contracts I see don't include these notices, but should!

If you contract directly with a homeowner who uses the house as his personal residence, consider yourself a prime contractor who needs to include any and all required notices.

The purpose of these notices is primarily to inform the homeowner of his rights and legal exposure when contracting for residential home improvements.

These notices can be changed by state legislatures on a regular basis. Also, these required notices vary from state to state. Consult your construction attorney and your state agency that governs Contractors to find out what requirements apply in your state for your type of contracting business.

O. ADDITIONAL TERMS AND CONDITIONS

See page attached: _____ Yes __X__ No

■ The clause above has been placed in the Agreement to remind you that you are acting as your own attorney in using these agreements and to remind you that the forms as presented are a starting point and not necessarily entirely suited in their present form to your purpose without some additional contract language or modifications.

INITIAL
CC
HH

This clause specifically indicates that you may need to add additional clauses to the Agreement based on the unique needs of your business, the disposition of the Owner, or the particular job that is the subject of the contract. The additional contract language you may need to add is beyond the scope of any form agreement in this book.

You may also decide to delete certain clauses from the Agreement depending upon the same factors mentioned above. If you work with a word processor, modifications to the Agreement will be fast and simple to make. Simply add (or delete) any clauses necessary to your Agreement and then delete the last clause, "ADDITIONAL TERMS AND CONDITIONS." Remember to consult an attorney familiar with construction before making significant changes to any agreement.

I have read and understood, and I agree to, all of the terms and conditions in the Agreement above.

■ You need a statement that indicates the parties have read and agree to the terms and conditions of the Agreement. Be sure to have the Agreement signed by the Owner prior to the time you commence work. Make sure you keep a signed copy of the Agreement in your records — an unsigned copy won't do you any good later on if you have a dispute. Finally, have the Owner initial every page of your Agreement, including any supplemental attachments, such as materials or Subcontractor bids.

Date: __5/22/01__   __CHARLIE CONTRACTOR, PRESIDENT__
CHARLIE CONTRACTOR, PRESIDENT
CHARLIE CONTRACTOR CONSTRUCTION, INC.

■ If your business is a corporation, be sure to sign the Agreement using your corporate title and place the word, "Inc.," after your company's name. If you fail to do this, you may have personal liability under the Agreement.

I then attach the Notice of Cancellation Form, any other notice required by state law, and all other bids listed in the Scope of Work as "enclosures" to this Agreement. Two copies of the Agreement should be given to the Owners, the original for signing, and a copy for their records. Each page of the original should be stamped with the initial stamp, and initialed by both you and the Owners. When the Owners have initialed and signed the Agreement, they should return it to you.

Date: __5/22/01__   __Harry Homeowner__
HARRY HOMEOWNER

Date: __5/22/01__   __Helen Homeowner__
HELEN HOMEOWNER

Remodelers are often called on by owners, architects, or designers to match existing finishes both inside and outside a building. On the exterior, the contractor may be required to match materials or textures such as roofing, siding (both materials and size), paint, stucco, brick, concrete finishes, etc. On the interior, he may be called on to match finish details such as paint, plaster, or drywall, wood flooring, wood trim details, or carpet.

The Pitfalls

It may seem obvious, but there are inherent pitfalls in trying to match new materials or finishes to existing ones. You can try to match something as closely as possible, but often an exact match is not possible. In other cases, an exact match is possible but prohibitively expensive. If this is not understood up front and addressed in some manner with the owner (namely, in your agreement), it can lead to a time-consuming and sometimes costly dispute between the contractor and the owner.

For example, even the most professional stucco patch will be noticeable unless the whole wall is painted a single color. Some materials, such as roofing shingles or carpeting fade over time and cannot be matched even if that color is still available. Other finishes, like varnish, darken over time making them difficult to match. Furthermore, in most pigmented products, dye lots change from one run to another. And in many cases a ceramic tile pattern, piece of hardware, type of brick, etc. simply is not available.

In other cases, the material is available, but at a big premium. For instance, let's say the plans call for matching the existing siding — a common scenario in remodeling work. After work begins, you find out that the existing siding is clear redwood that costs three times as much as the pine siding you factored into your bid and have already installed. However, the owner is demanding the redwood siding (even though it will be painted) and he wants you to remove the pine siding and install redwood siding solely at your own expense. The confusion in this area could easily cost you several thousand dollars.

Or let's say that the existing siding is close to a siding you thought was available, but after the contract is signed and the addition framed, you discover that to reach an exact match, you'll have to have knives cut to mill a custom batch of siding. This scenario also can cost you an additional thousand dollars or so (which wasn't factored into your bid).

Another example: you may assume it's fine with the owner to make small patches in the plaster walls using conventional drywall repair methods. The owner is terribly upset when he discovers you have done this.

Sometimes the construction agreement calls for the matching of existing finishes. Sometimes there is a simple note in the plans or specifications calling for the matching of an existing finish. Occasionally an owner will verbally ask you if you can match an existing texture, trim, or siding, and without thinking, or out of enthusiasm and a desire to please and secure the job, you'll blurt out, "No problem."

All of the scenarios above (including the verbal statement) can lead to a situation where you are legally obligated to "match an existing finish" at your own expense. Depending upon the competency of the matching job, the availability of stock materials which truly match existing materials in every way (type of material, size, etc.), and the degree to which the owner is reasonable about the matching efforts, you may or may not later have a serious dispute on your hands.

A small dispute in this area with a reasonable owner can often be quickly resolved. However, a large dispute in this area with an unreasonable owner can end up with the contractor either absorbing a large unanticipated expense for removal and replacement, or perhaps even with the cessation of the work and the filing of costly arbitration or litigation proceedings by the owner or contractor. Either situation erodes both profits and trust between the parties.

Clarify the Rules

The owner's discontent with matching finishes often shows up toward the end of the job when tensions are running high. The owner wants his castle back and the contractor wants his money. This is about the time the owner says that you should tear out an area and rework it because you didn't adequately "match the existing finish." It's then that you realize that this phrase was like a time bomb quietly ticking away.

What to do? Unfortunately, not much now, other than try to reach a reasonable agreement that doesn't cost you too much in lost profit. Next time, however, plan ahead. Because reasonable people can and do differ on the meaning of contract language — particularly on issues as vague as matching an existing finish — I always add clarifying language when I anticipate a potential matching problem (almost always the case on larger remodels, especially historical restoration projects).

This contract language serves to alert the owner to areas of work that may or may not meet their expectations and provides a road map for how to proceed. It puts the owner on notice that he needs to let the contractor know which areas of matching finishes are most critical to him. It also educates owners about the inherent difficulties in matching existing paint colors, textures, old-style wood trim, etc.

Another function of this type of contract language is to let the owner know there is no guarantee about matching certain finishes, It also reminds the contractor how important these details can be to the owner and forces a procedure on both the contractor and the owner which should help deter many future problems.

This contract clause also gives the owner two shots at test patches before he starts getting charged by the contractor for all the additional time involved in matching existing finishes. (Be sure also to communicate with the project architect in advance of providing test patches and sample materials to the owner. The architect will almost always want to have notice and the opportunity to be involved.) Two test patches, at no additional expense to the owner, should be enough to meet the reasonable expectations of most owners or architects in most situations. It certainly qualifies as a good faith effort on the part of the contractor. However, in the event that all this fails, it allows the contractor to say, "Of course I want you to be perfectly happy with the match of the finish and material. If this match or material isn't close enough, I'll rip it out and replace it as "extra work" on a time-and-materials change order basis, just as our agreement calls for."

If you clearly communicate your policy prior to signing the construction agreement, you'll avoid most potential disputes. On the other hand, if a problem arises when you're 70% through a job and relationships are strained (because everyone just wants the project to be finished), but you failed to install a procedure in your agreement to deal with "matching existing finishes," it can be difficult to come to an on-the-spot solution that seems fair to everyone. So when you have this matching requirement, plan ahead and avoid problems by better communication in your construction agreement.

How To Use the Addendum

If your agreements are on a word processor and you are writing up a job that involves the matching of existing finishes, it's better to simply incorporate the language as a clause in your agreement rather than present a separate addendum (see Form 1.3, Section III.G.1).

If you do not use a computer to generate your agreements, you may add this addendum to your agreements whenever you foresee a potential problem with matching existing finishes. Or, even if you use a computer to generate your primary agreement, a large change order could arise which involves the problematic matching of existing finishes. If this occurs, you can incorporate this addendum by reference into your change order in order to clarify and set the terms.

In order to avoid potential problems, be sure to have the addendum signed at the same time as the change order. The two documents must be signed concurrently because there must be new and current legal "consideration" (i.e., payment) supporting the obligations they contain. The legal requirement of "consideration" or a "bargained for exchange" must ordinarily be a current part of every agreement between parties. Agreements or addenda to contracts that lack current legal consideration may later be considered invalid and non-binding on the parties, even though they were signed by the parties.

For instance, if you have no match existing finishes clause in your contract, then sign a change order with the owner which involves matching existing finishes but you don't give the owner the addendum until two or three weeks after he signed the change order, the addendum could later be voided — even though the owner signed it — because it was not supported by current consideration. The lesson: don't spring the addendum on the owner after the agreement being amended has been signed. As a practical matter, most owners wouldn't sign the addendum after the fact anyway, but even if they did, you could still have a problem.

You should include explanatory language in the text of your change order, along with your description of the work to be performed: e.g., "This work will be performed according to the terms and conditions in this Change Order and the terms and conditions contained in the match existing finishes addendum attached which is hereby incorporated into and made a part of this Change Order." And it's a good idea to reference the addendum in your change order because the addendum sets the standards (or limitations) by which the work will be performed.

Finally, it's a good idea to state in your scope of work description (whether in the agreement or the change order) the areas where you anticipate potential difficulty in matching existing finishes.

However, much of this extra work is unnecessary when the language of the addendum is included in the original agreement as a contract clause.

As always, have an attorney familiar with construction review any changes you make to your construction agreements.

1.4 ■ ADDENDUM FOR MATCHING EXISTING FINISHES

PARTIES: This Addendum is made and entered into on this _____ day of _____, 19__, between _____, hereinafter referred to as "Contractor"; and _____, hereinafter referred to as "Owner."

This is an addendum to the Agreement between Contractor and Owner dated: _____.
The project is located at: _____.

In consideration of the mutual promises and obligations contained in the Agreement between Owner and Contractor referred to above which is hereby incorporated by reference, the parties agree as follows:

MATCHING EXISTING FINISHES AND MATERIALS: Where Contractor's work involves the "matching of existing finishes or materials," Contractor will use his best efforts to match existing finishes and materials. However, an exact match is *not* guaranteed by Contractor due to such factors as discoloration from aging, a difference in dye lots, discontinuation of product lines, and the difficulty of exactly matching certain finishes, colors, and planes.

Custom milling of materials has not been included in this Agreement, unless specifically stated in the Scope of Work section above. Unless *custom milling of materials* is specifically called for in the plans, specifications, or Scope of Work description above, any material not readily available at local lumberyards or suppliers is *not* included in this Agreement.

If Owner requires an exact match of materials or textures in a particular area, Owner must inform Contractor of this requirement in writing within seven (7) days of signing this Agreement. Contractor will then provide Owner with either a materials sample or test patch prior to the commencement of work involving the matching of existing finishes.

Owner must then approve or disapprove of the suitability of the match within 24 hours. After that time, or after Contractor has provided Owner with two or more test patches that have been rejected by Owner, all further test patches, materials submittals, or any removal and replacement of materials already installed in accordance with the terms of this section will be performed strictly as Extra Work on a time-and-materials basis by Contractor.

I have read and understood, and I agree to, all of the terms and conditions contained in this Addendum.

Date: _____   _____
                    CONTRACTOR'S SIGNATURE

Date: _____   _____
                    OWNER'S SIGNATURE

State and federal consumer protection laws have given the owner the right to cancel or rescind many residential construction contracts within three days of entering into those contracts. The contractor should check with his state contractor's license board or his local attorney to find out which rules apply to his projects and obtain the necessary forms and notices. Your state may have adopted a different more detailed form than is required under federal law.

The contractor must inform the owner of these rights or he technically faces the possibility that the owner will not be legally obligated to pay him for labor and materials.

Under federal law, the owner must be informed that he has the right of rescission when the contractor provides the construction funds for the work (unlikely with most builders). However, this right of rescission is also triggered when the contractor has the right to place a mechanic's lien on the property. Since this is a right residential contractors have on nearly all private projects, the right of rescission form is usually required. The law requires that each owner of the property be given two copies of the notice. Have the owners initial one copy for your files so you have a record of their receipt of the notice.

The Notice of Right of Rescission (see sample on page 92) should tell the owner that he can rescind the contract within three days from entering into the transaction. It must tell the owner exactly how to rescind the contract. It must also tell the owner that the contractor has the right to place a lien on the home and foreclose on the lien. Some states have modified this form so be sure to determine if your state has a particular format to follow.

The right of cancellation is similar to the right of rescission. The right of cancellation similarly informs the owner that he has three days to cancel the transaction. The owner is entitled to this right of cancellation under federal law when he enters into a contract in a location other than the contractor's customary place of business (typically his office) or the payment schedule calls for more than four payments. Since change orders may be considered additional payments, and since many contracts are signed in the customer's home versus the builder's office, it's safe to assume that the owner is entitled to receive the notice of cancellation on most residential construction projects.

The Notice of Cancellation form (see sample, facing page) explains that, in the event of cancellation by the owner, the owner's deposit will be returned by the contractor, and the owner must make available to the contractor any goods stored at the owner's premises. It also states exactly how and until what date the transaction can be canceled by the owner.

The contract you provide to the owner must also state in boldface 10-point type that he has the right to cancel the transaction.

The contractor should make sure that he verbally informs the owner of any rights he has of cancellation and rescission. The contractor should also be sure he has copies of these forms in his file acknowledging delivery to and receipt by the owner of all notices required by state and federal law.

Again, check with your state contractor's license board or local construction attorney to obtain the proper forms for these notices, and inquire whether your state places any additional requirements on the types of contracts that your firm enters into on a regular basis.

In the event of an emergency such as a fire, flood, hurricane, or earthquake, a waiver of this right can be obtained in writing from the owner. When you obtain the proper forms referred to above for use in your state, also obtain information on how to obtain the waivers in the event emergency work must be performed.

This notice procedure can seem burdensome if you haven't yet implemented it, but once you make it a standard practice, you'll realize how little time it takes. You'll also appreciate knowing that you are conforming to state and federal law and that you are not exposing yourself to the harsh consequences of failing to provide these notices.

As a practical matter, an unbelievably small percentage of owners ever cancel or rescind their contracts. I've made it a practice to simply include these forms with every residential contract I send out. If an owner comments on how much paperwork you've given him compared to other contractors he has worked with in the past, tell him to read these forms because they are designed to protect the consumer and are required by law.

■ Notice of Cancellation ■
(Notice Required by Federal Law)

DATE OF TRANSACTION: _____

You may cancel this transaction, without any penalty or obligation, within three business days from the above date.

If you cancel, any property traded in, any payments made by you under the contract or sale, and any negotiable instrument executed by you will be returned within 10 days following receipt by the seller of your cancellation notice, and any security interest arising out of the transaction will be canceled.

If you cancel, you must make available to the seller at your residence, in substantially as good condition as when received, any goods delivered to you under this contract or sale, or you may, if you wish, comply with the instructions of the seller regarding the return shipment of the goods at the seller's expense and risk.

If you do make the goods available to the seller and the seller does not pick them up within 20 days of the date of your notice of cancellation, you may retain or dispose of the goods without any further obligation. If you fail to make the goods available to the seller, or if you agree to return the goods to the seller and fail to do so, then you remain liable for performance of all obligations under the contract.

To cancel this transaction, mail or deliver a signed and dated copy of this cancellation notice, or any other written notice, or send a telegram to:

_____

NAME OF SELLER

at:

_____

ADDRESS OF SELLER'S PLACE OF BUSINESS

not later than midnight of: _____

DATE

I hereby cancel this transaction.

_____   _____

BUYER'S SIGNATURE   DATE

■ **NOTICE OF RIGHT OF RESCISSION** ■
(Notice Required by Federal Law)

You have entered into a transaction on _____ that could result in a lien, mortgage, or other security interest being placed on your property. You have a legal right under federal law to cancel this transaction, without penalty or cost within three business days from the date above or any later date on which all material disclosures required under the Truth in Lending Act have been given to you.

If you cancel the transaction, any lien, mortgage, or other security interest on your property arising from this transaction will automatically be void. Upon this cancellation, you will also be entitled to receive a refund of any down payment or other property or consideration.

If you decide to cancel this transaction, you may do so by simply notifying us of your decision to cancel this transaction at the address below by mail or telegram. Your decision to cancel must be in writing and must be sent on or before: _____ .

_____
CONTRACTOR'S NAME

_____
STREET

_____
CITY, STATE, ZIP CODE

You may use any form of written notice to cancel this transaction. You may also use this notice to cancel the transaction by dating and signing below:

I hereby cancel this transaction.

_____   _____
DATE   CUSTOMER

Customer hereby acknowledges receiving this notice: _____
(Initials)

CHAPTER 2

COST-PLUS AND LABOR-ONLY AGREEMENTS

2.1 ▪ COST-PLUS-FIXED-FEE AGREEMENT

2.2 ▪ COST-PLUS-PERCENTAGE-FEE AGREEMENT

2.3 ▪ LABOR-ONLY AGREEMENT

With typical cost-plus and labor-only agreements, the owner agrees to pay the contractor the contractually defined "cost of the work," plus the contractor's profit and overhead or a fixed fee. These types of agreements are based on a more cooperative or "partnership" style of relationship between the owner and contractor than are traditional fixed price agreements.

Cost-plus agreements are commonly referred to as time-and-materials agreements. Many contractors think of T&M agreements as very brief agreements which may not even include any profit and overhead and only very briefly define the contractor's compensation, e.g., "$35 per hour plus reimbursement for the cost of materials purchased by contractor for this project."

The cost-plus agreements below may be similar to the T&M agreements you are familiar with. However, cost-plus agreements do allow the contractor to charge profit and overhead. Also, the agreements I use go into more detail in defining the "cost of the work" for which the contractor is to be reimbursed because this is an area of typical dispute in T&M contracts.

Nevertheless, cost-plus and labor-only agreements have advantages and disadvantages for both the owner and the contractor.

Benefits for the Contractor

The primary advantage for the contractor with cost-plus and labor-only agreements is that the risk of losing money through inaccurate bidding or "Murphy's Law"

types of occurrences is greatly reduced because the contract defines what is and is not a direct, billable job cost. This helps prevent disputes over payments. The contractor's profit and overhead, depending on the type of agreement, is either calculated as a fixed fee or a percentage of the cost of the work.

Another advantage is that you don't need to have all the construction details and finishes completely spelled out at the time the contract is signed because the owner is paying the contractor the "cost of the work" however that work (within reason) is finally defined by the owner and the contractor.

As long as the project is built properly, the contractor is practically guaranteed to make a profit.

Drawbacks for the Contractor

What are some of the drawbacks of these types of contracts for the contractor?

One common disadvantage can be lower profits. Because the contractor's inherent risk of losing money on the project is reduced, the owner is often able to negotiate a lower profit and overhead rate than he might with a lump sum contract.

Another possible drawback for the contractor is that any money he saves through ingenuity, special discounts from suppliers, or other good fortune are passed directly on to the owner rather than being retained as profit because the owner is, by definition, paying the "cost of the work," plus the contractor's markup.

Another common risk the contractor faces is being

blamed for cost overruns which the owner may claim occurred because the contractor gave an unrealistically low ballpark estimate of the eventual final project cost or mismanaged the project. In fact, the costs may have risen mostly due to the *owner's* decisions and changes of mind through no fault or miscalculation of the contractor.

In this sense, a cost-plus or labor-only contract has a lot in common with a time-and-materials change order. In general, the contractor and the owner need to have a reasonably high level of faith in each other's sense of honesty and integrity or else the relationship is subject to rapid erosion at the first sign of cost overruns, or when the owner starts down the all-too-common path of upgrading finishes without realizing the significant impact this will have on his typically tight budget.

For this reason, cost-plus and labor-only agreements place the contractor in more of a *fiduciary* relationship (or legal position of trust) with the owner. The owner is relying on the contractor's representation that he will diligently complete the project in a workmanlike manner and won't "milk" the owner for more money through inefficient production or repeated field mistakes which are billed to the owner as direct job costs.

With a cost-plus-fixed-fee agreement or a labor-only-plus-fixed-fee agreement, this is not a concern because the contractor has no real incentive to have the job take longer or cost more. In fact, the less time the job takes, the sooner the contractor can collect his final draw and begin another project.

However, with cost-plus-percentage agreements and labor-only-plus-percentage agreements, the longer a job takes and the more it costs, the more money the contractor makes. Every owner is keenly aware of this fact and will let you know it sooner or later. This is where the trust comes in. This is also where a good detailed agreement helps.

If and when the owner starts to worry about his budget or the type of agreement he has entered into, you'll find it very helpful to have the exact scope of the "cost of the work" defined in your agreement.

Finally, if a contractor has trouble with record keeping and tracking receipts, he'll probably have a hard time with cost-plus agreements due to the fact that they are based on "actual costs incurred" and should require him to provide the owner with clear documentation for all expenses. If your record keeping is lax, you'll have a hard time collecting all the money you are owed for job costs.

Beware of Guaranteed Maximums

One control the owner may place on the contractor, which is not reflected in the agreements below, is a guaranteed maximum price clause (or not-to-exceed final project cost) in the construction agreement. This is not an unreasonable demand on the contractor as long as the cap price is realistic. It is appealing to owners because it gives them the security of a locked-in price with the opportunity to save money if the project costs less than expected.

If you don't have to provide a not-to-exceed price or guaranteed maximum price, don't. It can potentially hurt you, and ordinarily can't do you much good. For this reason, a not-to-exceed clause does not appear in the agreements below.

Nevertheless, if you are required to provide a not-to-exceed or guaranteed maximum price, and if the project is completed for less than the not-to-exceed or guaranteed maximum price, you can insist that the owner contractually agree to split the saved money with you. For example, if your not-to-exceed price is $200,000, you can draft a guaranteed maximum price clause like this:

"Contractor agrees the not-to-exceed, guaranteed maximum price for the work described in this agreement is $200,000, subject to an increase in this guaranteed maximum price if Owner increases either the Scope of Work or the quality of the finishes described in the construction documents. (All such increases will be put in writing and signed by both Contractor and Owner prior to the commencement of this additional work.)

Any costs exceeding the guaranteed maximum price or adjusted guaranteed maximum price will be paid for by Contractor. However, if the final cost of the work as defined by this Agreement is less than the guaranteed maximum price or adjusted guaranteed maximum price. Contractor will be paid 50% of the difference between the guaranteed maximum price or adjusted guaranteed maximum price and the final cost of the work under this Agreement."

For example, the guaranteed maximum price for this agreement is $200,000. The owner may order several upgrades to finishes such as cabinets and floor coverings that were not included in the original scope of work on which the contractor based his original $200,000 guaranteed maximum price. The cost of these owner upgrades amounts to $10,000.

Accordingly, the adjusted guaranteed maximum price is $210,000. If the final cost of the work under this agreement is $215,000, the contractor pays the $5,000 in excess of the adjusted guaranteed price. However, if the final cost of the work is $205,000, the owner will pay the contractor an additional $2,500 (50% of the $5,000 savings).

You may or may not be successful with this request, but it can't hurt to negotiate. If you do have to provide a guaranteed maximum price, make sure it is not unrealistically low and allows for Murphy's Law types of occurrences.

If you go 10% over the guaranteed number, you may end up paying for the owner's final 10% of the project out of your 10% profit on the job. Also, if you have a guaranteed maximum price, *you must keep track of changes in the scope of work and issue change orders for increases in the scope of work* so that your maximum price is constantly adjusted upwards as the owner increases the scope of the work or as legitimate change orders arise.

Remember, if your cost-plus agreement has a not-to-exceed price built into it, the scope of work and quality of the finishes *must be completely defined!* With vague plans and finishes, you'll leave yourself open to never-ending battles about what really was included in the guaranteed maximum price.

Nearly every owner has a budget. If the owner asks for a not-to-exceed price, you may be able to suggest instead that you will provide the owner with non-contractual bid sheets showing your "ballpark" estimates for the different phases of work on the project. This carries much less risk than a contractually binding maximum price.

When you provide "ballpark" estimates for the owner, make sure they are reasonably accurate and not unrealistically low. The owner will be relying on your estimate and if you knowingly are grossly negligent in estimating it too low, you are almost guaranteed to have a dispute on your hands later when the owner runs out of money and the project is only 80% complete. For this reason, avoid signing a contract and starting a project based on a sloppy "shoot from the hip" estimate.

Also, even though you may sign a cost-plus agreement, generally define with the owner the scope of work and quality of finishes anticipated so that if the owner runs out of money because of his constant upgrades rather than the cost of your work, you'll have a letter or plans or some type of written documentation which draws this distinction.

Advantages to the Owner

Cost-plus and labor-only agreements based on actual labor or project costs incurred can be beneficial to the owner because the owner pays only the actual costs incurred for the project work, plus the contractor's markup.

The owner does not need to pay for the contractor's contingency or "padding" which is sometimes factored into a lump sum bid to compensate the contractor for the inherent risks of the project. The owner also may be able to negotiate a slightly lower profit and overhead rate with the contractor because the contractor's risk with this type of contract is lower.

In addition, if the contractor — or the owner — is able to reduce the anticipated cost of certain phases of the work through his ingenuity, these savings will be passed on to the owner rather than being retained by the contractor as they might be with a fixed price contract.

When cost-plus agreements are executed well the owner pays no more or less than the actual cost to build his project, plus a negotiated markup. This seems inherently fair to both parties as long as both live up to their promises, and have a high level of integrity and competence.

Cost-Plus-Fixed-Fee Agreement

The cost-plus-fixed-fee agreement is inherently fair to both the contractor and owner. It is fair to the owner because it provides him with the opportunity to pay only for the actual cost of the work, plus a "fixed fee" covering the contractor's profit and overhead, and because the contractor has no financial incentive to increase project costs since he is not paid a percentage of these higher costs.

Rather, the contractor has an incentive to perform the work quickly and well because the sooner he completes the job, the sooner he collects the balance of his fixed fee and can begin another project. If he can save the owner money through more efficient methods which require less labor or use materials more efficiently, this will ordinarily shorten the amount of time spent on the project and the amount of money the owner pays for project labor.

Also, if the owner doesn't have a fully detailed set of plans or wants the contractor to spend more time on value engineering — that is, common sense, cost-saving ideas — the contractor is better able to provide these because he is not roped into a fixed price and fixed scope of work where every change requires the preparation and issuance of a change order.

This agreement is fair to the contractor because he can negotiate a reasonable fee for overhead and profit for the entire project without exposing himself to all the risks associated with fixed price contracts.

With cost-plus-fixed-fee agreements, you have the possibility of a win-win relationship. If the contractor comes up with ways to get the work done just as well for less money, he earns the same fee in less time and the owner gets the project completed for less money.

However, once the work begins, the contractor must take care to track any significant increases to the scope of work because he will normally want to increase his fixed fee accordingly.

2.1 ■ Cost-Plus-Fixed-Fee Agreement

CONTRACTOR'S NAME: _____

ADDRESS: _____

_____

PHONE: _____

FAX: _____

LIC #: _____

DATE: _____

OWNER'S NAME: _____

ADDRESS: _____

_____

PROJECT ADDRESS: _____

_____

I. PARTIES

This contract (hereinafter referred to as "Agreement") is made and entered into on this _____ day of
_____ , 19_____ , by and between _____ ,
(hereinafter referred to as "Owner"); and _____ ,
(hereinafter referred to as "Contractor"). In consideration of the mutual promises contained herein,
Contractor agrees to perform the following work:

II. GENERAL SCOPE OF WORK DESCRIPTION

_____

_____

_____

_____

_____

(Additional Scope of Work page(s) attached: _____ Yes _____ No)

III. GENERAL CONDITIONS FOR THE AGREEMENT ABOVE

A. CONTRACTOR'S DUTIES

Contractor acknowledges and accepts the relationship of trust implicit in this Construction Agreement.
Contractor agrees to use good efforts, judgment, and skills to complete the work according to the Contract

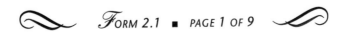

Documents referred to in this Agreement. Contractor agrees to furnish competent construction management and administration and to adequately supervise the work in progress. Contractor agrees to complete the work in a timely and workmanlike manner.

Contractor represents and warrants the following to Owner:

1. Contractor is financially solvent.

2. Contractor is able to furnish the tools, materials, supplies, equipment, and labor required to complete the work and perform his obligations hereunder and has sufficient experience and skills to do so.

3. Contractor will employ only skilled and properly trained staff for the performance of the work. Contractor will submit a "Rate Schedule for Contractor's Personnel" (see Section H below) which states the name and total rate charged for each of his employees who works on this project. Adjustments to the personnel on this list will be made by Contractor on an as-needed basis and Owner will be informed of all such changes.

B. CONTRACT DOCUMENTS
The Contract Documents consist of the following documents which are hereby incorporated by reference into this Agreement:

1. This Agreement.
2. Any plans, specifications, or addenda referred to in the General Scope of Work section above.
3. Other: _____

_____

_____

C. EXCLUSIONS
This Agreement does *not* include labor or materials for the following work:

1. PROJECT SPECIFIC EXCLUSIONS:

_____

_____

_____

_____

_____

_____

2. STANDARD EXCLUSIONS: Unless specifically included in the "General Scope of Work" section above, this Agreement does *not* include *labor or materials* for the following work (any Exclusions in this paragraph which have been lined out and initialed by the parties do not apply to this Agreement): Removal and disposal of any materials containing asbestos (or any other hazardous material as defined by the EPA).

Custom milling of any wood for use in project. Moving Owner's property around the site. Labor or materials required to repair or replace any Owner-supplied materials. Repair of concealed underground utilities not located on prints or physically staked out by Owner which are damaged during construction. Surveying that may be required to establish accurate property boundaries for setback purposes (fences and old stakes may not be located on actual property lines). Final construction cleaning (Contractor will leave site in "broom swept" condition). Landscaping and irrigation work of any kind. Temporary sanitation, power, or fencing. Removal of soils under house in order to obtain 18 inches (or code-required height) of clear space between bottom of joists and soil. Removal of filled ground or rock or any other materials not removable by ordinary hand tools (unless heavy equipment is specified in Scope of Work section above), correction of existing out-of-plumb or out-of-level conditions in existing structure. Correction of concealed substandard framing. Rerouting/removal of vents, pipes, ducts, structural members, wiring or conduits, steel mesh which may be discovered in the removal of walls or the cutting of openings in walls. Removal and replacement of existing rot or insect infestation. Failure of surrounding part of existing structure, despite Contractor's good faith efforts to minimize damage, such as plaster or drywall cracking and popped nails in adjacent rooms, or blockage of pipes or plumbing fixtures caused by loosened rust within pipes. Construction of continuously level foundation around structure (if lot is sloped more than 6 inches from front to back or side to side, Contractor will step the foundation in accordance with the slope of the lot). Exact matching of existing finishes. Repair of damage to existing roads, sidewalks, and driveways that could occur when construction equipment and vehicles are being used in the normal course of construction.

3. FEES FOR BUILDING PERMITS, PLANS, ENGINEERING & ARCHITECTURAL SERVICES, UTILITY CONNECTIONS, AND SPECIAL TESTING: This Agreement does not include the cost of coordinating or submitting for the permits, fees, and services referred to above. If Owner requests Contractor to coordinate any of these services or obtain any of the permits above, Contractor will perform this work on an hourly basis at the hourly rate of: $_____

Owner (not Contractor) is to enter into contracts for all of the above-mentioned services and provide direct payment to the people or agencies contracted with for all of the services and permit fees in the paragraph above.

If Owner requests that Contractor meet with Owner and architect or other design professionals to review the construction plans and specifications prior to completion of the final design documents, Contractor will perform this work on an hourly basis at the hourly rate of: $_____

D. DATE OF WORK COMMENCEMENT AND SUBSTANTIAL COMPLETION

Commence work:_____. Construction time through substantial completion: Approximately ___ to ___ weeks/months, *not* including delays and adjustments for delays caused by: holidays, inclement weather, accidents, shortage of labor or material, additional time required for performance of Change Order work (as specified in each Change Order), delays caused by Owner, and other delays unavoidable or beyond the control of the Contractor.

E. EXPIRATION OF THIS AGREEMENT

This Agreement will expire 30 days after the date at the top of page one of this Agreement if not accepted in writing by Owner and returned to Contractor within that time.

F. CONTRACTOR'S FEE

Owner will pay Contractor the Contract Sum consisting of the Cost of the Work as defined in Section H of this Agreement, plus, as compensation for Contractor's profit and overhead, a fixed fee of: $_____.

G. PROGRESS PAYMENTS

Based upon applications for payment *and all supporting documentation* submitted to Owner by Contractor on Thursday of: ___ every week / ___every other week, Owner shall make a progress payment to Contractor as provided below on the following Friday of that week (the next day). The amount of each progress payment shall be calculated as follows and paid on or before the Friday following the date on which Contractor submitted the payment request/invoice:

Add up the total Cost of the Work as defined in Section H below, which has been performed during the payment period, add the appropriate percentage of Contractor's Fee, and the total of these two amounts will be due each Friday to Contractor.

The percentage of Contractor's Fee due with each payment by Owner will be based on a "Schedule of Values" submitted by Contractor with this Agreement which shows the relative value of each phase of the work, e.g., foundation complete: 15% of Contractor's Fee; framing complete: 30% of Contractor's Fee; rough mechanical, rough electrical, rough plumbing complete: 15% of Contractor's Fee; etc.

H. COSTS TO BE REIMBURSED

Owner shall reimburse Contractor the Cost of the Work. The term "Cost of the Work" shall mean costs necessarily incurred by Contractor in good faith and in the proper performance of the work. The Cost of the Work shall include the items set forth in this section.

    1. LABOR COSTS: Wages of construction workers directly employed by Contractor to perform the construction work ("In-House Labor") will be paid as established by the Rate Schedule for Contractor's Personnel set forth below. This rate schedule is the gross amount to be charged for each worker (any and all applicable labor burden, medical and retirement benefits, bonuses, etc. have been factored into these rates).

RATE SCHEDULE FOR CONTRACTOR'S PERSONNEL

WORKER   RATE

A. _____ : $_____ PER HR.

B. _____ : $_____ PER HR.

C. _____ : $_____ PER HR.

D. _____ : $_____ PER HR.

E. _____ : $_____ PER HR.

F. _____ : $_____ PER HR.*

* Clerical time spent preparing payment applications.

2. CONTRACTOR'S SUPERVISORY PERSONNEL: When Contractor or Contractor's employee is performing both carpentry work and supervisory work, there shall be no duplication of payment for such labor (i.e., payment for both carpentry work and supervisory work at the same time).

Owner will be billed for Contractor or Contractor's supervisory personnel performing off-site coordination activities or off-site job-related meetings directly related to the progress of the work. This off-site time billed to Owner shall not exceed __ hours per week unless the off-site meeting is requested by Owner, or otherwise as agreed to in writing by Owner and Contractor.

All accounting work and documentation preparation in connection with payment applications is a direct job cost which will be performed at the rate of $____ per hour. Accounting and document preparation work is guaranteed not to exceed __ hours per payment application.

3. COST OF TIME SPENT PICKING UP MATERIALS AND MOBILIZING JOB: Time spent by Contractor and his employees at lumberyards and material supply houses (including travel time to and from) to pick up materials, and time required to move tools and equipment onto the job site at the start of the project and away from the site at the end of the project, is part of the Cost of Work.

4. SUBCONTRACT COSTS: Payments made to Subcontractors in accordance with the requirements of the Subcontracts is part of the Cost of Work for the project.

5. COST OF MATERIALS INCORPORATED INTO THE PROJECT: The cost of materials and equipment (and applicable sales tax, freight, or delivery charges) incorporated into the completed project is part of the Cost of the Work for the project. Any unused, excess materials shall be returned to the supplier for a credit. This credit will be issued to Owner along with the next billing by Contractor.

6. COSTS OF OTHER MATERIALS, EQUIPMENT, TEMPORARY EQUIPMENT, TAXES, SECURITY, AND RELATED ITEMS:
a. The cost of temporary fencing, temporary sanitation, monthly utility fees paid directly by Contractor, and rental equipment, including the costs of transporting and installing the equipment (if required).
b. The cost of consumable supplies which are consumed during the course of the project, e.g., circular saw blades, chalk, string line, reciprocating saw blades, wood stakes, forming lumber, pencils, etc.). No power tools or capital equipment will be paid for by Owner. Upon request by Owner, leftover consumable supplies will become the property of Owner at the end of the project.
c. Costs of removal of debris from the site, and hauling and dump fees.
d. Phone company charges for business-related phone calls.
e. The cost of all taxes on the project itself imposed by local, state, or federal agencies related to the work (not including taxes on employees and subcontractors which have already been factored by Contractor into the labor rates above).
f. Security costs required by Owner or deemed essential by Contractor.

7. EMERGENCY REPAIRS AND PRECAUTIONS: The Cost of the Work shall also include any actions taken in case of an emergency to prevent threatened damage, injury or loss to persons and property on the job site.

I. COSTS NOT TO BE REIMBURSED
The Cost of Work shall *not* include:

1. Any general insurance costs and state and federal taxes of Contractor (e.g., worker's compensation, comprehensive general liability insurance, auto insurance, health insurance, or labor burden expenses such as state and federal employer taxes, etc.). Contractor has factored these costs into the Rate Schedule for Contractor's Personnel in Section H.1 above, or these costs will be paid out of Contractor's profit and overhead percentage.

2. Travel time to and from the job site for Contractor and his employees. Costs associated with travel time such as: gas, vehicle maintenance, mileage payments, vehicle insurance, etc.

3. Costs to purchase, repair, and maintain Contractor's tools, vehicles, and equipment.

4. Cellular phone charges (unless specifically agreed to in writing by Owner and Contractor).

J. SUBCONTRACTS AND OTHER AGREEMENTS
Any portions of the work that Contractor chooses to subcontract shall be performed under appropriate subcontracts with Contractor.

Contractor will allow only skilled Subcontractors who are properly licensed, bonded, and insured in accordance with the terms of this Agreement to bid and perform work on this project.

K. ACCOUNTING RECORDS
Contractor shall keep full and detailed accounts and exercise such controls as may be necessary for detailed and responsible financial management of all aspects of this Agreement.

L. WORK PERFORMED BY OWNER OR OWNER'S SEPARATE CONTRACTORS AND MATERIALS FURNISHED BY OWNER
Prior to the time Contractor has entered into subcontracts, Owner may designate his own Subcontractors for Contractor to work with on the project. However, Contractor has the right to prequalify and approve Owner's Subcontractors and require evidence of work experience, proper licensing, and insurance.

Contractor has the right to refuse job-site access to Subcontractors who are not first prequalified by Contractor. Contractor's warranty will not extend to any work performed by Owner or Subcontractors who are not prequalified by Contractor. If Owner furnishes materials to the job site, Owner is responsible for verifying suitability of these materials prior to their delivery to the job site.

M. CHANGES IN THE WORK AND PERSONS AUTHORIZED TO ORDER CHANGES

1. PEOPLE AUTHORIZED TO MAKE DESIGN DECISIONS AFFECTING THE COST OF THE WORK: The following people are authorized to make design decisions which affect the Cost of the Work:

_____

_____

(Please fill in line(s) above at time of signing Agreement)

2. CHANGES IN THE WORK AND ADDITIONAL CONTRACTOR'S FEE: During the course of the work, Owner may request Contractor to perform additional work. Owner may also alter the selection of products or building design. Any significant decrease in the Scope of Work will reduce Contractor's Fee in an amount to be determined by Contractor.

Any significant increase in the Scope of Work (an increase that extends the time required to complete the work under normal circumstances by more than __%, or causes Contractor to hire additional labor to complete the work) will result in a proportional increase in Contractor's Fee.

N. MISCELLANEOUS CONDITIONS

1. OWNER COORDINATION WITH CONTRACTOR: Owner agrees to promptly furnish to Contractor all details and decisions about unspecified construction finishes and to consent to or deny changes in the Scope of Work that may arise so as not to delay the progress of the Work. Owner agrees to furnish Contractor with continual access to the job site.

2. WORK STOPPAGE AND TERMINATION OF AGREEMENT FOR DEFAULT: Contractor shall have the right to stop all work on the project and keep the job idle if payments are not made to Contractor in accordance with the Payment Schedule in this Agreement, or if Owner repeatedly fails or refuses to furnish Contractor with access to the job site and /or product selections or information necessary for the advancement of Contractor's work. Simultaneous with stopping work on the project, the Contractor must give Owner written notice of the nature of Owner's default and must also give the Owner a 14-day period in which to cure this default.

If work is stopped due to any of the above reasons (or for any other material breach of contract by Owner) for a period of 14 days, and the Owner has failed to take significant steps to cure his default, then Contractor may, without prejudicing any other remedies Contractor may have, give written notice of termination of the Agreement to Owner and demand payment for all completed work and materials ordered through the date of work stoppage, and any other loss sustained by Contractor, including Contractor's Profit and Overhead at the rate of ____% on the balance of the incomplete work under the Agreement. Thereafter, Contractor is relieved from all other contractual duties, including all Punch List and warranty work.

3. INTEREST CHARGES: Interest in the amount of ____% per month will be charged on all late payments under this Agreement. "Late Payments" are defined as any payment not received within _____ days of receipt of invoice from Contractor.

4. CONTRACTOR NOT TO BE RELIED UPON AS ARCHITECT, ENGINEER, OR DESIGNER: The Contractor is *not* an architect, engineer, or designer. Contractor is not being hired to perform any of these services. To the extent that Contractor makes any suggestions in these areas, the Owner acknowledges and agrees that Contractor's suggestions are merely options that the Owner may want to review with the appropriate design professional. Contractor's suggestions are not a substitute for professional engineering, architectural, or design services, and are not to be relied on as such by Owner.

5. LIEN RELEASES: Upon request of Owner, Contractor and Subcontractors will issue appropriate lien releases prior to receiving final payment from Owner.

O. DISPUTE RESOLUTION AND ATTORNEY'S FEES

Any controversy or claim arising out of or related to this Agreement involving an amount of *less* than $5,000 (or the maximum limit of the court) must be heard in the Small Claims Division of the Municipal Court in the county where the Contractor's office is located. Any controversy or claim arising out of or related to this Agreement which is over the dollar limit of the Small Claims Court must be settled by binding arbitration administered by the American Arbitration Association in accordance with the Construction Industry Arbitration Rules. Judgment upon the award may be entered in any Court having jurisdiction thereof.

The prevailing party in any legal proceeding related to this Agreement shall be entitled to payment of reasonable attorney's fees, costs, and expenses.

P. INSURANCE

Owner shall pay for and maintain "Course of Construction" or "Builder's Risk" or any other insurance that provides the same type of coverage to the Contractor's work in progress during the course of the project. It is Owner's express responsibility to insure dwelling and all work in progress against all damage caused by fire and Acts of God such as earthquakes, floods, etc.

Q. WARRANTY

Contractor provides a limited warranty on all Contractor- and Subcontractor-supplied labor and materials used in this project for a period of one year following substantial completion of all work.

No warranty is provided by Contractor on any materials furnished by the Owner for installation. No warranty is provided on any existing materials that are moved and/or reinstalled by the Contractor within the dwelling (including any warranty that existing/used materials will not be damaged during the removal and reinstallation process). One year after substantial completion of the project, the Owner's sole remedy (for materials and labor) on all materials that are covered by a manufacturer's warranty is strictly with the manufacturer, not with the Contractor.

Repair of the following items is specifically excluded from Contractor's warranty: Damages resulting from lack of Owner maintenance; damages resulting from Owner abuse or ordinary wear and tear; deviations that arise such as the minor cracking of concrete, stucco and plaster; minor stress fractures in drywall due to the curing of lumber; warping and deflection of wood; shrinking/cracking of grouts and caulking; fading of paints and finishes exposed to sunlight.

THE EXPRESS WARRANTIES CONTAINED HEREIN ARE IN LIEU OF ALL OTHER WARRANTIES, EXPRESS OR IMPLIED, INCLUDING ANY WARRANTIES OF MERCHANTABILITY, HABITABILITY, OR FITNESS FOR A PARTICULAR USE OR PURPOSE. THIS LIMITED WARRANTY EXCLUDES CONSEQUENTIAL AND INCIDENTAL DAMAGES AND LIMITS THE DURATION OF IMPLIED WARRANTIES TO THE FULLEST EXTENT PERMISSIBLE UNDER STATE AND FEDERAL LAW.

R. ENTIRE AGREEMENT, SEVERABILITY, AND MODIFICATION

This Agreement represents and contains the entire agreement between the parties. Prior discussions or verbal representations by Contractor or Owner that are not contained in this Agreement are *not* a part of this Agreement. In the event that any provision of this Agreement is at any time held by a Court to be invalid or unenforceable, the parties agree that all other provisions of this Agreement will remain in full force and effect. Any future modification of this Agreement must be made in writing and executed by Owner and Contractor in order to be valid and binding upon the parties.

S. ADDITIONAL LEGAL NOTICES REQUIRED BY STATE OR FEDERAL LAW

See page(s) attached: _____ Yes; _____ No

T. ADDITIONAL TERMS AND CONDITIONS

See page(s) attached: _____ Yes; _____ No

I have read and understood, and I agree to, all of the terms and conditions in the Agreement above.

Date: _____   _____
                         CONTRACTOR'S SIGNATURE

Date: _____   _____
                         OWNER'S SIGNATURE

Date: _____   _____
                         OWNER'S SIGNATURE

2.1 ■ Cost-Plus-Fixed-Fee Agreement

ANNOTATED

Charlie Contractor Construction, Inc.
123 Hammer Lane
Anywhere, USA 33333
Phone: (123) 456-7890
Fax: (123) 456-7899
Lic#: 11111

DATE: **May 22, 2001**

OWNER'S NAME: **Mr. & Mrs. Harry Homeowner**
ADDRESS: **333 Swift St.**
Anywhere, USA 33333

PROJECT ADDRESS: **same**

I. PARTIES

This contract (hereinafter referred to as "Agreement") is made and entered into on this **22nd** day of **May** , 20 **01** , by and between **Harry and Helen Homeowner** , (hereinafter referred to as "Owner"); and **Charlie Contractor Construction, Inc.** , (hereinafter referred to as "Contractor"). In consideration of the mutual promises contained herein, Contractor agrees to perform the following work:

> ■ See annotation, Form 1.3: Section I. Parties.

II. GENERAL SCOPE OF WORK DESCRIPTION

1. Furnish all labor, materials, tools, equipment, and supervision to construct the addition to Owner's residence according to the plans by **Art Architect** , dated **August 1, 2001, six pages** .

(Additional Scope of Work page(s) attached: _____ Yes __X__ No)

> ■ Provide general details of the Scope of Work here. Refer to the latest edition of the plans and specifications on which you have based your bid (including any addenda). Many disputes occur because the Contractor does

INITIAL
CC
HH

not accurately define the Scope of Work. The Contractor's fixed fee in this type of agreement is based on the Contractor's agreement with Owner that Contractor will generally be performing no more than a specified quantity of work.

If the Scope of Work is significantly increased, Contractor has a basis for issuing a Change Order and increasing his Contractor's Fee because the project will take longer to complete, requiring more of his time and overhead. Therefore, within reason, be sure to accurately define the general Scope of Work in this section of the Agreement so that there is no future confusion over just how much work Contractor is expected to perform.

III. GENERAL CONDITIONS FOR THE AGREEMENT ABOVE

A. CONTRACTOR'S DUTIES

Contractor acknowledges and accepts the relationship of trust implicit in this Construction Agreement. Contractor agrees to use good efforts, judgment, and skills to complete the work according to the Contract Documents referred to in this Agreement. Contractor agrees to furnish competent construction management and administration and to adequately supervise the work in progress. Contractor agrees to complete the work in a timely and workmanlike manner.

Contractor represents and warrants the following to Owner:

1. Contractor is financially solvent.
2. Contractor is able to furnish the tools, materials, supplies, equipment, and labor required to complete the work and perform its obligations hereunder and has sufficient experience and skills to do so.
3. Contractor will employ only skilled and properly trained staff for the performance of the work. Contractor will submit a "Rate Schedule for Contractor's Personnel" (see Section H below) which states the name and total rate charged for each of his employees who works on this project. Adjustments to the personnel on this list will be made by Contractor on an as-needed basis and Owner will be informed of all such changes.

■ The clause above forms the basis of the foundation of trust which the Owner is relying on in signing this type of agreement with the Contractor. By generally enumerating the Contractor's duties to the Owner in the contract, the Owner has a contractual basis for relying on the Contractor's sense of good faith and fair dealing.

The Contractor is agreeing to complete the work in the amount of time specified by the Contract Documents. He is also agreeing to perform the work in a workmanlike manner, i.e., a manner that is consistent with the level of skill and judgment expected from a Contractor who regularly performs the class of work being described by the Contract Documents.

B. CONTRACT DOCUMENTS

The Contract Documents consist of the following documents which are hereby incorporated by reference into this Agreement:

INITIAL
CC
HH

1. This Agreement.
2. Any plans, specifications, or addenda referred to in the General Scope of Work section above.
3. Other: __N/A__

C. EXCLUSIONS

This Agreement does *not* include labor or materials for the following work:

1. PROJECT SPECIFIC EXCLUSIONS:

Electrical service panel upgrade, if required; any work on plumbing pipes that are above the bottom of the floor joists; exterior concrete flatwork of any kind; furnishing and installing towel bars, medicine chests, t.p. holders, mirrors, shower doors, or tub enclosures; fire sprinkler system (if required by local building department during plan check process).

> ■ Be sure to state what work you are *not* performing. If you are asked later to perform this work, you may have a claim for an increase in the Contractor's Fee because the excluded work will be considered "additional work" not within the scope of the original Agreement. It is very important to monitor this area with "fixed fee" types of cost-plus agreements so that you can be paid an additional Contractor's Fee if you perform additional work.

2. STANDARD EXCLUSIONS:

2. STANDARD EXCLUSIONS: Unless specifically included in the "General Scope of Work" section above, this Agreement does *not* include *labor or materials* for the following work (any Exclusions in this paragraph which have been lined out and initialed by the parties do not apply to this Agreement): Removal and disposal of any materials containing asbestos (or any other hazardous material as defined by the EPA). Custom milling of any wood for use in project. Moving Owner's property around the site. Labor or materials required to repair or replace any Owner-supplied materials. Repair of concealed underground utilities not located on prints or physically staked out by Owner which are damaged during construction. Surveying that may be required to establish accurate property boundaries for setback purposes (fences and old stakes may not be located on actual property lines). Final construction cleaning (Contractor will leave site in "broom swept" condition). Landscaping and irrigation work of any kind. Temporary sanitation, power, or fencing. Removal of soils under house in order to obtain 18 inches (or code-required height) of clear space between bottom of joists and soil. Removal of filled ground or rock or any other materials not removable by ordinary hand tools (unless heavy equipment is specified in Scope of Work section above), correction of existing out-of-plumb or out-of-level conditions in existing structure. Correction of concealed substandard framing. Rerouting/removal of vents, pipes, ducts, structural members, wiring or conduits, steel mesh which may be discovered in the removal of walls or the cutting of openings in walls. Removal and replacement of existing rot or insect infestation. Failure of surrounding part of existing structure, despite Contractor's good faith efforts to minimize damage, such as plaster or drywall cracking and popped nails in adjacent rooms, or blockage of pipes or plumbing fixtures caused by loosened rust within pipes. Construction of continuously level foundation around structure

INITIAL
CC
HH

(if lot is sloped more than 6 inches from front to back or side to side, Contractor will step the foundation in accordance with the slope of the lot). Exact matching of existing finishes. Repair of damage to existing roads, sidewalks, and driveways that could occur when construction equipment and vehicles are being used in the normal course of construction.

> ■ See annotation, Form 1.3: Section III.A.2. Standard Exclusions.

3. FEES FOR BUILDING PERMITS, PLANS, ENGINEERING & ARCHITECTURAL SERVICES, UTILITY CONNECTIONS, AND SPECIAL TESTING: This Agreement does not include the cost of coordinating or submitting for the permits, fees, and services referred to above. If Owner requests Contractor to coordinate any of these services or obtain any of the permits above, Contractor will perform this work on an hourly basis at the hourly rate of: $ **35.00**

Owner (not Contractor) is to enter into contracts for all of the above-mentioned services and provide direct payment to the people or agencies contracted with for all of the services and permit fees in the paragraph above.

If Owner requests that Contractor meet with Owner and architect or other design professionals to review the construction plans and specifications prior to completion of the final design documents, Contractor will perform this work on an hourly basis at the hourly rate of: $ **35.00**

> ■ Contractors have different preferences for handling the types of work described above. With some clients you may want to charge for this work; with others you may not. If you don't want to charge for this work, either remove the clause or write in "0" for the labor rate. Or, if you don't want involvement in the permit/design review phase, simply delete the clause or line it out and initial it.

D. DATE OF WORK COMMENCEMENT AND SUBSTANTIAL COMPLETION
Commence work: **June 15, 2001** . Construction time through substantial completion: Approximately **12** to **16** weeks, *not* including delays and adjustments for delays caused by: holidays, inclement weather, accidents, shortage of labor or material, additional time required for performance of Change Order work (as specified in each Change Order), delays caused by Owner, and other delays unavoidable or beyond the control of the Contractor.

> ■ See annotation, Form 1.3: Section III.B. Date of Work Commencement and Substantial Completion.

E. EXPIRATION OF THIS AGREEMENT
This Agreement will expire 30 days after the date at the top of page one of this Agreement if not accepted in writing by Owner and returned to Contractor within that time.

INITIAL
CC
HH

> ■ See annotation, Form 1.3: Section III.C. Expiration of This Agreement.

F. CONTRACTOR'S FEE

Owner will pay Contractor the Contract Sum consisting of the Cost of the Work as defined in Section H of this Agreement, plus, as compensation for Contractor's profit and overhead, a fixed fee of: $ __22,500.00__.

> ■ This is your profit and overhead for performing the work, and is usually based on the duration and cost of the job. This amount is often negotiated between the Owner and the Contractor. The amount should approximate your normal profit and overhead rate if this was a fixed price job.

G. PROGRESS PAYMENTS

Based upon applications for payment *and all supporting documentation* submitted to Owner by Contractor on Thursday of every week, Owner shall make a progress payment to Contractor as provided below on the following Friday of every week (the next day). The amount of each progress payment shall be calculated as follows and paid on or before the Friday following the date on which Contractor submitted the payment request/invoice:

Add up the total Cost of the Work as defined in Section H below, which has been performed during the payment period, add the appropriate percentage of Contractor's Fee, and the total of these two amounts will be due each Friday to Contractor.

The percentage of Contractor's Fee due with each payment by Owner will be based on a "Schedule of Values" submitted by Contractor with this Agreement which shows the relative value of each phase of the work, e.g., foundation complete: 15% of Contractor's Fee; framing complete: 30% of Contractor's Fee; rough mechanical, rough electrical, rough plumbing complete: 15% of Contractor's Fee; etc.

> ■ If you are being paid weekly or biweekly based on the amount of work completed plus a percentage of your fee, you need a "measuring stick" to determine how much of the Contractor's Fee is due. This measuring stick is the "Schedule of Values." By creating one and incorporating it into the contract, you can simply indicate what percentage of a particular phase of work has been completed during the billing period and invoice for that amount.
>
>   For example, if you have completed approximately one third of your framing during the weekly billing period and your Schedule of Values shows that the framing phase represents 30% of your $22,500 fee, then your Contractor's Fee would be calculated as follows:
>
>   .30 (framing phase) x $22,500 = $6,750 x .33 (percent completed) = $2,227.50

INITIAL
CC
HH

H. COSTS TO BE REIMBURSED

Owner shall reimburse Contractor the Cost of the Work. The term "Cost of the Work" shall mean costs necessarily incurred by Contractor in good faith and in the proper performance of the work. The Cost of the Work shall include the items set forth in this section.

1. LABOR COSTS: Wages of construction workers directly employed by Contractor to perform the construction work ("In-House Labor") will be paid as established by the Rate Schedule for Contractor's Personnel set forth below. This rate schedule is the gross amount to be charged for each worker (any and all applicable labor burden, medical and retirement benefits, bonuses, etc. have been factored into these rates).

RATE SCHEDULE FOR CONTRACTOR'S PERSONNEL

WORKER	RATE
A. __Charlie Contractor__ :	$ __35.00__ PER HR.
B. __Chuck Carpenter__ :	$ __28.00__ PER HR.
C. __Larry Laborer__ :	$ __18.00__ PER HR.
D. __Susan Secretary__ :	$ __18.00__ PER HR. *

* Clerical time spent preparing payment applications.

> ■ Be sure to make the labor rate you charge high enough to cover all taxes, insurance, and all other benefits and contributions you must pay on your workers.

2. CONTRACTOR'S SUPERVISORY PERSONNEL: When Contractor or Contractor's employee is performing both carpentry work and supervisory work, there shall be no duplication of payment for such labor (i.e., payment for both carpentry work and supervisory work at the same time).

Owner will be billed for Contractor or Contractor's supervisory personnel performing off-site coordination activities or off-site job-related meetings directly related to the progress of the work. This off-site time billed to Owner shall not exceed **6** hours per week unless the off-site meeting is requested by Owner, or otherwise as agreed to in writing by Owner and Contractor.

> ■ The off-site billable hours have been limited to six hours per week in this sample contract, unless the off-site meeting has been requested by the Owner. Change this off-site cap to suit your needs. Just keep in mind that the Owner's perception — especially with residential remodeling — is that most of the work is done at the job site. If the Owner is billed for hours and hours of off-site work each week, he is likely to feel he is being taken advantage of by the Contractor.

INITIAL
CC
HH

All accounting work and documentation preparation in connection with payment applications is a direct job cost which will be performed at the rate of ___**$18.00**___ per hour. Accounting and document preparation work is guaranteed not to exceed _**3**_ hours per payment application.

> ■ Similar to off-site work by the Contractor, accounting work has been clearly defined and limited so that the Contractor is paid something for this work, but the Owner has a cap on what the Contractor can charge for this invoice documentation. Again, change this figure to suit your needs and the size of the job.

3. COST OF TIME SPENT PICKING UP MATERIALS AND MOBILIZING JOB: Time spent by Contractor and his employees at lumberyards and material supply houses (including travel time to and from) to pick up materials, and time required to move tools and equipment onto the job site at the start of the project and away from the site at the end of the project, is part of the Cost of the Work for the project.

4. SUBCONTRACT COSTS: Payments made to Subcontractors in accordance with the requirements of the Subcontracts is part of the Cost of the Work for the project.

5. COST OF MATERIALS INCORPORATED INTO THE PROJECT: The cost of materials and equipment (and applicable sales tax, freight, or delivery charges) incorporated into the completed project is part of the Cost of the Work for the project. Any unused, excess materials shall be returned to the supplier for a credit. This credit will be issued to Owner along with the next billing by Contractor.

> ■ The credit for unused materials is only fair and bolsters the Owner's trust and confidence in the Contractor.

6. COSTS OF OTHER MATERIALS, EQUIPMENT, TEMPORARY EQUIPMENT, TAXES, SECURITY, AND RELATED ITEMS:

a. The cost of temporary fencing, temporary sanitation, monthly utility fees paid directly by Contractor, and rental equipment, including the costs of transporting and installing the equipment (if required).
b. The cost of consumable supplies which are consumed during the course of the project (e.g., circular saw blades, chalk, string line, reciprocating saw blades, wood stakes, forming lumber, pencils, etc.). No power tools or capital equipment will be paid for by Owner. Upon request by Owner, leftover consumable supplies will become the property of Owner at the end of the project.

> ■ It is very important to list consumable supplies as a direct job cost. Add to this list to suit your job. This can be a gray area with some Owners who expect the Contractor to supply these items. However, since they are literally "burned up" or fully consumed on the project, they are properly charged as direct job costs.

INITIAL
_CC_
_H H_

c. Costs of removal of debris from the site, and hauling and dump fees.

d. Phone company charges for business-related phone calls.

e. The cost of all taxes on the project itself imposed by local, state, or federal agencies related to the work (not including taxes on employees and subcontractors which have already been factored by Contractor into the labor rates above).

f. Security costs required by Owner or deemed essential by Contractor.

> ■ If you anticipate the need for fencing or other job-site security, let the Owner know this in writing so that he is not surprised when he sees this as a direct job cost.

7. EMERGENCY REPAIRS AND PRECAUTIONS: The Cost of the Work shall also include any actions taken in case of an emergency to prevent threatened damage, injury, or loss to persons and property on the job site.

> ■ This category could include tarping roofs or building temporary pedestrian barricades or other such activities designed to protect persons and property. If you know such work will be required, let the Owner know this in writing prior to signing the contract. Also, be sure to take these emergency precautions whenever reasonably necessary: tarping an exposed roof will cost everyone a lot less than replacing insulation, ceilings, and furniture if a rainstorm blasts your unprotected roof.

I. COSTS NOT TO BE REIMBURSED

The Cost of the Work shall *not* include:

1. Any general insurance costs and state and federal taxes of Contractor (e.g., worker's compensation, comprehensive general liability insurance, auto insurance, health insurance, or labor burden expenses such as state and federal employer taxes, etc.). Contractor has factored these costs into the Rate Schedule for Contractor's Personnel in Section H.1 above, or these costs will be paid out of Contractor's profit and overhead percentage.

> ■ The expenses in this category should be factored into either your hourly rates or your Contractor's Fee.

2. Travel time to and from the job site for Contractor and his employees. Costs associated with travel time such as: gas, vehicle maintenance, mileage payments, vehicle insurance, etc.

> ■ Travel time is not normally reimbursable. Gas, vehicle maintenance, mileage payments, auto insurance, etc. should be factored into your Contractor's Fee or profit and overhead rate.

INITIAL
CC
HH

3. Costs to purchase, repair, and maintain Contractor's tools, vehicles, and equipment.

> ■ Again, this should be factored into your Contractor's Fee or profit and overhead rate.

4. Cellular phone charges (unless specifically agreed to in writing by Owner and Contractor).

J. SUBCONTRACTS AND OTHER AGREEMENTS

Any portions of the work that Contractor chooses to subcontract shall be performed under appropriate subcontracts with Contractor.

Contractor will allow only skilled Subcontractors who are properly licensed, bonded, and insured in accordance with the terms of this Agreement to bid and perform work on this project.

K. ACCOUNTING RECORDS

Contractor shall keep full and detailed accounts and exercise such controls as may be necessary for detailed and responsible financial management of all aspects of this Agreement.

> ■ This clause makes it the Contractor's responsibility to maintain good recordkeeping and accounting practices. This is a reasonable expectation because the Owner will pay for the work based on these receipts. If the Contractor is sloppy about tracking receipts and job costs, it is most likely the Contractor who will suffer if the Owner refuses to pay.
>
> You'll need to devise a systematic approach to track the work performed by your in-house crew on a daily basis. This can be done either on the time sheets themselves, or in a daily job log. You will need to rely on this documentation if the Owner questions what your crew did during the 220 billable work hours listed on your invoice. Be sure to provide plenty of back-up paperwork with your invoices.

L. WORK PERFORMED BY OWNER OR OWNER'S SEPARATE CONTRACTORS AND MATERIALS FURNISHED BY OWNER

Prior to the time Contractor has entered into subcontracts, Owner may designate his own Subcontractors for Contractor to work with on the project. However, Contractor has the right to prequalify and approve Owner's Subcontractors and require evidence of work experience, proper licensing, and insurance.

Contractor has the right to refuse job-site access to Subcontractors who are not first prequalified by Contractor. Contractor's warranty will not extend to any work performed by Owner or Subcontractors who are not prequalified by Contractor. If Owner furnishes materials to the job site, Owner is responsible for verifying suitability of these materials prior to their delivery to the job site.

INITIAL
CC
HH

■ It is important to maintain the right to qualify the Owner's Subcontractors because you will be contractually liable for their performance just as if they were your own subs (unless you draft a clause relieving you of this liability and the Owner is willing to sign it).

M. CHANGES IN THE WORK AND PERSONS AUTHORIZED TO ORDER CHANGES

1. PEOPLE AUTHORIZED TO MAKE DESIGN DECISIONS AFFECTING THE COST OF THE WORK: The following people are authorized to make design decisions which affect the Cost of the Work:

Harry Homeowner

Helen Homeowner

(Please fill in line(s) above at time of signing Agreement)

2. CHANGES IN THE WORK AND ADDITIONAL CONTRACTOR'S FEE: During the course of the work, Owner may request Contractor to perform additional work. Owner may also alter the selection of products or building design. Any significant decrease in the Scope of Work will reduce Contractor's Fee in an amount to be determined by Contractor.

Any significant increase in the Scope of Work (an increase that extends the time required to complete the work under normal circumstances by more than **5**%, or causes Contractor to hire additional labor to complete the work) will result in a proportional increase in Contractor's Fee.

■ This clause is important because it entitles the Contractor to charge an additional fee if the Owner increases the Scope of Work significantly. Your fixed fee is based on an anticipated quantity of work. If the Owner significantly increases the Scope of Work, you should write up a Change Order increasing your fixed fee accordingly. Significant decreases in the Scope of Work will result in a decrease in your fee which you should determine based on how much less time the project will take you to complete. This is a break from the normal policy of the lump sum agreements which ordinarily do not credit the Contractor's overhead and profit back to the Owner with deductive Change Orders. Some credit for deductive Change Orders is due the Owner with a cost-plus agreement because the Contractor ordinarily has not invested the same amount of work bidding a T&M type of job as he has with a fixed price project.

N. MISCELLANEOUS CONDITIONS

1. OWNER COORDINATION WITH CONTRACTOR: Owner agrees to promptly furnish to Contractor all details and decisions about unspecified construction finishes and to consent to

INITIAL
CC
HH

or deny changes in the Scope of Work that may arise so as not to delay the progress of the Work. Owner agrees to furnish Contractor with continual access to the job site.

2. WORK STOPPAGE AND TERMINATION OF AGREEMENT FOR DEFAULT: Contractor shall have the right to stop all work on the project and keep the job idle if payments are not made to Contractor in accordance with the Payment Schedule in this Agreement, or if Owner repeatedly fails or refuses to furnish Contractor with access to the job site and /or product selections or information necessary for the advancement of Contractor's work. Simultaneous with stopping work on the project, the Contractor must give Owner written notice of the nature of Owner's default and must also give the Owner a 14-day period in which to cure this default.

If work is stopped due to any of the above reasons (or for any other material breach of contract by Owner) for a period of 14 days, and the Owner has failed to take significant steps to cure his default, then Contractor may, without prejudicing any other remedies Contractor may have, give written notice of termination of the Agreement to Owner and demand payment for all completed work and materials ordered through the date of work stoppage, and any other loss sustained by Contractor, including Contractor's Profit and Overhead at the rate of __20__% on the balance of the incomplete work under the Agreement. Thereafter, Contractor is relieved from all other contractual duties, including all Punch List and warranty work.

> ■ See annotation, Form 1.3: Section III.H. Work Stoppage and Termination of Agreement for Default.

3. INTEREST CHARGES: Interest in the amount of __1.5__% per month will be charged on all late payments under this Agreement. "Late Payments" are defined as any payment not received within __5__ days of receipt of invoice from Contractor.

> ■ See annotation, Form 1.3: Section III.F.7. Interest Charges.

4. CONTRACTOR NOT TO BE RELIED UPON AS ARCHITECT, ENGINEER, OR DESIGNER: The Contractor is *not* an architect, engineer, or designer. Contractor is not being hired to perform any of these services. To the extent that Contractor makes any suggestions in these areas, the Owner acknowledges and agrees that Contractor's suggestions are merely options that the Owner may want to review with the appropriate design professional. Contractor's suggestions are not a substitute for professional engineering, architectural, or design services, and are not to be relied on as such by Owner.

> ■ See annotation, Form 1.3: Section III.G.6. Contractor Not To Be Relied Upon as Architect, Engineer or Designer.

INITIAL
CC
HH

5. LIEN RELEASES: Upon request of Owner, Contractor and Subcontractors will issue appropriate lien releases prior to receiving final payment from Owner.

■ See annotation, Form 1.3: Section III.I. Lien Releases.

O. DISPUTE RESOLUTION AND ATTORNEY'S FEES

Any controversy or claim arising out of or related to this Agreement involving an amount of *less* than $5,000 (or the maximum limit of the court) must be heard in the Small Claims Division of the Municipal Court in the county where the Contractor's office is located. Any controversy or claim arising out of or related to this Agreement which is over the dollar limit of the Small Claims Court must be settled by binding arbitration administered by the American Arbitration Association in accordance with the Construction Industry Arbitration Rules. Judgment upon the award may be entered in any Court having jurisdiction thereof.

The prevailing party in any legal proceeding related to this Agreement shall be entitled to payment of reasonable attorney's fees, costs, and expenses.

■ See annotation, Form 1.3: Section III.J. Dispute Resolution and Attorney's Fees.

P. INSURANCE

Owner shall pay for and maintain "Course of Construction" or "Builder's Risk" or any other insurance that provides the same type of coverage to the Contractor's work in progress during the course of the project. It is Owner's express responsibility to insure dwelling and all work in progress against all damage caused by fire and Acts of God such as earthquakes, floods, etc.

■ See annotation, Form 1.3: Section III.K. Insurance.

Q. WARRANTY

Contractor provides a limited warranty on all Contractor- and Subcontractor-supplied labor and materials used in this project for a period of one year following substantial completion of all work.

No warranty is provided by Contractor on any materials furnished by the Owner for installation. No warranty is provided on any existing materials that are moved and/or reinstalled by the Contractor within the dwelling (including any warranty that existing/used materials will not be damaged during the removal and reinstallation process). One year after substantial completion of the project, the Owner's sole remedy (for materials and labor) on all materials that are covered by a manufacturer's warranty is strictly with the manufacturer, not with the Contractor.

Repair of the following items is specifically excluded from Contractor's warranty: Damages resulting from lack of Owner maintenance; damages resulting from Owner abuse or ordinary wear and tear; deviations that arise such as the minor cracking of concrete, stucco and plaster; minor stress

INITIAL
CC
HH

fractures in drywall due to the curing of lumber; warping and deflection of wood; shrinking/cracking of grouts and caulking; fading of paints and finishes exposed to sunlight.

THE EXPRESS WARRANTIES CONTAINED HEREIN ARE IN LIEU OF ALL OTHER WARRANTIES, EXPRESS OR IMPLIED, INCLUDING ANY WARRANTIES OF MERCHANTABILITY, HABITABILITY, OR FITNESS FOR A PARTICULAR USE OR PURPOSE. THIS LIMITED WARRANTY EXCLUDES CONSEQUENTIAL AND INCIDENTAL DAMAGES AND LIMITS THE DURATION OF IMPLIED WARRANTIES TO THE FULLEST EXTENT PERMISSIBLE UNDER STATE AND FEDERAL LAW.

> ■ See annotation, Form 1.3: Section III.L. Warranty.

R. ENTIRE AGREEMENT, SEVERABILITY, AND MODIFICATION

This Agreement represents and contains the entire agreement between the parties. Prior discussions or verbal representations by Contractor or Owner that are not contained in this Agreement are *not* a part of this Agreement. In the event that any provision of this Agreement is at any time held by a Court to be invalid or unenforceable, the parties agree that all other provisions of this Agreement will remain in full force and effect. Any future modification of this Agreement must be made in writing and executed by Owner and Contractor in order to be valid and binding upon the parties.

> ■ See annotation, Form 1.3: Section III.M. Entire Agreement, Severability, and Modification.

S. ADDITIONAL LEGAL NOTICES REQUIRED BY STATE OR FEDERAL LAW

See page(s) attached: __**X**__ Yes _____ No

> ■ See annotation, Form 1.3: Section III.N. Additional Legal Notices Required by State or Federal Law.

T. ADDITIONAL TERMS AND CONDITIONS

See page(s) attached: _____ Yes __**X**__ No

> ■ See annotation, Form 1.3: Section III.O. Additional Terms and Conditions.

I have read and understood, and I agree to, all of the terms and conditions in the Agreement above.

INITIAL
CC
HH

■ You need a statement that indicates the parties have read and agree to the terms and conditions of the Agreement. Be sure to have the Agreement signed by the Owner *prior* to the time you commence work. Make sure you keep a signed copy of the Agreement in your records — an unsigned copy won't do you any good later on if you have a dispute. Finally, have the Owner initial every page of your Agreement, including any supplemental attachments, such as materials or Subcontractor bids.

Date: __5/22/01__   **CHARLIE CONTRACTOR, PRESIDENT**
CHARLIE CONTRACTOR, PRESIDENT
CHARLIE CONTRACTOR CONSTRUCTION, INC.

■ If your business is a corporation, be sure to sign the Agreement using your corporate title and place the word, "Inc.," after your company's name. If you fail to do this, you may have personal liability under the Agreement.

I then attach the Notice of Cancellation Form, any other notice required by state law, and all other bids listed in the Scope of Work as "enclosures" to this Agreement. Two copies of the Agreement should be given to the Owners, the original for signing, and a copy for their records. Each page of the original should be stamped with the initial stamp, and initialed by both you and the Owners. When the Owners have initialed and signed the Agreement, they should return it to you.

Date: __5/22/01__   *Harry Homeowner*
HARRY HOMEOWNER

Date: __5/22/01__   *Helen Homeowner*
HELEN HOMEOWNER

The cost-plus-percentage-fee agreement is identical to the preceding cost-plus-fixed-fee agreement except that the contractor's fee is not fixed in the contract at a predetermined amount by the parties.

Rather, in a cost-plus-percentage-fee agreement the contractor's fee is calculated as a *percentage* of the contractually defined job costs. This percentage is typically in the 10% to 25% range depending on the size of the job, your company's overhead, and how much you want the work.

One advantage of this type of agreement is that you are somewhat less at risk if the owner significantly increases the scope of work. You still collect the profit and overhead rate you've set without having to renegotiate your fee. For that reason, this type of contract can work well for jobs that have many unknowns at bid time.

Otherwise, this type of cost-plus agreement is subject to all the advantages and disadvantages of the cost-plus-fixed fee agreement (Form 2.1) discussed at the beginning of this chapter.

Because the two types of contracts are so similar, this section does not contain the cost-plus-percentage-fee agreement in its full form. Rather, what follows are the three clauses you need to change to convert Form 2.1 to Form 2.2. Simply substitute these three clauses for their respective clauses in Form 2.1. The names and letter designations of the clauses remain the same in both contracts.

F. CONTRACTOR'S FEE

Owner will pay Contractor the Contract Sum consisting of the Cost of the Work as defined in Section H of this Agreement, plus a fixed percentage fee of ___% of the cost of all work as compensation for Contractor's profit and overhead.

G. PROGRESS PAYMENTS

Based upon applications for payment *and all supporting documentation* submitted to Owner by Contractor on Thursday of: ___ every week / ___every other week, Owner shall make a progress payment to Contractor as provided below on the following Friday of every week (the next day). The amount of each progress payment shall be calculated as follows and paid on or before the Friday following the date on which Contractor submitted the payment request/invoice:

Add up the total Cost of the Work as defined in Section H below, which has been performed during the payment period, add the appropriate percentage of Contractor's Fee, and the total of these two amounts will be due each Friday to Contractor.

M. CHANGES IN WORK AND ADDITIONAL CONTRACTOR'S FEE

During the course of the work, Owner may request Contractor to perform Additional Work. Owner may also alter the selection of products or building design. All such changes in the work will be performed by Contractor according to the terms and conditions in this Agreement.

In addition, Contractor will charge profit and overhead — either at the rate set in this Agreement or at the rate of ___% — on *all* Subcontractors designated by Owner to work on the project up until the point in time when Contractor has completed all of his work under this Agreement.

Labor-Only Agreement

The labor-only agreement is one in which the contractor furnishes only labor, no materials. There are numerous variations on how the contractor can charge for his labor. There are also variations on how the contractor can charge for his markup or contractor's fee (profit and overhead).

In one approach, the contractor charges a fixed price for his labor. For instance, the contractor can agree to furnish a fixed labor price of $28,500 to install the foundation and framing on a project. By providing a fixed price for a fixed scope of work, this type of contract is essentially like a standard lump sum or fixed price contract in which the contractor furnishes no materials. With this type of contract, the contractor ordinarily adds his profit and overhead into the lump sum labor price.

In one variation on this, the contractor provides no fixed price, but rather sets forth his labor rates and the general scope of work he'll perform. Then, the owner pays whatever labor is incurred on the project, plus the contractor's fee which is defined in the contract as either a fixed fee or a percentage of the total labor costs.

With either of these agreements, you can set a guaranteed maximum price if the owner requires it. However, this type of clause can lead to disputes if the scope of work is not clearly defined. See the discussion of the guaranteed maximum clause in the introduction to this chapter. The sample agreement that follows is a labor-only agreement on a cost-plus-percentage-fee basis. The contractor is paid the cost of the labor, plus a percentage fee as compensation for his profit and overhead.

2.3 ■ Labor-Only Agreement
(Cost-Plus-Percentage-Fee)

CONTRACTOR'S NAME: _____

ADDRESS: _____

_____

PHONE: _____

FAX: _____

LIC #: _____

DATE: _____

OWNER'S NAME: _____

ADDRESS: _____

_____

PROJECT ADDRESS: _____

_____

I. PARTIES

This contract (hereinafter referred to as "Agreement") is made and entered into on this _____ day of
_____ , 19_____ , by and between _____ ,
(hereinafter referred to as "Owner"); and _____ ,
(hereinafter referred to as "Contractor"). In consideration of the mutual promises contained herein,
Contractor agrees to perform the following work:

II. GENERAL SCOPE OF WORK DESCRIPTION

Contractor agrees to furnish labor *only* (no materials of any kind whatsoever) to complete the following work:

_____

_____

_____

_____

_____

_____

_____

(Additional Scope of Work page(s) attached: _____ Yes _____ No)

III. GENERAL CONDITIONS FOR THE AGREEMENT ABOVE

A. CONTRACTOR'S DUTIES

Contractor acknowledges and accepts the relationship of trust implicit in this Construction Agreement. The Contractor agrees to use good efforts, judgment, and skills to complete the work according to the Contract Documents referred to in this Agreement. Contractor agrees to furnish competent construction management and administration and to adequately supervise the work in progress. Contractor agrees to complete the work in a timely and workmanlike manner.

Contractor represents and warrants the following to Owner:
1. Contractor is financially solvent.
2. Contractor is able to furnish the tools and labor required to complete the work and perform its obligations hereunder and has sufficient experience and skills to do so.
3. Contractor shall furnish only skilled and properly trained staff for the performance of the work. Contractor will submit a "Rate Schedule for Contractor's Personnel" (see Section H. below) which states the name and total hourly rate charged for each worker who will work on this project as an employee of Contractor.

B. CONTRACT DOCUMENTS

The Contract Documents consist of the following documents which are hereby incorporated by reference into this Agreement:

1. This Agreement.
2. Any plans, specifications, or addenda referred to in the General Scope of Work section above.
3. Other:_____

C. EXCLUSIONS

This Agreement does *not* include materials of any kind whatsoever. This Agreement does *not* include labor for the following work:

1. PROJECT SPECIFIC EXCLUSIONS:

_____
_____
_____
_____
_____
_____

2. STANDARD EXCLUSIONS: Unless specifically included in the "General Scope of Work" section above, this Agreement does *not* include *labor or materials* for the following work (any Exclusions in this paragraph which have been lined out and initialed by the parties do not apply to this Agreement): Removal and disposal of any materials containing asbestos (or any other hazardous material as defined by the EPA). Custom milling of any wood for use in project. Moving Owner's property around the site. Labor or materials required to repair or replace any Owner-supplied materials. Repair of concealed underground utilities not located on prints or physically staked out by Owner which are damaged during construction. Surveying that may be required to establish accurate property boundaries for setback purposes (fences and old stakes may not be located on actual property lines). Final construction cleaning (Contractor will leave site in "broom swept" condition). Landscaping and irrigation work of any kind. Temporary sanitation, power, or fencing. Removal of soils under house in order to obtain 18 inches (or code-required height) of clear space between bottom of joists and soil. Removal of filled ground or rock or any other materials not removable by ordinary hand tools (unless heavy equipment is specified in Scope of Work section above), correction of existing out-of-plumb or out-of-level conditions in existing structure. Correction of concealed substandard framing. Rerouting/removal of vents, pipes, ducts, structural members, wiring or conduits, steel mesh which may be discovered in the removal of walls or the cutting of openings in walls. Removal and replacement of existing rot or insect infestation. Failure of surrounding part of existing structure, despite Contractor's good faith efforts to minimize damage, such as plaster or drywall cracking and popped nails in adjacent rooms, or blockage of pipes or plumbing fixtures caused by loosened rust within pipes. Construction of continuously level foundation around structure (if lot is sloped more than 6 inches from front to back or side to side, Contractor will step the foundation in accordance with the slope of the lot). Exact matching of existing finishes. Public or private utility connection fees. Repair of damage to existing roads, sidewalks, and driveways that could occur when construction equipment and vehicles are being used in the normal course of construction.

3. SUPPLEMENTAL OWNER OBLIGATIONS: Owner agrees to purchase all materials, rental equipment, and incidental services required for the completion of Contractor's work under this Agreement. No materials of any kind shall be purchased on the accounts of the Contractor. Owner will purchase both materials that will be fastened into the project (e.g., concrete and lumber, if applicable) and "consumable" materials required by Contractor to perform his work (e.g., saw blades, chalk, etc.).

Owner will pay for all debris removal (Contractor to leave debris in one pile on site), utilities, job-site sanitation (porta-potty), job-site power and water sufficient to perform Contractor's work, and job-site telephone for local calls if the work will take more than two weeks to complete. Owner will pay for any rental equipment, incidental Subcontractors (e.g., concrete pumpers and site contractors), and security measures reasonably required by Contractor to complete the work. At this time, Contractor expects the following security items, incidental Subcontractors, and rental equipment to be required:

a._____

b._____

c._____

4. BUILDING PERMITS, PLANS, ENGINEERING & ARCHITECTURAL FEES, UTILITY CONNECTION FEES AND SPECIAL TESTING FEES: This Agreement does not include the cost of coordinating, paying for or submitting for, the permits, fees, and services referred to above. If Owner requests Contractor to coordinate any of these services or obtain any of the permits above, Contractor will perform this work on an hourly basis at the hourly rate of: $_____.

Owner (not Contractor) is to enter into contracts for all of the above-mentioned services and provide direct payment to the people or agencies contracted with for all of the services and permit fees in the paragraph above.

If Owner requests that Contractor meet with Owner and architect or other design professionals to review the construction plans and engineering details prior to completion of the final design documents, Contractor will perform this work on an hourly basis at the hourly rate of: $_____.

D. DATE OF WORK COMMENCEMENT AND SUBSTANTIAL COMPLETION

Commence work:_____. Construction time through substantial completion: Approximately _____ to _____ weeks/months, *not* including delays and adjustments for delays caused by: holidays, inclement weather, accidents, shortage of labor or material, additional time required for performance of Change Order work (as specified in each Change Order), delays caused by Owner, and other delays unavoidable or beyond the control of the Contractor.

E. EXPIRATION OF THIS AGREEMENT

This Agreement will expire 30 days after the date at the top of page one of this Agreement if not accepted in writing by Owner and returned to Contractor within that time.

F. CONTRACTOR'S FEE

Owner will pay Contractor the Contract Sum consisting of the Cost of the Work as defined in Section H of this Agreement, plus a fixed percentage fee of ___% of the cost of all work as compensation for Contractor's profit and overhead.

G. PROGRESS PAYMENTS

Based upon applications for payment *and all supporting documentation* submitted to Owner by Contractor on Thursday of: ___ every week / ___every other week, Owner shall make a progress payment to Contractor as provided below on the following Friday of that week (the next day). The amount of each progress payment shall be calculated as follows and paid on or before the Friday following the date on which Contractor submitted the payment request/invoice:

Add up the total Cost of the Work as defined in Section H below, which has been performed during the payment period, add the appropriate percentage of Contractor's Fee, and the total of these two amounts will be due each Friday to Contractor.

H. COSTS TO BE REIMBURSED

Owner shall reimburse Contractor the Cost of the Work. The term "Cost of the Work" shall mean costs necessarily incurred by Contractor in good faith and in the proper performance of the work. The Cost of the Work shall include the items set forth in this section.

   1. LABOR COSTS: Wages of construction workers directly employed by Contractor to perform the construction work ("In-House Labor") will be paid as established by the Rate Schedule for Contractor's Personnel set forth below. This rate schedule is the gross amount to be charged for each worker and the Contractor (any and all applicable labor burden, medical and retirement benefits, bonuses, etc. have been factored into these rates).

RATE SCHEDULE FOR CONTRACTOR'S PERSONNEL

WORKER   RATE

A. _____ : $_____ PER HR.

B. _____ : $_____ PER HR.

C. _____ : $_____ PER HR.

D. _____ : $_____ PER HR.

E. _____ : $_____ PER HR.

2. CONTRACTOR'S SUPERVISORY PERSONNEL: When Contractor or Contractor's employee is performing both carpentry work and supervisory work, there shall be no duplication of payment for such labor (i.e., payment for both carpentry work and supervisory work at the same time).

Owner will be billed for Contractor or Contractor's supervisory personnel performing off-site coordination activities or off-site job-related meetings directly related to the progress of the work. This off-site time billed to Owner shall not exceed __ hours per week unless the off-site meeting is requested by Owner, or otherwise as agreed to in writing by Owner and Contractor.

All accounting work and documentation preparation with payment applications is a direct job cost which will be performed at the rate of $_____ per hour. Accounting and documentation preparation work is guaranteed not to exceed _____ hours per payment application.

3. COST OF TIME SPENT PICKING UP MATERIALS AND MOBILIZING JOB: Time spent by Contractor and his employees at lumberyards and material supply houses (including travel time to and from) to pick up materials, and time required to move tools and equipment onto the job site at the start of the project and away from the site at the end of the project, is part of the Cost of the Work.

4. PROFIT AND OVERHEAD: Contractor's profit and overhead at the rate of __% will be charged on all labor included in the Rate Schedule for Contractor's Personnel, above. The sum of the labor expenses in the above Rate Schedule and the profit and overhead on this labor amount will be the amount charged to Owner with each invoice from Contractor.

5. COSTS OF MATERIALS INCORPORATED INTO THE PROJECT: If Contractor purchases materials for the project due to Owner's inability to have materials at the site when needed by Contractor, Owner agrees to immediately reimburse Contractor for the cost of these materials, plus profit and overhead on these materials at the rate of __% if Contractor must purchase these materials on his own accounts.

6. EMERGENCY REPAIRS AND PRECAUTIONS: Taking action to prevent threatened damage, injury, or loss in case of an emergency which could affect the safety of persons and property on the site is part of the Cost of the Work.

I. COSTS NOT TO BE REIMBURSED BY OWNER

The following expenses shall *not* be reimbursed by owner:

1. Any general insurance costs and state and federal taxes of Contractor (e.g., worker's compensation, comprehensive general liability insurance, auto insurance, health insurance, or labor burden expenses such as state and federal employer taxes, etc.). Contractor has factored these costs into the Rate Schedule for Contractor's Personnel in Section H.1 above, or these costs will be paid out of Contractor's profit and overhead percentage.

2. Travel time to and from the job site for Contractor and his employees. Costs associated with travel time such as: gas, vehicle maintenance, mileage payments, vehicle insurance, etc.

3. Costs to purchase, repair, and maintain Contractor's tools, vehicles, and equipment.

4. Cellular phone charges (unless specifically agreed to in writing by Owner and Contractor).

J. WORK PERFORMED BY OWNER OR OWNER'S SEPARATE CONTRACTORS AND MATERIALS FURNISHED BY OWNER

Owner is responsible for supervising all of Owner's separate contractors. Contractor has no duty to supervise or coordinate Owner's separate contractors. Owner is responsible for verifying suitability and conformity of all materials he furnishes prior to their delivery to the job site.

K. CHANGES IN THE WORK AND ADDITIONAL CONTRACTOR'S FEE

During the course of the work, Owner may request Contractor to perform Additional Work. Owner may also alter the selection of products or building design. All such changes in the work will be performed by Contractor according to the terms and conditions in this Agreement.

L. MISCELLANEOUS CONDITIONS

1. OWNER COORDINATION WITH CONTRACTOR: Owner agrees to promptly furnish to Contractor all details and decisions about unspecified construction finishes and to consent to or deny changes in the Scope of Work that may arise so as not to delay the progress of the Work. Owner agrees to furnish Contractor with continual access to the job site.

2. WORK STOPPAGE AND TERMINATION OF AGREEMENT FOR DEFAULT: Contractor shall have the right to stop all work on the project and keep the job idle if payments are not made to Contractor in accordance with the Payment Schedule in this Agreement, or if Owner repeatedly fails or refuses to furnish Contractor with access to the job site and /or product selections or information necessary for the advancement of Contractor's work. Simultaneous with stopping work on the project, the Contractor must give Owner written notice of the nature of Owner's default and must also give the Owner a 14-day period in which to cure this default.

If work is stopped due to any of the above reasons (or for any other material breach of contract by Owner) for a period of 14 days, and the Owner has failed to take significant steps to cure his default, then Contractor may, without prejudicing any other remedies Contractor may have, give written notice of termination of the Agreement to Owner and demand payment for all completed work and materials ordered through the date of work stoppage, and any other loss sustained by Contractor, including

Contractor's Profit and Overhead at the rate of ___% on the balance of the incomplete work under the Agreement. Thereafter, Contractor is relieved from all other contractual duties, including all Punch List and warranty work.

3. INTEREST CHARGES: Interest in the amount of ___% per month will be charged on all late payments under this Agreement. "Late Payments" are defined as any payment not received within ___ days of receipt of invoice from Contractor.

4. CONTRACTOR NOT TO BE RELIED UPON AS ARCHITECT, ENGINEER, OR DESIGNER: The Contractor is *not* an architect, engineer, or designer. Contractor is not being hired to perform any of these services. To the extent that Contractor makes any suggestions in these areas, the Owner acknowledges and agrees that Contractor's suggestions are merely options that the Owner may want to review with the appropriate design professional. Contractor's suggestions are not a substitute for professional engineering, architectural, or design services, and are not to be relied on as such by Owner.

5. LIEN RELEASES: Upon request by Owner, Contractor and Subcontractors will issue appropriate lien releases prior to receiving final payment from Owner.

M. DISPUTE RESOLUTION AND ATTORNEY'S FEES

Any controversy or claim arising out of or related to this Agreement involving an amount of *less* than $5,000 (or the maximum limit of the court) must be heard in the Small Claims Division of the Municipal Court in the county where the Contractor's office is located. Any controversy or claim arising out of or related to this Agreement which is over the dollar limit of the Small Claims Court must be settled by binding arbitration administered by the American Arbitration Association in accordance with the Construction Industry Arbitration Rules. Judgment upon the award may be entered in any Court having jurisdiction thereof.

The prevailing party in any legal proceeding related to this Agreement shall be entitled to payment of reasonable attorney's fees, costs, and expenses.

N. WARRANTY

Contractor provides a limited warranty on his workmanship for a period of one year following substantial completion of all work.

No warranty is provided by Contractor on any materials furnished by the Owner for installation. No warranty is provided on any existing materials that are moved and/or reinstalled by the Contractor within the dwelling (including any warranty that existing/used materials will not be damaged during the removal and reinstallation process). One year after substantial completion of the project, the Owner's sole remedy (for materials and labor) on all materials that are covered by a manufacturer's warranty is strictly with the manufacturer, not with the Contractor.

Repair of the following items is specifically excluded from Contractor's warranty: Damages resulting from lack of Owner maintenance; damages resulting from Owner abuse or ordinary wear and tear; deviations that arise such as the minor cracking of concrete, stucco and plaster; minor stress fractures in drywall due to the curing of lumber; warping and deflection of wood; shrinking/cracking of grouts and caulking; fading of paints and finishes exposed to sunlight.

Because this is a Labor-Only Agreement, Contractor provides no warranty of any kind whatsoever on any of the materials used on this project or on the work performed by Owner's separate subcontractors.

THE EXPRESS WARRANTIES CONTAINED HEREIN ARE IN LIEU OF ALL OTHER WARRANTIES, EXPRESS OR IMPLIED, INCLUDING ANY WARRANTIES OF MERCHANTABILITY, HABITABILITY, OR FITNESS FOR A PARTICULAR USE OR PURPOSE. THIS LIMITED WARRANTY EXCLUDES CONSEQUENTIAL AND INCIDENTAL DAMAGES AND LIMITS THE DURATION OF IMPLIED WARRANTIES TO THE FULLEST EXTENT PERMISSIBLE UNDER STATE AND FEDERAL LAW.

O. ENTIRE AGREEMENT, SEVERABILITY, AND MODIFICATION

This Agreement represents and contains the entire agreement between the parties. Prior discussions or verbal representations by Contractor or Owner that are not contained in this Agreement are *not* a part of this Agreement. In the event that any provision of this Agreement is at any time held by a Court to be invalid or unenforceable, the parties agree that all other provisions of this Agreement will remain in full force and effect. Any future modification of this Agreement must be made in writing and executed by Owner and Contractor in order to be valid and binding upon the parties.

P. ADDITIONAL LEGAL NOTICES REQUIRED BY STATE OR FEDERAL LAW

See page(s) attached: _____ Yes; _____ No

Q. ADDITIONAL TERMS AND CONDITIONS

See page(s) attached: _____ Yes; _____ No

I have read and understood, and I agree to, all the terms and conditions contained in the Agreement above.

Date: _____   _____
CONTRACTOR'S SIGNATURE

Date: _____   _____
OWNER'S SIGNATURE

2.3 ■ LABOR-ONLY AGREEMENT
(COST-PLUS-PERCENTAGE-FEE)

ANNOTATED

Charlie Contractor Construction, Inc.
123 Hammer Lane
Anywhere, USA 33333
Phone: (123) 456-7890
Fax: (123) 456-7899
Lic#: 11111

DATE: **May 22, 2001**

OWNER'S NAME: **Mr. & Mrs. Harry Homeowner**
ADDRESS: **333 Swift St.**
Anywhere, USA 33333

PROJECT ADDRESS: **same**

I. PARTIES

This contract (hereinafter referred to as "Agreement") is made and entered into on this **22nd** day of **May**, 20**01**, by and between **Harry and Helen Homeowner**, (hereinafter referred to as "Owner"); and **Charlie Contractor Construction, Inc.**, (hereinafter referred to as "Contractor"). In consideration of the mutual promises contained herein, Contractor agrees to perform the following work:

> ■ See annotation, Form 1.3: Section I. Parties.

II. GENERAL SCOPE OF WORK DESCRIPTION

Contractor agrees to furnish labor *only* (no materials of any kind whatsoever) to complete the following work:

Furnish LABOR ONLY and supervision of Contractor's employees to complete the foundation and framing portion only of Owner's residence according to the plans by Art Architect, dated August 1, 2001, six pages.

INITIAL
CC
HH

Work to include: Contractor to make up materials list for framing and foundation work and place material orders on Owner's accounts. Provide labor (labor *only*, no materials) to set foundation forms (including required hand excavation which can not be performed by heavy equipment), set steel in forms, set all wet-set foundation hardware, pour concrete, strip forms, clean forms for reuse, frame all portions of residence, apply vapor barrier under wood siding to residence, install siding, install all exterior doors and windows, install all exterior door and window trim, install all exterior trim, stand required inspections by building department for foundation and rough framing.

Note To Owner: Contractor recommends Owner have all exterior doors pre-primed on all four sides prior to installation by Contractor, or primed within 24 hours of installation by Contractor in order to maintain manufacturer's warranty on doors. Owner should review installation instructions for Masonite siding to determine whether the joints need to be caulked immediately after installation of siding. Owner may want to have exterior trim boards back-primed and sealed prior to installation by Contractor in order to avoid bleeding of resins.

(Additional Scope of Work page(s) attached: _____ Yes __**X**__ No)

> ■ Provide general details of the Scope of Work here. Refer to the latest edition of the plans and specifications. Many disputes occur because the Contractor does not accurately define the Scope of Work.

III. GENERAL CONDITIONS FOR THE AGREEMENT ABOVE

A. CONTRACTOR'S DUTIES
Contractor acknowledges and accepts the relationship of trust implicit in this Construction Agreement. The Contractor agrees to use good efforts, judgment, and skills to complete the work according to the Contract Documents referred to in this Agreement. Contractor agrees to furnish competent construction management and administration and to adequately supervise the work in progress. Contractor agrees to complete the work in a timely and workmanlike manner.

Contractor represents and warrants the following to Owner:
1. Contractor is financially solvent.
2. Contractor is able to furnish the tools and labor required to complete the work and perform its obligations hereunder and has sufficient experience and skills to do so.
3. Contractor shall furnish only skilled and properly trained staff for the performance of the work. Contractor will submit a "Rate Schedule for Contractor's Personnel" (see Section H below) which states the name and total hourly rate charged for each worker who will work on this project as an employee of Contractor.

INITIAL
CC
HH

■ The clause above forms the basis of the foundation of trust which the Owner is relying on in signing this type of agreement with the Contractor. By generally enumerating the Contractor's duties to the Owner in the contract, the Owner has a contractual basis for relying on the Contractor's sense of good faith and fair dealing.

The Contractor is agreeing to complete the work in the amount of time specified by the Contract Documents. He is also agreeing to perform the work in a workmanlike manner, i.e., a manner that is consistent with the level of skill and judgment expected from a Contractor who regularly performs the class of work described by the Contract Documents.

B. CONTRACT DOCUMENTS

The Contract Documents consist of the following documents which are hereby incorporated by reference into this Agreement:

1. This Agreement.
2. Any plans, specifications, or addenda referred to in the General Scope of Work section above.
3. Other: **N/A**

C. EXCLUSIONS

This Agreement does *not* include materials of any kind whatsoever. This Agreement does *not* include labor for the following work:

1. PROJECT SPECIFIC EXCLUSIONS:

Materials of any kind to be incorporated into the residence; decks, exterior concrete flatwork, fences, driveways, any work outside the footprint of the house, installation of garage door, door hardware, weatherstripping or door bottoms. Concrete materials, concrete pumper. Consumable materials ordinarily purchased and consumed during the course of Contractor's work. Temporary power, phone, water. Locating buck outs, conduits or ufer grounds in foundation for other subtrades. Supervising other subtrade work. Cutting, patching, and blocking work for subtrades who are not on site when Contractor is performing his work on site. Painting, caulking, or patching work of any kind. Sheet metal work. Dumpsters, hauling of debris or dump fees of any kind. Excavator for foundation or site work. Scraping or grading any part of the site for foundation work, including finish grading. French drains. Moving any dirt except for the final cleaning of footings prior to the concrete pour.

■ Be sure to state work that you are *not* performing. While generally this is not as important in a labor-only percentage-fee agreement as it is with fixed price contracts, it is still a good idea and avoids future confusion.

INITIAL
CC
HH

2. STANDARD EXCLUSIONS: Unless specifically included in the "General Scope of Work" section above, this Agreement does *not* include *labor or materials* for the following work (any Exclusions in this paragraph which have been lined out and initialed by the parties do not apply to this Agreement): Removal and disposal of any materials containing asbestos (or any other hazardous material as defined by the EPA). Custom milling of any wood for use in project. Moving Owner's property around the site. Labor or materials required to repair or replace any Owner-supplied materials. Repair of concealed underground utilities not located on prints or physically staked out by Owner which are damaged during construction. Surveying that may be required to establish accurate property boundaries for setback purposes (fences and old stakes may not be located on actual property lines). Final construction cleaning (Contractor will leave site in "broom swept" condition). Landscaping and irrigation work of any kind. Temporary sanitation, power, or fencing. Removal of soils under house in order to obtain 18 inches (or code-required height) of clear space between bottom of joists and soil. Removal of filled ground or rock or any other materials not removable by ordinary hand tools (unless heavy equipment is specified in Scope of Work section above), correction of existing out-of-plumb or out-of-level conditions in existing structure. Correction of concealed substandard framing. Rerouting/removal of vents, pipes, ducts, structural members, wiring or conduits, steel mesh which may be discovered in the removal of walls or the cutting of openings in walls. Removal and replacement of existing rot or insect infestation. Failure of surrounding part of existing structure, despite Contractor's good faith efforts to minimize damage, such as plaster or drywall cracking and popped nails in adjacent rooms, or blockage of pipes or plumbing fixtures caused by loosened rust within pipes. Construction of continuously level foundation around structure (if lot is sloped more than 6 inches from front to back or side to side, Contractor will step the foundation in accordance with the slope of the lot). Exact matching of existing finishes. Repair of damage to existing roads, sidewalks, and driveways that could occur when construction equipment and vehicles are being used in the normal course of construction.

■ See annotation, Form 1.3: Section III.A.2. Standard Exclusions.

3. SUPPLEMENTAL OWNER OBLIGATIONS: Owner agrees to purchase all materials, rental equipment, and incidental services required for the completion of Contractor's work under this Agreement. No materials of any kind shall be purchased on the accounts of the Contractor. Owner will purchase both materials that will be fastened into the project (e.g., concrete and lumber, if applicable) and "consumable" materials required by Contractor to perform his work (e.g., saw blades, chalk, etc.).

Owner will pay for all debris removal (Contractor to leave debris in one pile on site), utilities, job-site sanitation (porta-potty), job-site power and water sufficient to perform Contractor's work, and job-site telephone for local calls if the work will take more than two weeks to complete. Owner will pay for any rental equipment, incidental Subcontractors (e.g., concrete pumpers and site contractors), and security measures reasonably required by Contractor to complete the work. At this time, Contractor expects the following security items, incidental Subcontractors, and rental equipment to be required:

a. Concrete pumper.

b. Site work contractor.

c. Fencing and gate around residence.

INITIAL
CC
HH

> ■ This exclusion covers work that the Owner agrees to undertake and should be included so that it is clear that this work is necessary and that the Owner will pay for this work.

4. BUILDING PERMITS, PLANS, ENGINEERING & ARCHITECTURAL FEES, UTILITY CONNECTION FEES AND SPECIAL TESTING FEES: This Agreement does not include the cost of coordinating, paying for or submitting for, the permits, fees, and services referred to above. If Owner requests Contractor to coordinate any of these services or obtain any of the permits above, Contractor will perform this work on an hourly basis at the hourly rate of: $ _35.00_

Owner (not Contractor) is to enter into contracts for all of the above-mentioned services and provide direct payment to the people or agencies contracted with for all of the services and permit fees in the paragraph above.

If Owner requests that Contractor meet with Owner and architect or other design professionals to review the construction plans and engineering details prior to completion of the final design documents, Contractor will perform this work on an hourly basis at the hourly rate of: $ _35.00_

> ■ Contractors have different preferences for handling the types of work described above. With some clients you may want to charge for this work; with others you may not. If you don't want to charge for this work, either remove the clause or write in "0" for the labor rate. Or, if you don't want involvement in the permit/design review phase, simply delete the clause or line it out and initial it.

D. DATE OF WORK COMMENCEMENT AND SUBSTANTIAL COMPLETION

Commence work: **June 15, 2001** . Construction time through substantial completion: Approximately _10_ to _14_ weeks, *not* including delays and adjustments for delays caused by: holidays, inclement weather, accidents, shortage of labor or material, additional time required for performance of Change Order work (as specified in each Change Order), delays caused by Owner, and other delays unavoidable or beyond the control of the Contractor.

> ■ See annotation, Form 1.3: Section III.B. Date of Work Commencement and Substantial Completion.

E. EXPIRATION OF THIS AGREEMENT

This Agreement will expire 30 days after the date at the top of page one of this Agreement if not accepted in writing by Owner and returned to Contractor within that time.

> ■ See annotation, Form 1.3: Section III.C. Expiration of This Agreement.

INITIAL
CC
HH

F. CONTRACTOR'S FEE

Owner will pay Contractor the Contract Sum consisting of the Cost of the Work as defined in Section H of this Agreement, plus a fixed percentage fee of __20__ % of the cost of all work as compensation for Contractor's profit and overhead.

> ■ Designate your profit and overhead rate above. This is the main way which this agreement is different from a fixed price labor-only agreement. The Contractor is paid a percentage of the cost of his labor, rather than a predetermined fixed fee.

G. PROGRESS PAYMENTS

Based upon applications for payment *and all supporting documentation* submitted to Owner by Contractor on Thursday of every week, Owner shall make a progress payment to Contractor as provided below on the following Friday of every week (the next day). The amount of each progress payment shall be calculated as follows and paid on or before the Friday following the date on which Contractor submitted the payment request/invoice:

Add up the total cost of the labor as defined in Section H below, which has been performed during the payment period, add the appropriate percentage of Contractor's Fee, and the total of these two amounts will be due each Friday to Contractor.

> ■ Add up the total cost of the labor as defined in Section H below, which has been performed during the payment period, add the appropriate percentage of Contractor's Fee, and the total of these two amounts will be due each Friday to Contractor.

H. COSTS TO BE REIMBURSED

Owner shall reimburse Contractor the Cost of the Work. The term "Cost of the Work" shall mean costs necessarily incurred by Contractor in good faith and in the proper performance of the work. The Cost of the Work shall include the items set forth in this section.

1. LABOR COSTS: Wages of construction workers directly employed by Contractor to perform the construction work ("In-House Labor") will be paid as established by the Rate Schedule for Contractor's Personnel set forth below. This rate schedule is the gross amount to be charged for each worker and the Contractor (any and all applicable labor burden, medical and retirement benefits, bonuses, etc. have been factored into these rates).

RATE SCHEDULE FOR CONTRACTOR'S PERSONNEL

WORKER	RATE
A. __Charlie Contractor__ :	$ __35.00__ PER HR.
B. __Chuck Carpenter__ :	$ __28.00__ PER HR.
C. __Larry Laborer__ :	$ __28.00__ PER HR.

INITIAL
CC
HH

■ Be sure to make the labor rate you charge high enough to cover all taxes, insurance, and all other benefits and contributions you must pay on your workers.

2. CONTRACTOR'S SUPERVISORY PERSONNEL: When Contractor or Contractor's employee is performing both carpentry work and supervisory work, there shall be no duplication of payment for such labor (i.e., payment for both carpentry work and supervisory work at the same time).

Owner will be billed for Contractor or Contractor's supervisory personnel performing off-site coordination activities or off-site job-related meetings directly related to the progress of the work. This off-site time billed to Owner shall not exceed __4__ hours per week unless the off-site meeting is requested by Owner, or otherwise as agreed to in writing by Owner and Contractor.

All accounting work and documentation preparation with payment applications is a direct job cost which will be performed at the rate of $__18__ per hour. Accounting and documentation preparation work is guaranteed not to exceed __2__ hours per payment application.

■ The off-site billable hours have been limited to two hours per week in this sample contract, unless the off-site meeting has been requested by the Owner. Change this off-site cap to suit your needs. Just keep in mind that the Owner's perception is that most of the work is done at the job site. If the Owner is billed for hours and hours of off-site work each week, he is likely to feel that the Contractor is taking advantage of him.

3. COST OF TIME SPENT PICKING UP MATERIALS AND MOBILIZING JOB: Time spent by Contractor and his employees at lumberyards and material supply houses (including travel time to and from) to pick up materials, and time required to move tools and equipment onto the job site at the start of the project and away from the site of the end at the project, is part of the Cost of the Work.

4. PROFIT AND OVERHEAD: Contractor's profit and overhead at the rate of __20__ % will be charged on all labor included in the Rate Schedule for Contractor's Personnel, above. The sum of the labor expenses in the above Rate Schedule and the profit and overhead on this labor amount will be the amount charged to Owner with each invoice from Contractor.

■ Rather than charging profit and overhead, you can instead charge a fixed fee. This is similar to the cost-plus-fixed-fee agreement shown earlier in this chapter, except in this case, you provide labor only, not materials.

   If you do this, your fixed fee is equivalent to your profit and overhead for performing the work. Usually this is roughly based on the size of the job. This amount is often negotiated between the Owner and the Contractor. This amount should approximate your normal profit and overhead rate if this was a fixed price job.

INITIAL
CC
HH

5. COSTS OF MATERIALS INCORPORATED INTO THE PROJECT: If Contractor purchases materials for the project due to Owner's inability to have materials at the site when needed by Contractor, Owner agrees to immediately reimburse Contractor for the cost of these materials, plus profit and overhead on these materials at the rate of __20__ % if Contractor must purchase these materials on his own accounts.

■ This clause lets the Owner know that the project requires consumable supplies and that these are a direct job cost. This can be a gray area with the Owner. Some Owners expect the Contractor to supply these items, however, these items are literally burned up or fully consumed on the project so they are more properly charged as direct job costs.

This clause also lets the Owner know that materials need to be charged on the Owner's account, not the Contractor's account. If they *are* purchased on the Contractor's account the Contractor will charge overhead and profit. Remember, this is a "Labor-Only" contract. Why be responsible for ordering the materials and having them on your account and not be paid profit and overhead on them? If that's what the Owner wants, use one of the cost-plus agreements from earlier in this chapter where you are making a profit and overhead markup on both the labor *and* the materials on the project.

If you have a dispute with the Owner over a small item and the Owner has say, $3,000 to $6,000 worth of materials charged on your accounts around town, you'll find yourself in a very poor negotiating position over that small item the Owner owes you for.

Also, if the Owner develops money problems and he has charged materials on your account which he can't pay for, either you'll end up paying for his materials, or you'll ruin your credit with your supplier while the supplier is deciding whether to sue just the Owner or the Owner *and* you (after all, it is your account).

If you anticipate the need for fencing or other job-site security let the owner know this in writing so that he is not surprised when you inform him of this requirement after the contract has been signed.

6. EMERGENCY REPAIRS AND PRECAUTIONS: Taking action to prevent threatened damage, injury, or loss in case of an emergency which could affect the safety of persons and property on the site is part of the Cost of the Work.

■ This category could include tarping roofs or building temporary pedestrian barricades or other such activities designed to protect persons and property. If you know such work will be required, let the Owner know this in writing prior to signing the contract. Also, be sure to take these emergency precautions whenever reasonably necessary: tarping an exposed roof will cost everyone a lot less than replacing insulation, ceilings, and furniture if a rainstorm blasts your unprotected roof.

INITIAL
CC
HH

I. COSTS NOT TO BE REIMBURSED BY OWNER

The following expenses shall *not* be reimbursed by owner:

1. Any general insurance costs and state and federal taxes of Contractor (e.g., worker's compensation, comprehensive general liability insurance, auto insurance, health insurance, or labor burden expenses such as state and federal employer taxes, etc.). Contractor has factored these costs into the Rate Schedule for Contractor's Personnel in Section H.1 above, or these costs will be paid out of Contractor's profit and overhead percentage.

> ■ The expenses in this category should be factored into either your hourly rates or your Contractor's Fee.

2. Travel time to and from the job site for Contractor and his employees. Costs associated with travel time such as: gas, vehicle maintenance, mileage payments, vehicle insurance, etc.

> ■ These costs are either not normally reimbursable, or, in the case of gas, vehicle maintenance, mileage payments, auto insurance, etc. should be factored into your Contractor's Fee or profit and overhead rate.

3. Costs to purchase, repair, and maintain Contractor's tools, vehicles, and equipment.

> ■ This should be factored into your Contractor's Fee or profit and overhead rate.

4. Cellular phone charges (unless specifically agreed to in writing by Contractor and Owner).

J. WORK PERFORMED BY OWNER OR OWNER'S SEPARATE CONTRACTORS AND MATERIALS FURNISHED BY OWNER

Owner is responsible for supervising all of Owner's separate contractors. Contractor has no duty to supervise or coordinate Owner's separate contractors. Owner is responsible for verifying suitability and conformity of all materials he furnishes prior to their delivery to the job site.

K. CHANGES IN THE WORK AND ADDITIONAL CONTRACTOR'S FEE

During the course of the work, Owner may request Contractor to perform Additional Work. Owner may also alter the selection of products or building design. All such changes in the work will be performed by Contractor according to the terms and conditions in this Agreement.

> ■ If you are working on a contract based on actual labor costs incurred plus a *fixed* Contractor's Fee, this clause is important because it entitles the Contractor to charge an additional fee if the Owner increases the Scope of Work significantly. Your fixed fee is based on an anticipated quantity of

INITIAL
CC
HH

work. If the Owner significantly increases the scope of this work, you should write up a Change Order increasing your fixed fee accordingly.

L. MISCELLANEOUS CONDITIONS

1. OWNER COORDINATION WITH CONTRACTOR: Owner agrees to promptly furnish to Contractor all details and decisions about unspecified construction finishes and to consent to or deny changes in the Scope of Work that may arise so as not to delay the progress of the Work. Owner agrees to furnish Contractor with continual access to the job site.

2. WORK STOPPAGE AND TERMINATION OF AGREEMENT FOR DEFAULT: Contractor shall have the right to stop all work on the project and keep the job idle if payments are not made to Contractor in accordance with the Payment Schedule in this Agreement, or if Owner repeatedly fails or refuses to furnish Contractor with access to the job site and /or product selections or information necessary for the advancement of Contractor's work. Simultaneous with stopping work on the project, the Contractor must give Owner written notice of the nature of Owner's default and must also give the Owner a 14-day period in which to cure this default.

If work is stopped due to any of the above reasons (or for any other material breach of contract by Owner) for a period of 14 days, and the Owner has failed to take significant steps to cure his default, then Contractor may, without prejudicing any other remedies Contractor may have, give written notice of termination of the Agreement to Owner and demand payment for all completed work and materials ordered through the date of work stoppage, and any other loss sustained by Contractor, including Contractor's Profit and Overhead at the rate of __20__ % on the balance of the incomplete work under the Agreement. Thereafter, Contractor is relieved from all other contractual duties, including all Punch List and warranty work.

> ■ See annotation, Form 1.3: Section III.H. Work Stoppage and Termination of Agreement for Default.

3. INTEREST CHARGES: Interest in the amount of __1.5__% per month will be charged on all late payments under this Agreement. "Late Payments" are defined as any payment not received within __5__ days of receipt of invoice from Contractor.

> ■ See annotation, Form 1.3: Section III.F.7. Interest Charges.

4. CONTRACTOR NOT TO BE RELIED UPON AS ARCHITECT, ENGINEER, OR DESIGNER: The Contractor is *not* an architect, engineer, or designer. Contractor is not being hired to perform any of these services. To the extent that Contractor makes any suggestions in these areas, the Owner acknowledges and agrees that Contractor's suggestions are merely options that the Owner may want to review with the appropriate design professional. Contractor's suggestions are

INITIAL
CC
HH

not a substitute for professional engineering, architectural, or design services, and are not to be relied on as such by Owner.

> ■ See annotation, Form 1.3: Section III.G.6 Contractor Not To Be Relied Upon as Architect, Engineer, or Designer.

5. LIEN RELEASES: Upon request by Owner, Contractor and Subcontractors will issue appropriate lien releases prior to receiving final payment from Owner.

> ■ See annotation, Form 1.3: Section III.I. Lien Releases.

M. DISPUTE RESOLUTION AND ATTORNEY'S FEES:

Any controversy or claim arising out of or related to this Agreement involving an amount of *less* than $5,000 (or the maximum limit of the court) must be heard in the Small Claims Division of the Municipal Court in the county where the Contractor's office is located. Any controversy or claim arising out of or related to this Agreement which is over the dollar limit of the Small Claims Court must be settled by binding arbitration administered by the American Arbitration Association in accordance with the Construction Industry Arbitration Rules. Judgment upon the award may be entered in any Court having jurisdiction thereof.

The prevailing party in any legal proceeding related to this Agreement shall be entitled to payment of reasonable attorney's fees, costs, and expenses.

> ■ For annotation, see Form 1.3 Annotated: Section III.J Dispute Resolution and Attorney's Fees.

N. WARRANTY

Contractor provides a limited warranty on his workmanship for a period of one year following substantial completion of all work.

No warranty is provided by Contractor on any materials furnished by the Owner for installation. No warranty is provided on any existing materials that are moved and/or reinstalled by the Contractor within the dwelling (including any warranty that existing/used materials will not be damaged during the removal and reinstallation process). One year after substantial completion of the project, the Owner's sole remedy (for materials and labor) on all materials that are covered by a manufacturer's warranty is strictly with the manufacturer, not with the Contractor.

Repair of the following items is specifically excluded from Contractor's warranty: Damages resulting from lack of Owner maintenance; damages resulting from Owner abuse or ordinary wear and tear; deviations that arise such as the minor cracking of concrete, stucco and plaster; minor stress fractures in drywall due to the curing of lumber; warping and deflection of wood; shrinking/cracking of grouts and caulking; fading of paints and finishes exposed to sunlight.

INITIAL
CC
HH

Because this is a Labor-Only Agreement, Contractor provides no warranty of any kind whatsoever on any of the materials used on this project or on the work performed by Owner's separate subcontractors.

THE EXPRESS WARRANTIES CONTAINED HEREIN ARE IN LIEU OF ALL OTHER WARRANTIES, EXPRESS OR IMPLIED, INCLUDING ANY WARRANTIES OF MERCHANTABILITY, HABITABILITY, OR FITNESS FOR A PARTICULAR USE OR PURPOSE. THIS LIMITED WARRANTY EXCLUDES CONSEQUENTIAL AND INCIDENTAL DAMAGES AND LIMITS THE DURATION OF IMPLIED WARRANTIES TO THE FULLEST EXTENT PERMISSIBLE UNDER STATE AND FEDERAL LAW.

■ See annotation, Form 1.3: Section III.L. Warranty.

O. ENTIRE AGREEMENT, SEVERABILITY, AND MODIFICATION

This Agreement represents and contains the entire agreement between the parties. Prior discussions or verbal representations by Contractor or Owner that are not contained in this Agreement are *not* a part of this Agreement. In the event that any provision of this Agreement is at any time held by a Court to be invalid or unenforceable, the parties agree that all other provisions of this Agreement will remain in full force and effect. Any future modification of this Agreement must be made in writing and executed by Owner and Contractor in order to be valid and binding upon the parties.

■ See annotation, Form 1.3: Section III.M. Entire Agreement, Severability, and Modification.

P. ADDITIONAL LEGAL NOTICES REQUIRED BY STATE OR FEDERAL LAW

See page(s) attached: __X__ Yes; _____ No

■ See annotation, Form 1.3: Section III.N. Additional Legal Notices Required by State or Federal Law.

Q. ADDITIONAL TERMS AND CONDITIONS

See page(s) attached: _____ Yes; __X__ No

■ See annotation, Form 1.3: Section III.O. Additional Terms and Conditions.

I have read and understood, and I agree to, all of the terms and conditions in the Agreement above.

INITIAL
CC
HH

■ You need a statement that indicates the parties have read and agree to the terms and conditions of the Agreement. Be sure to have the Agreement signed by the Owner prior to the time you commence work. Make sure you keep a signed copy of the Agreement in your records — an unsigned copy won't do you any good later on if you have a dispute. Finally, have the Owner initial every page of your Agreement, including any supplemental attachments, such as materials or Subcontractor bids.

Date: _5/22/01_   **_CHARLIE CONTRACTOR, PRESIDENT_**
                     CHARLIE CONTRACTOR, PRESIDENT
                     CHARLIE CONTRACTOR CONSTRUCTION, INC.

■ If your business is a corporation, be sure to sign the Agreement using your corporate title and place the word, "Inc.," after your company's name. If you fail to do this, you may have personal liability under the Agreement.
    I then attach the Notice of Cancellation Form, any other notice required by state law, and all other bids listed in the Scope of Work as "enclosures" to this Agreement. Two copies of the Agreement should be given to the Owners, the original for signing, and a copy for their records. Each page of the original should be stamped with the initial stamp, and initialed by both you and the Owners. When the Owners have initialed and signed the Agreement, they should return it to you.

Date: _5/22/01_   _Harry Homeowner_
                     HARRY HOMEOWNER

Date: _5/22/01_   _Helen Homeowner_
                     HELEN HOMEOWNER

CHAPTER 3

DESIGN-BUILD AGREEMENTS

3.1 ■ DESIGN-BUILD PRECONSTRUCTION SERVICES AGREEMENT

3.2 ■ DESIGN-BUILD AGREEMENT

Ordinarily, when a contractor builds a project, he has not designed the project and therefore he is not responsible for the implied accuracy or suitability of the plans. However, with the design-build approach, the contractor assumes part or all of the responsibility for the design work, as well as the usual responsibility for the construction of the project.

Along with this dual responsibility for both construction and design, the design-build contractor also inherits the potential conflict of interest that comes with wearing two hats, that of both designer and contractor.

If serious defects show up on the project, this places the owner in an unusually favorable position legally. The owner does not have to prove whether any patent (generally apparent) or latent (less apparent or concealed) defects were due to design errors/omissions or construction defects because the design-build contractor is legally responsible for all aspects of both design and construction.

The design-build contractor should take steps to reduce this potential conflict of interest and increased legal exposure by adhering to very high standards of ethics, construction, and communication in all dealings with the owner. The design-build contractor should also pay close attention to all written agreements he uses for this type of construction. Finally, contractors should not lightly venture into design-build construction as if it is merely a simple twist on the same old work. For the many reasons stated in this chapter, it's not!

One way to reduce some of the risk associated with design-build construction is to subcontract the design work to a separate entity. Or, have the owner contract separately with an outside architect or designer for all design work with you acting merely as a third party who puts the owner together with an architect or designer you have worked well with in the past.

Benefits of Design-Build

Whichever approach you select, the primary benefit of the design-build approach is that it dismantles some of the barriers in the sometimes conflicted relationship between the owner, architect, and contractor.

Traditionally, the owner hires the architect to interpret his ideas, design the project to suit his budget, and monitor the work of the contractor. The architect is also traditionally hired to make sure the contractor's work follows the plans and is of acceptable quality.

The owner/architect/contractor relationship can feel like a hierarchy with the owner on the top, the architect in the middle, and the contractor at the bottom. With reasonable people, this hierarchy works; with unreasonable people, this division between the players can cause endless friction as a result of the divergent interests that each party brings to the table.

With the design-build approach, on the other hand, all the parties potentially benefit from having the construction and design work proceed more cooperatively and seamlessly than in the traditional building scheme.

If the contractor and the designer are one and the same entity, or are working closely together, the contractor is likely to have more influence and control over the design and final cost of the project, and there is apt to be less job-site friction. The contractor may also have an advantage in marketing his business to owners who are in the preliminary stages of planning their project.

Construction costs may also be reduced since the contractor is often in a better position to design a project in line with the owner's target budget. Many a project has not been built because the owner's architect inadvertently designed a project that was 25% more expensive to build than the owner's budget allowed for.

Another major advantage with the design-build approach is that the contractor will no doubt find it easier to resolve design and construction problems that regularly arise in the field. In this case, the designer and builder are focused on solving the problem, not on who caused it.

This type of "one-stop shopping" can be attractive to an owner who has not yet hired an architect or firmed up his idea of how to proceed with a project. Presenting this cooperative approach to a project — where finding solutions rather finger-pointing is the goal — is also a relatively easy sell for the contractor.

The design-build approach to construction elicits a myriad of responses from residential contractors. Most contractors I know have never formally entered into the design-build arena. Informally, however, most of them have at one time or another drawn a simple sketch for a deck or garage, or slightly revised a floor plan or construction detail and watched the owner shake his head and say, "That's just what I had in mind — this will work. Build it for me."

In this sense, most contractors have worked on the fringe of design-build, perhaps without knowing it, and without directly being paid for their design. Depending on the specific situation, this can be relatively harmless or rather risky.

Assess Your Skills

There are two major concerns in the design of a project. First, there are the structural, engineering, and mechanical concerns: sizing of foundations and lumber to carry loads and resist forces, sizing of wires and pipes to carry electrical current, gases, and liquids, matching of products and materials to unique site conditions, weatherproofing, etc. And then there are the aesthetic

concerns such as layout of the floor plan, architectural details, selection of finishes and colors, effect of lighting, etc.

Of course, some of the items in these two categories overlap. And regardless of the category, the way each of these items is designed will affect the cost, aesthetics, and functionality of the project. Furthermore, whoever designs these items will be held responsible to one degree or another for the suitability, accuracy, and functionality of the design.

Some contractors are stronger in the structural area, others are much better in the aesthetic design area. Know your strong areas and weak areas if you plan on doing design work yourself, whether it's for a fee in conjunction with your building or merely "free" advice. The bottom line is this: If you design it, it had better work as represented! Defects in the structural design of the building will carry the greatest potential for future legal liability.

The Risks of "Free" Design

If you design it, it had better work because, if any property damage or personal injury occurs due to your design error or omission, you will likely be found liable for any damages the owner suffers.

This could well be the outcome whether you are paid for the design, or are merely offering your sage advice free of charge — as part of a fixed price agreement, for example — on how the owner can change that crazy structural detail by Art Architect to make it "just the way you did it at your own house." If it turns out your design doesn't work at the owner's house and he suffers a loss, you may or may not have insurance that covers this type of claim by the owner. In many areas it is difficult and expensive, if not impossible, for a builder to get insured against design errors and omissions.

Be aware that if you even briefly unbuckle your tool belt and step into the shoes of the designer or architect, even with the best of intentions, and it turns out your suggested design failed under the circumstances (whether due to your errors or omissions), you may be found to have induced the owner into accepting your negligent design and later be held liable by an arbitrator or a judge for the resulting damages.

What's the solution? Do you have to bury your head in the sand every time you see a ridiculous construction detail you know won't work? Absolutely not! To the contrary, this approach also has its own potential legal pitfalls. If you recognize a design that you know will fail and yet you refuse to raise this issue in writing with the owner and architect, you may share in the liability.

For example, let's say you write up a quick bid for a garage job using the owner's preliminary plans, give the owner a lump-sum bid, and get the job. Then you notice some problems when you pick up the permit and approved job copy of the plans (which were submitted by the owner). You see that the roof for the garage has 3-foot parapet walls, is framed without slope, has a small primary drain, and no secondary overflow drain or scuppers.

You know from experience that some type of overflow drain or scupper is required and that the primary drain should be larger than designed. You also know it's better if the roof slopes toward the drains. You also notice that the nearby cluster of elm trees will shed their leaves onto the roof and clog the one undersized drain on the roof — which likely will result in a large "swimming pool" on top of the garage which could collapse the entire roof system.

What do you do? Keep your mouth shut or just design and implement a fix on your own? I wouldn't suggest doing either. I would write a quick letter to the architect or designer bringing this situation to his attention. Give the architect or the designer the chance to address this error, omission, or concern *first*.

If the architect or designer changes the detail and it results in extra work, issue a change order to the owner referencing the new detail for the roof as drawn by the architect or designer.

If the architect or designer refuses to change the detail, however, I would send the same memo to the owner and let him know that you will build according to the plans given to you if instructed in writing to do so — as long as there is no code violation. But also state in the memo that it's your professional opinion that the details provided won't work well (briefly explain why) and will likely lead to premature failure and significant property loss or even possible injury to people.

Tell the owner to again review the design with his architect or designer and the local building department, and then to sign and return your letter acknowledging he has read it and indicating how he wants you to proceed. Finally, if you think the error or omission on the plans involves a violation of a building code, you may want to send a letter to the building department requesting clarification of the code as it relates to the detail in question.

Having done all of the above, you may have not only avoided a potential building disaster, you have also avoided legal liability for the fix of the problem. If you just fix it on your own, you may be assuming *all* the liability for the fix of the problem and consequently your fix has to be 100% right, or you may still be liable for future failures.

For instance, let's say you point out the problem of the flat roof, the undersized primary drain, the missing secondary drain, and the likelihood of roof collapse to the owner, but *not* to the architect or designer. On your own, you suggest to the owner some design deviations to correct all the problems you are aware of. While you're cursing that "crazy detail" by the designer, the owner praises you for saving his garage and the '56 Mercedes Gull Wing that he will soon store in the garage. Suddenly you're smugly thinking, "Chalk up a few goodwill points for the contractor!"

But your goodwill may be short-lived. Suppose the following week you fail to place the overflow drain at the right height off the roof and the building inspector also misses the error. Unfortunately for you (and the Gull Wing), the architect has no site inspection duties, nor did you bother to involve him in the design, so he also does not discover your "design" error.

The following winter, months after collecting your final check, the primary roof drain clogs, the roof collects water like a swimming pool and collapses before the water is high enough to reach the overflow drain that you placed too high off the roof surface. Result: You may be found liable for the loss of the Gull Wing and the garage because you negligently specified the height of the overflow drain, even though you corrected numerous other problems with the design of the roof.

The simple way to handle any details you disagree with or know will fail (or are simply suspicious about) is to make any suggestions you want to the owner, architect, or designer, but make sure you tell the designer or architect that they must review and approve all design suggestions you offer.

In some cases, your suggested deviation from the plans may also require building department review and a detail submittal in writing. If you simply make the change on your own in the field without submitting the detail to the building department, and it later turns out (after a costly failure) that your deviation from the plans was the source of the failure, you'll be up the proverbial creek without a paddle.

Also, understand that if you deviate from the manufacturer's recommended installation instructions when installing products that this deviation will generally *void* any manufacturer's warranty.

Why take on legal liability for the design when you are not paid for it, perhaps not licensed for it, and probably not insured for it in the event that it fails? Don't seek the "I'm the good guy who fixed your bad design" spotlight. Have a procedure in place that you follow when you deviate from approved plans and specifications. Involve the owner and architect in this procedure. In all but trivial cases, leave a paper trail. Educate your employees in this procedure.

If you are working as a contractor who is *not* being paid to design, adopt the approach that you are the "mechanic" or craftsman paid to construct on the basis of plans furnished by the owner. If you encounter a detail you think won't work and it exposes you to future liability (which includes any detail that involves structural work or waterproofing), have the owner's designer or architect draft the new detail. Whether you offer alternate design possibilities or merely point out the possibility of a problem is up to you.

Unless you are hired as the design-build contractor, it's important to recognize that you are part of a team and that your role is not the same as that of the designer or architect. Resist the urge to be a know-it-all contractor who interprets and changes plans at will, for better or worse.

But, if you like the idea of designing and building, and recognize and accept the greater liability inherent with design-build contracting, read on.

Preparing for Design-Build

With design-build contracting, wearing the hat of both designer and contractor solves some problems and creates others. These new problems stem from the fact that you are doing two separate jobs at once for the owner. Here is a brief list of the key problems:

Potential conflict of interest. Traditionally, the architect/designer prepares the plans and specifications, and also acts as the owner's unofficial strong man when it comes to making sure the contractor is conforming to the plans and specifications and not cutting corners. Especially according to the A.I.A. documents, the role

of the architect also involves monitoring certain contract administration procedures for the owner — e.g., approval of payment requests, schedules of value, input on who should pay for disputed change orders, informal dispute resolution, etc.

If the contractor both designs and builds the project and is sloppy about conforming to the plans or maintaining high work quality, the owner will suffer a loss, in part, simply because he failed to hire a separate designer who had no financial interest in cutting corners either during design or construction.

A design-build contractor who is both the builder and the "in-house" designer may be perceived by a skeptical owner as deriving a direct financial gain by cutting corners during the construction phase. Whether or not he is cutting corners, just the fact that the design-build contractor has no third party (like an architect or designer) familiar with construction looking over his shoulder to monitor his work sets him up for a higher standard of care — at least in the owner's mind — than is typical in the traditional building scheme where the builder simply follows someone else's plans.

In this area, minor problems can rapidly escalate into major misunderstandings between the owner and the contractor. Be sure to draft language into your agreement that tempers the owner's unrealistic expectations that the job will go off perfectly, without a hitch. Whether or not the design-build approach or the traditional approach is employed, rarely does any project get completed without a few misunderstandings or problems arising.

Proper licensing and legal relationships. Licensing requirements for designers and design-build contractors vary from state to state. Without proper licensing, a court might even consider your design-build contract to be void and unenforceable. Without proper licensing, some states may consider the contractor to have automatically forfeited his mechanic's lien rights. These are major areas of concern.

Some states have laws that specifically address design-build contracting; others do not. Be sure to check with an attorney in your state about proper licensing prior to offering design-build services of any kind. Questions you should address which could affect licensing and different types of design-build relationships include:

• Does your state law allow a properly licensed building

contractor to perform both the construction and the design work of a project without the need for a separately licensed architect or designer? Does a state statute specifically address design-build licensing and contracting?

- If no separate design license is required, what is the limit on the size or type of construction that the contractor can design without having a design license? Residential wood-frame through three stories? Commercial? Residential multi-unit through four units?

- Does the builder plan to subcontract the design function out as he would any other subtrade? If so, how does this affect any licensing requirement for the design-build contractor?

- Will the builder hire a properly licensed architect as an "in-house" employee to perform the design work? If so, how will this affect any licensing requirement for the design-build contractor?

- If a separate design license is required, will the properly licensed design professional become a partner or joint venturer with the design-build contractor (more common on commercial projects)? If so, how will this affect any licensing requirement for the design-build contractor?

- Does your state require a special "design-build" contracting license (other than separate contracting and architectural licenses) for design-build contractors? If a separate "design-build" license is required in your state, is the failure to have this license an absolute bar to any kind of monetary recovery (or mechanic's lien) in a legal action?

Bonding and insurance. The design-build contractor will typically need errors and omissions insurance for his design work as well as the comprehensive general liability insurance that contractors ordinarily carry. Insurance that covers errors and omissions for the design of construction projects is ordinarily very costly and is hard to obtain in some states by someone other than a licensed architect.

If your state requires minimum bonding of every contractor, then every design-build contractor in that state will also require bonding.

The problem is, you can't assume that your existing insurance and your existing bond will automatically cover your operations as a design-build contractor. It may not. If you currently do design-build work and aren't sure about the extent of your insurance or bond

coverage, make a note *now* to call your insurance agent to verify that you are properly insured for design errors and omissions.

Insurance companies view design-build contractors as carrying greater risk than their traditional counterparts. For one thing, there are fewer design-build firms around. So insurance companies don't have a well-established baseline for assessing their risk.

In addition, when you enter into a design-build contract, even if you subcontract out the design function to a separately licensed architectural firm, you will still be legally responsible to the owner for the design. Accordingly, it's important to verify that your designer has proper licensing and insurance just as you would for any other subcontractor.

Suffice it to say that prior to entering into design-build contracts, you should meet with your attorney and your insurance agent and carefully review your insurance policy, licensing requirements in your state, and bonding to make sure that you are properly prepared for the greater risks of design-build construction.

Allocation of risk with the design entity. If the design aspect of your work is done by in-house employees or principals in the design-build business, the design-build contractor will "own" nearly all of the various risks associated with design-build contracting.

However, if the design-build contractor subcontracts the design work or enters into a joint venture agreement with the designer, then he should carefully consider having his agreement with the designer (or even the owner) allocate the various types of risk he will face on the project.

The issues you should consider when allocating risk include the following: design errors, design delays, construction delays, cost overruns, breach of contract by owner, indemnification clauses, construction defects, and legal fees associated with the project. Review these issues with your construction attorney if you are either subcontracting out the design services or entering into a joint venture with the designer.

Additional Protection

In addition to verifying that you and/or your designer have proper licensing, insurance, and bonding, there are other strategies you should consider to lower the risks associated with design-build work when it is done in-house.

Consider placing a *"Not for Construction"* note on all editions of your plans (except the one that's headed for plan check at the building department) and/or a stamp that reads *"These plans are the property of XYZ Construction and are intended and suitable for use only by XYZ Construction."* You could also add, *"These plans may not be used, copied, or reproduced without the written permission of the designer."*

This type of notice offers you some protection in the event that your plans get into the hands of another contractor who misinterprets them and then blames a future construction defect on a perceived or actual design error or omission.

If you design in-house, don't be cheap about bringing in an engineer to review structural aspects of your plans or an experienced architect to review construction details which you are uncertain about.

Another approach that will help you steer clear of most of the legal pitfalls of design-build is to not contract directly with the owner for design services. Instead, simply recommend that the owner work closely with either XYZ Architects or ABC Architects (and your company) to prepare the plans and budgets. You can still work cooperatively with the designer in this case, but on an informal — not a contractual — basis. This is like design-build "lite." Tell the owner that you have good past working relationships with both architects presented for consideration. But also tell him to interview both architects and make up his own mind.

Under this scenario you may want an agreement with the owner charging so much per hour for "Coordination of Preconstruction Services." This will cover your time spent helping prepare plans and budgets. The amount paid to you by the owner for these services might then be partially or completely credited back to the owner once he signs a contract with you to build the work. See Form 3.1, Section III.C "Future Construction Work and Credit for Design Work," and Form 2.1, Section III.C.3 "Fees for Building Permits, Plans, Engineering & Architectural Services, Utility Connections, and Special Testing." However, do not use the design-build preconstruction services agreement because this agreement assumes the contractor will be doing the design work.

By employing this approach, many of the benefits of a close working relationship between designer and contractor can be offered to the owner without the contractor taking on the legal liabilities of the design.

The Design-Build Relationship With the Owner

With commercial construction, the old maxim that "time is money" can be very true. The carrying costs on a large piece of land that cannot be used during construction can run up the owner's expenses in a hurry. As a way of reducing overall project costs, the design-build approach has worked well with larger commercial projects that are on a "fast track" schedule where plans for later phases of work have not been fully developed when construction begins.

However, the agreements and design-build information in this book are *not* suitable for commercial projects. The complexities and risks of commercial design-build projects are simply too great and are far beyond the scope of the agreements in this book.

When building homes or residential additions, the design-build contractor needs to be sure that he doesn't shoot himself in the foot during the initial client meetings by promising more than he can deliver in the area of price, time to perform, and scope of work, or by failing to ascertain the owner's primary needs and concerns right from the start.

Nearly every owner wants to know early on about one or more of the following:

• How long will it take to design the project?
• How much will the design work cost?
• What scope of work is included in the design cost?
• How long will it take to build the project?
• How much will the construction work cost?
• What is the scope of work included in the construction cost?

The design-build contractor needs to realistically temper the owner's expectations during the first one or two meetings until he has had enough time to rough out the general scope of work for the project and the project's budget range. If the design-build contractor doesn't understand the general scope of work and quality of finishes desired by the owner, he may provide an initial budget range that is too low and have problems down the road when the final project price is given to the owner.

At the same time, if the design-build contractor ascertains early on that the owner's primary concern is cost, he may be able to adjust the scope of work and choice of finishes in order to match the owner's budget.

Likewise, if the owner's primary concern is getting the project "weathered in" prior to the winter rains or snow, this is important to know because it may affect the design-build contractor's construction costs and schedule for preparation of drawings.

Accordingly, cost, scope of work, quality of finishes, and time allowed to perform the work (both design and construction) are all issues that should be explored in the first one or two meetings — *before* the design-build contractor invests much time in the project.

Design-Build Preconstruction Services Agreement 3.1

There are many ways to approach a design-build customer. Some contractors may want to start off with a "Preconstruction Services Agreement" that covers the design aspects of the project, then follow up this agreement with a fixed price agreement after the scope of work is completely detailed and the owner is ready to sign a contract.

Other contractors, however, may want to provide a design-build agreement to the owner before the project has a fully detailed set of plans. If the contractor submits a lump sum price for the work before the plans and specifications have been fully developed, he must be careful to accurately describe the design criteria on which the lump sum price is based.

After the first meeting with the owner where the contractor explains the company's design-build approach and determines the general scope of the project, the contractor can give the owner the preconstruction services agreement, assuring payment for his design and/or bidding time.

The sample preconstruction services agreement (and the design-build agreement shown in Section 3.2) is based on a scenario where the builder has "in-house" design capabilities that he is offering the owner. This is the most typical type of residential design-build contracting.

The preconstruction services agreement can also place limitations on the owner's use of the drawings made by the contractor should the contractor not be awarded the construction phase of the work.

The sample agreements — in this section and the next — may be adapted for use by contractors who satisfy all the state licensing, insurance, bonding, and legal concerns that apply to design-build contractors.

3.1 ■ DESIGN-BUILD PRECONSTRUCTION SERVICES AGREEMENT

CONTRACTOR'S NAME: _____

ADDRESS: _____

_____

PHONE: _____

FAX: _____

LIC #: _____

DATE: _____

OWNER'S NAME: _____

ADDRESS: _____

_____

PROJECT ADDRESS: _____

_____

I. PARTIES

This contract (hereinafter referred to as "Agreement") is made and entered into on this _____ day of
_____ , 19_____ , by and between _____ ,
(hereinafter referred to as "Owner"); and _____ ,
(hereinafter referred to as "Contractor"). In consideration of the mutual promises contained herein,
Contractor agrees to perform the following work:

II. GENERAL BACKGROUND INFORMATION AND ANTICIPATED SCOPE OF WORK

Owner and Contractor have been discussing Owner's project referred to above. Owner does not yet have a
fully detailed set of plans for this project which are generally suitable for bidding or construction by a
Contractor.

Owner agrees to meet with Contractor from time to time in order to provide information to Contractor
regarding design, cost, function, and aesthetics of the project.

Owner would like Contractor to develop plans and certain specifications which are generally suitable for
construction by Contractor.

A. ANTICIPATED SCOPE OF WORK

The scope of the plans and specifications to be provided by Contractor is strictly limited to the following
(both parties initial the appropriate items):

_____ Conceptual Design Sketches
_____ Floor Plan
_____ Foundation Plan
_____ Framing Plan
_____ Exterior Elevations
_____ Interior Elevations (including elevations for kitchen and baths)
_____ Interior Finish Schedule For Doors, Hardware and Trim
_____ Other:_____
_____ Other:_____
_____ Engineering Details and Outside Consultants Listed Below (Owner will directly pay for and enter into separate agreements with these outside consultants and services):

1. _____
2. _____
3. _____
4. _____

Owner expressly authorizes Contractor to rely on the accuracy of any existing reports (e.g., surveys, engineering reports, etc.) furnished to Contractor by Owner.

1. TARGET BUDGET RANGE: The target budget range for the construction phase of the project is: $_____ to $_____, *exclusive of the following:* plans, permits, outside consultant fees, Change Orders after the initial plans have been prepared, land costs, finance fees, public and private utility company hook-up fees, any work inconsistent with the design criteria set forth below.

2. DESIGN CRITERIA FOR ESTABLISHING TARGET BUDGET RANGE: Due to the fact that the plans and specifications are not fully developed at the time this Agreement is entered into, the target budget price for construction is based on the following design criteria:

_____
_____
_____

III. GENERAL CONDITIONS

A. TIME TO COMPLETE INITIAL DESIGN WORK ABOVE
The date when plans suitable for Contractor's bidding purposes will be ready is estimated to be:
_____, 19__.

B. CONTRACTOR'S COMPENSATION AND BILLING
Owner agrees to pay Contractor in one of the methods set forth below:

1. LUMP SUM AMOUNT: Owner agrees to pay the Lump Sum Amount of $_____ (exclusive of the cost of outside consultants, engineers, permit fees, and any work not specifically included above). However, if the scope of the design work is increased by Owner, then Contractor will adjust the Lump Sum Amount of this Agreement with a written Change Order.

2. HOURLY RATE: Owner agrees to pay Contractor at the hourly rate of $_____ per hour to perform the Preconstruction Services noted above.

Contractor will bill for his time on a periodic basis. Payment is due within seven (7) days of receipt of Contractor's invoice.

C. FUTURE CONSTRUCTION WORK AND CREDIT FOR DESIGN WORK

Owner has indicated a willingness to work with Contractor in the construction phase of the project, if the project can be built for a cost that is within Owner's budget. However, Owner is under absolutely no obligation to have the Contractor build the project. Owner and Contractor will enter into a new and separate agreement covering the construction phase of the project if Contractor is selected to perform the construction work for Owner on this project.

If Owner selects Contractor to perform the construction phase of the work, Contractor agrees to credit back to Owner ___% of the total fees paid to Contractor for his time spent performing work under this preconstruction services agreement.

D. USE OF PLANS AND SPECIFICATIONS

The plans and specifications are being developed for the sole use of the Contractor named in this Agreement. If the Contractor named in this agreement is not awarded the construction agreement, Owner is *strictly prohibited* from using these plans and specifications for *construction*. However, Owner may furnish these plans and specifications to an architect or other competent design professional to be used strictly for conceptual design ideas, *not for construction*. Owner's new architect or design professional will then take complete responsibility for the accuracy and suitability of *all* plans.

E. INDEMNIFICATION

If the Contractor who is a party to this Agreement does not build this project, then he is fully released, held harmless, and indemnified by Owner from liability and claims of every kind (including attorney's fees) related to the use of these drawings and specifications by any and all persons subsequently engaged by Owner or a future Owner of this property to design or build the project.

The Contractor named in this Agreement is hereby released, held harmless, and indemnified from all claims of every kind whatsoever (including attorney's fees) brought by any person or entity that result from alleged errors or omissions existing in the plans or specifications of Contractor *if* these plans are relied on in any way by anyone other than this Contractor and his designated agents in building or designing the project.

F. TERMINATION

Owner may terminate this Agreement at any time during the design phase by giving Contractor written notice of the termination and providing payment to Contractor for all services rendered through the date of

termination. Upon termination and payment for all work performed to date, all drawings, details, and estimates performed through the termination date by Contractor will be delivered to Owner, subject to the limitations in the "Use of Plans and Specifications" clause above. Contractor may terminate this Agreement if payments are not made to Contractor or outside consultants in accordance with the terms of this Agreement.

G. DISPUTE RESOLUTION

Any controversy or claim arising out of or related to this Agreement involving an amount of *less* than $5,000 (or the maximum limit of the court) must be heard in the Small Claims Division of the Municipal Court in the county where the Contractor's office is located. Any controversy or claim arising out of or related to this Agreement which is over the dollar limit of the Small Claims Court must be settled by binding arbitration administered by the American Arbitration Association in accordance with the Construction Industry Arbitration Rules. Judgment upon the award may be entered in any Court having jurisdiction thereof.

The prevailing party in any legal proceeding related to this Agreement shall be entitled to payment of reasonable attorney's fees, costs, and expenses.

H. ENTIRE AGREEMENT

This Agreement represents and contains the entire agreement between the parties. Prior discussions or verbal representations by the parties that are not contained in this Agreement are *not* a part of this Agreement.

I. ADDITIONAL LEGAL NOTICES REQUIRED BY STATE OR FEDERAL LAW

See page(s) attached: _____Yes _____No

J. ADDITIONAL TERMS AND CONDITIONS

See page(s) attached: _____Yes _____No

I have read and understood, and I agree to, all the terms and conditions contained in the Agreement above.

Date: _____   _____
                              CONTRACTOR'S SIGNATURE

Date: _____   _____
                              OWNER'S SIGNATURE

3.1 ■ Design-Build Preconstruction Services Agreement

ANNOTATED

Charlie Contractor Construction, Inc.
123 Hammer Lane
Anywhere, USA 33333
Phone: (123) 456-7890
Fax: (123) 456-7899
Lic#: 11111

DATE: **May 22, 2001**

OWNER'S NAME: **Mr. & Mrs. Harry Homeowner**
ADDRESS: **333 Swift St.**
Anywhere, USA 33333

PROJECT ADDRESS: **same**

I. PARTIES

This contract (hereinafter referred to as "Agreement") is made and entered into on this **22nd** day of **May**, 20**01**, by and between **Harry and Helen Homeowner**, (hereinafter referred to as "Owner"); and **Charlie Contractor Construction, Inc.**, (hereinafter referred to as "Contractor"). In consideration of the mutual promises contained herein, Contractor agrees to perform the following work:

■ See annotation, Form 1.3: Section I. Parties.

II. GENERAL BACKGROUND INFORMATION AND ANTICIPATED SCOPE OF WORK

Owner and Contractor have been discussing Owner's project referred to above. Owner does not yet have a fully detailed set of plans for this project which are generally suitable for bidding or construction by a Contractor.

Owner agrees to meet with Contractor from time to time in order to provide information to Contractor regarding design, cost, function, and aesthetics of the project.

Owner would like Contractor to develop plans and certain specifications which are generally suitable for construction by Contractor.

INITIAL
CC
HH

■ The clause above is a type of recital clause — a clause that establishes the general purpose of the contract, the duty of the Owner to make himself available to the Contractor for design meetings, and the expectation that the design-build Contractor will bid and possibly perform the construction work.

A. ANTICIPATED SCOPE OF WORK

The scope of the plans and specifications to be provided by Contractor is strictly limited to the following (both parties initial the appropriate items):

_HH_ _CC_   Conceptual Design Sketches
_HH_ _CC_   Floor Plan
_HH_ _CC_   Foundation Plan
_HH_ _CC_   Framing Plan
_HH_ _CC_   Exterior Elevations
_____   Interior Elevations (including elevations for kitchen and baths)
_____   Interior Finish Schedule For Doors, Hardware and Trim
_HH_ _CC_   Other: **Landscaping and Irrigation plans**
_____   Other:_____
_HH_ _CC_   Engineering Details and Outside Consultants Listed Below (Owner will directly pay for and enter into separate agreements with these outside consultants and services)

1. Energy Consultant for Title 24 Energy Compliance
2. Soils Engineer to review, approve, and stamp foundation details and to visit and test site, as required.

Owner expressly authorizes Contractor to rely on the accuracy of any existing reports (e.g., surveys, engineering reports, etc.) furnished to Contractor by Owner.

■ The clause above is very important because it establishes the scope of the design-build Contractor's design work. As with construction work, clearly establishing the scope of work is critical to avoiding future misunderstandings. Pay particular attention to this if the design work is being performed on a fixed price basis.

   The statement indicating that the Contractor can rely on the Owner's existing surveys, engineering reports, etc. is important because if one of the reports later turns out to be inaccurate, the design-build Contractor stands a better chance of not being held responsible for subsequent problems.

1. TARGET BUDGET RANGE: The target budget range for the construction phase of the project is: $ _125,000_ to $ _150,000_ , *exclusive of the following:* plans, permits, outside consultant fees, Change Orders after the initial plans have been prepared, land costs, finance

INITIAL
_CC_
_HH_

fees, public and private utility company hook-up fees, any work inconsistent with the design criteria set forth below.

> ■ The clause above indicates a "target budget range" rather than a fixed price for Owner's desired construction budget. As with fixed price construction contracts, this Agreement is careful to exclude difficult-to-predict costs from the "target budget range." Be sure to list exclusions which are specific to the project. As with other construction contracts, stating *excluded* work is just as important as stating what work is included in the Agreement.
>
> If the Owner goes over budget, it will be very helpful to refer back to items which were excluded in the Agreement, but which the Owner later incorporated into his design and construction.
>
> Without these indicators, the Owner may assume that either your budget number was unrealistically low or your construction costs were too high, or that you were negligent in providing him with a realistic estimate of the true "soft" and "hard" costs associated with his project.

2. DESIGN CRITERIA FOR ESTABLISHING TARGET BUDGET RANGE: Due to the fact that the plans and specifications are not fully developed at the time this Agreement is entered into, the target budget price for construction is based on the following design criteria:

Materials and labor to construct an approximately 1,200-square-foot, two-story addition to Owner's residence. Target range based on allowances for finishes consistent with finish quality of existing residence. New siding, windows, doors, and trim to match existing as closely as possible. Target range also based on painting entire exterior of house, but painting only new areas of interior of house. Target range based on no work being performed inside any part of existing house except the remodeling of the existing kitchen and the upgrading of the electrical service panel and furnace. Target range based on new dimensional shingle, composition roof on new and existing roof. Target range based on the assumption that no unusual or hazardous existing site conditions will be discovered during design or construction work.

> ■ The section above describes the general scope of the anticipated project. This should generally include the size of the project, the quality of the finishes, and any areas of the existing residence that will be worked on in conjunction with the new addition.

III. GENERAL CONDITIONS

A. TIME TO COMPLETE INITIAL DESIGN WORK ABOVE

The date when plans suitable for Contractor's bidding purposes will be ready is estimated to be: **July 1**, 20**01**.

■ The clause above simply establishes a reasonable amount of time for you to complete the basic design work. Be sure to give yourself adequate time to perform the work. If the Scope of Work changes after this Agreement is signed and work has begun, you should change this date in your Change Order.

B. CONTRACTOR'S COMPENSATION AND BILLING

Owner agrees to pay Contractor in one of the methods set forth below:

1. LUMP SUM AMOUNT: Owner agrees to pay the Lump Sum Amount of ___$4,500___ (exclusive of the cost of outside consultants, engineers, permit fees, and any work not specifically included above). However, if the scope of the design work is increased by Owner, then Contractor will adjust the Lump Sum Amount of this Agreement with a written Change Order.

2. HOURLY RATE: Owner agrees to pay Contractor at the hourly rate of $__N/A__ per hour to perform the Preconstruction Services noted above.

Contractor will bill for his time on a periodic basis. Payment is due within seven (7) days of receipt of Contractor's invoice.

■ Some design-build Contractors prefer to bill hourly for their design work, others prefer to charge a fixed price. Both methods have their pros and cons similar to those pointed out for construction agreements. When billing is hourly, many Owners will still want a not-to-exceed price for the design work.

C. FUTURE CONSTRUCTION WORK AND CREDIT FOR DESIGN WORK

Owner has indicated a willingness to work with Contractor in the construction phase of the project, if the project can be built for a cost that is within Owner's budget. However, Owner is under absolutely no obligation to have the Contractor build the project. Owner and Contractor will enter into a new and separate agreement covering the construction phase of the project if Contractor is selected to perform the construction work for Owner on this project.

If Owner selects Contractor to perform the construction phase of the work, Contractor agrees to credit back to Owner __50__% of the total fees paid to Contractor for his time spent performing work under this preconstruction services agreement.

■ Whether you choose to credit back some, all, or none of the design fee will not affect your legal status or liability regarding your design work. It is strictly a matter of your company policy and typically reflects the amount of the design contract relative to the amount of the construction contract. However, crediting back to the Owner all or part of the design fee can

INITIAL
CC
HH

> be a powerful incentive for the Owner to award the construction contract to you. Also, the positive relationship that hopefully has developed between you and the Owner during the design phase and your intimate knowledge of the construction aspects of the project — since you have done the design work — are additional factors that usually place the design-build Contractor in the lead position to perform the construction work.

D. USE OF PLANS AND SPECIFICATIONS

The plans and specifications are being developed for the sole use of the Contractor named in this Agreement. If the Contractor named in this agreement is not awarded the construction agreement, Owner is *strictly prohibited* from using these plans and specifications for *construction*. However, Owner may furnish these plans and specifications to an architect or other competent design professional to be used strictly for conceptual design ideas, *not for construction*. Owner's new architect or design professional will then take complete responsibility for the accuracy and suitability of *all* plans.

> ■ The clause above is intended to limit the design-build Contractor's liability in the event he does not perform the construction work and the Owner's future Contractor improperly performs the work and then alleges faulty design work on the part of the original design-build Contractor. The clause states that the design-build Contractor's plans may *not* be used for construction by a future Contractor. This is a very important clause that may help to limit your liability if another Contractor performs work based on your plans and then alleges design errors.
>
> At the same time, the clause above allows the Owner to use the design-build Contractor's plans as *conceptual design ideas only* so that the Owner can obtain at least a minimal benefit from these plans if he decides not to have the design-build Contractor perform the construction phase of the project.
>
> Even if the Owner doesn't want the design-build Contractor to perform the work, he is free to use the Contractor's plans (after he pays for the preconstruction design work), subject to the plans being used strictly as *conceptual design ideas* when having *new* plans and specifications prepared.
>
> Then, if another Contractor performs the work, all actual construction, in theory, must necessarily be performed on the basis of the Owner's *new* plans, *not* on the basis of the original design-build Contractor's plans.

E. INDEMNIFICATION

If the Contractor who is a party to this Agreement does not build this project, then he is fully released, held harmless, and indemnified by Owner from liability and claims of every kind (including attorney's fees) related to the use of these drawings and specifications by any and all persons subsequently engaged by Owner or a future Owner of this property to design or build the project.

INITIAL
CC
HH

The Contractor named in this Agreement is hereby released, held harmless, and indemnified from all claims of every kind whatsoever (including attorney's fees) brought by any person or entity that result from alleged errors or omissions existing in the plans or specifications of Contractor *if* these plans are relied on in any way by anyone other than this Contractor and his designated agents in building or designing the project.

> ■ The indemnification clause above is critical and should be used in conjunction with the "Use of Plans and Specifications" clause. The indemnification clause states that the Owner releases the design-build Contractor from liability that results from another contractor working off the design-build Contractor's plans. Although indemnification clauses can be overruled in court in some instances, this is a very important clause that will help protect the interests of the design-build Contractor.

F. TERMINATION

Owner may terminate this Agreement at any time during the design phase by giving Contractor written notice of the termination and providing payment to Contractor for all services rendered through the date of termination. Upon termination and payment for all work performed to date, all drawings, details, and estimates performed through the termination date by Contractor will be delivered to Owner subject to the limitations in the "Use of Plans and Specifications" clause above. Contractor may terminate this Agreement if payments are not made to Contractor or outside consultants in accordance with the terms of this Agreement.

> ■ The termination clause makes it easy for the Owner to terminate the Contractor during the design phase if he is not satisfied with the Contractor's performance. If the Owner and design-build Contractor don't have a good relationship during the design phase, it's better to make it easy to end the relationship early on. Written notice is required if the Owner wants to terminate the design phase of the work and payment for all work performed to date is required.
>
> The Contractor may terminate the Agreement if the Owner fails to make payments in accordance with the payment schedule. The Contractor also should give written notice of the termination and the grounds for termination.

G. DISPUTE RESOLUTION

Any controversy or claim arising out of or related to this Agreement involving an amount of *less* than $5,000 (or the maximum limit of the court) must be heard in the Small Claims Division of the Municipal Court in the county where the Contractor's office is located. Any controversy or claim arising out of or related to this Agreement which is over the dollar limit of the Small Claims Court must be settled by binding arbitration administered by the American Arbitration Association in accordance with the Construction Industry Arbitration Rules. Judgment upon the award may be entered in any Court having jurisdiction thereof.

INITIAL
CC
HH

The prevailing party in any legal proceeding related to this Agreement shall be entitled to payment of reasonable attorney's fees, costs, and expenses.

H. ENTIRE AGREEMENT

This Agreement represents and contains the entire agreement between the parties. Prior discussions or verbal representations by the parties that are not contained in this Agreement are *not* a part of this Agreement.

I. ADDITIONAL LEGAL NOTICES REQUIRED BY STATE OR FEDERAL LAW

See page(s) attached: __**X**__ Yes; _____ No

J. ADDITIONAL TERMS AND CONDITIONS

See page(s) attached: _____ Yes; __**X**__ No

> ■ See annotation, Form 1.3: Section III.O. Additional Terms and Conditions.

I have read and understood, and I agree to, all the terms and conditions contained in the Agreement above.

> ■ You need a statement which indicates that the parties have read and agree to the terms and conditions of the Agreement. Be sure to have the Agreement signed by the Owner *prior* to the time you commence work. Make sure you keep a signed copy of the Agreement in your records — an unsigned copy won't do you any good later on if you have a dispute. Finally, have the Owner initial every page of your Agreement, including any supplemental attachments, such as materials or Subcontractor bids.

Date: _5/22/01_   *CHARLIE CONTRACTOR, PRESIDENT*
                            CHARLIE CONTRACTOR, PRESIDENT
                            CHARLIE CONTRACTOR CONSTRUCTION, INC.

> ■ If your business is a corporation, be sure to sign the Agreement using your corporate title and place the word, "Inc." after your company's name. If you fail to do this, you may have personal liability under the Agreement.

Date: _5/22/01_   *Harry Homeowner*
                            HARRY HOMEOWNER

Date: _5/22/01_   *Helen Homeowner*
                            HELEN HOMEOWNER

> ■ Be sure you have an Agreement in your files which is signed by the Owner and contains the Owner's initials on each page.

At the point in time when the design-build contractor has developed sufficiently detailed plans and specifications that have been approved by the owner, and the owner has given the contractor enough detail that he can *accurately* commit to a fixed price contract amount, the contractor can then provide the owner with a fixed price (lump sum) agreement. If, on the other hand, the design-build contractor wishes to establish a full construction agreement with the owner before all the construction details are known, the contractor may wish to use a design-build agreement, either on a fixed fee or cost-plus basis. The cost-plus type of design-build agreement is not shown in this book, but would also work well if agreeable to the owner.

If the design-build agreement seems more complicated than the other agreements in this book, it is because this type of contractual agreement between the parties is more complicated and subjects the contractor to certain additional risks. Because of these risks and the potential conflicts of interest, the design phase of the work and the construction phase of the work are treated separately in the agreement. Furthermore, the owner is given the right to terminate the agreement at any time during the design phase only. That way, if the relationship starts off rocky, it can be easily ended before the problems snowball.

You'll also find language specifying how the owner may and may not use the plans should the original contractor not end up doing the construction work. This is critical in design-build work, since the plans in the hands of the owner and/or other contractors may engender significant liability for the original "designer" for a long time to come.

Finally, the complexities and risks of design-build contracting are great enough that contractors are again strongly advised to *not* rely on any of the language in the sample agreements without the assistance and guidance of a good attorney who is familiar with design-build and with construction law in the contractor's state.

3.2 ■ Design-Build Agreement

CONTRACTOR'S NAME: _____

ADDRESS: _____

_____

PHONE: _____

FAX: _____

LIC #: _____

DATE: _____

OWNER'S NAME: _____

ADDRESS: _____

_____

PROJECT ADDRESS: _____

_____

I. PARTIES

This contract (hereinafter referred to as "Agreement") is made and entered into on this _____ day of _____ , 19_____ , by and between _____ , (hereinafter referred to as "Owner"); and _____ , (hereinafter referred to as "Contractor"). In consideration of the mutual promises contained herein, Contractor agrees to perform the following work:

II. GENERAL SCOPE OF WORK DESCRIPTION

Contractor agrees to perform the following work, in two phases, as set forth below. The price to be paid for each phase of work is indicated below.

A. PRECONSTRUCTION DESIGN SERVICES PHASE

This phase includes design, drawings, meetings with Owner, and bidding.

Owner agrees to meet with Contractor from time to time in order to provide information to Contractor regarding design, cost, function, and aesthetics of the project.

Contractor will develop plans and certain specifications which are generally suitable for construction by Contractor.

B. ANTICIPATED SCOPE OF WORK

The scope of the plans and specifications to be provided by Contractor is strictly limited to the following (both parties initial the appropriate items):

_____ Conceptual Design Sketches

_____ Floor Plan

_____ Foundation Plan

_____ Framing Plan

_____ Exterior Elevations

_____ Interior Elevations (including elevations for kitchen and baths)

_____ Interior Finish Schedule For Doors, Hardware and Trim

_____ Other:_____

_____ Other:_____

_____ Engineering Details and Outside Consultants Listed Below (Owner will directly pay for and enter into separate agreements with these outside consultants and services):

1. _____

2. _____

3. _____

4. _____

Owner expressly authorizes Contractor to rely on the accuracy of any existing reports (e.g., surveys, engineering reports, etc.) furnished to Contractor by Owner.

Contractor will inform Owner of the effects of Owner-requested changes which force the cost of work to exceed the Lump Sum Contract Amount set forth in this Agreement.

C. TIME TO COMPLETE INITIAL DESIGN WORK ABOVE

The date when plans will be substantially complete is estimated to be: _____, 19___.

D. CONTRACTOR'S COMPENSATION AND BILLING

Owner agrees to pay Contractor in one of the methods set forth below:

1. LUMP SUM PRICE FOR THE DESIGN PHASE OF PROJECT: Owner agrees to pay the Lump Sum Price of $_____ (exclusive of the cost of outside consultants, engineers, permit fees, and any work not specifically included above which will be contracted for and paid for directly by Owner). However, if the scope of the design work is increased by Owner, or involves design work in addition to that specified in the design criteria set forth below, then Contractor will adjust the Lump Sum Price of this Agreement with a written Change Order.

2. HOURLY RATE: Owner agrees to pay Contractor at the hourly rate of $_____ per hour to perform the Preconstruction Services noted above.

Contractor will bill for his time on a periodic basis. Payment is due within _____ days of receipt of Contractor's invoice.

E. FUTURE CONSTRUCTION WORK AND CREDIT FOR DESIGN WORK

Owner has indicated a willingness to work with Contractor during the construction phase of the project, but is under absolutely no obligation to do so.

However, if Owner selects Contractor to perform the construction phase of the work, Contractor agrees to credit back to Owner ____% of the total fees paid to Contractor for his time spent performing work under this Agreement.

F. USE OF PLANS AND SPECIFICATIONS

The plans and specifications are being developed for the sole use of the Contractor named in this Agreement. If the Contractor named in this Agreement is not awarded the Construction Agreement, Owner is *strictly prohibited* from using these plans and specifications for *construction*. However, Owner may furnish these plans and specifications to an architect or other competent design professional to be used strictly for conceptual design ideas, *not for construction*. Owner's new architect or design professional will then take complete responsibility for the accuracy and suitability of *all* plans.

G. OBTAINING BUILDING PERMIT

If construction of project is approved by Owner and the Contractor named in this Agreement is to perform the construction phase of the work, Contractor will assist Owner in obtaining building permit. Owner to directly pay the cost of all governmental permit fees.

H. CONSTRUCTION OF PROJECT AND DESIGN CRITERIA

Contractor agrees to furnish all labor, materials, and equipment to construct the project in accordance with the design criteria, plans, and specifications set forth in this Agreement.

    1. LUMP SUM PRICE FOR THE CONSTRUCTION PHASE OF PROJECT: This Lump Sum Price is subject to being increased or decreased by written Change Order if the Scope of Work as set forth by the design criteria below or the plans and specifications in existence at the time this Agreement is entered into is increased or decreased by Owner or any governmental agency: $_____

Allowances that are a part of the Lump Sum Price above are designated as follows:

Item	*Allowance Amount*
_____	$_____
_____	$_____
_____	$_____
_____	$_____

    2. DESIGN CRITERIA FOR ESTABLISHING LUMP SUM PRICE: Due to the fact the plans and specifications may not have been fully developed at the time this Agreement has been entered into, the basic design criteria which the Lump Sum Price for construction has been based on includes the following:

_____

_____

_____

_____

(See Additional Page Attached: ___Yes; ___No)

III. GENERAL CONDITIONS

A. EXCLUSIONS

This Agreement does *not* include *labor or materials* for the following work (unless Owner selects one of these items as an Additional Alternate):

1. PROJECT SPECIFIC EXCLUSIONS:

_____

_____

_____

_____

_____

_____

2. STANDARD EXCLUSIONS: Unless specifically included in the "General Scope of Work" section above, this Agreement does *not* include *labor or materials* for the following work (any Exclusions in this paragraph which have been lined out and initialed by the parties do not apply to this Agreement): Removal and disposal of any materials containing asbestos (or any other hazardous material as defined by the EPA). Custom milling of any wood for use in project. Moving Owner's property around the site. Labor or materials required to repair or replace any Owner-supplied materials. Repair of concealed underground utilities not located on prints or physically staked out by Owner which are damaged during construction. Surveying that may be required to establish accurate property boundaries for setback purposes (fences and old stakes may not be located on actual property lines). Final construction cleaning (Contractor will leave site in "broom swept" condition). Landscaping and irrigation work of any kind. Temporary sanitation, power, or fencing. Removal of soils under house in order to obtain 18 inches (or code-required height) of clear space between bottom of joists and soil. Removal of filled ground or rock or any other materials not removable by ordinary hand tools (unless heavy equipment is specified in Scope of Work section above), correction of existing out-of-plumb or out-of-level conditions in existing structure. Correction of concealed substandard framing. Rerouting/removal of vents, pipes, ducts, structural members, wiring or conduits, steel mesh which may be discovered in the removal of walls or the cutting of openings in walls. Removal and replacement of existing rot or insect infestation. Failure of surrounding part of existing structure, despite Contractor's good faith efforts to minimize damage, such as plaster or drywall cracking and popped nails in adjacent rooms, or blockage of pipes or plumbing fixtures caused by loosened rust within pipes. Construction of continuously level foundation around structure (if lot is sloped more than 6 inches from front to back or side to side, Contractor will step the foundation in accordance with the slope of the lot).

Exact matching of existing finishes. Public or private utility connection fees. Repair of damage to existing roads, sidewalks, and driveways that could occur when construction equipment and vehicles are being used in the normal course of construction.

B. DATE OF WORK COMMENCEMENT AND SUBSTANTIAL COMPLETION

Commence work:_____. Construction time through substantial completion: Approximately ___ to ___ weeks/months, *not* including delays and adjustments for delays caused by: holidays, inclement weather, accidents, shortage of labor or material, additional time required for performance of Change Order work (as specified in each Change Order), delays caused by Owner, and other delays unavoidable or beyond the control of the Contractor.

C. EXPIRATION OF THIS AGREEMENT

This Agreement will expire 30 days after the date at the top of page one of this Agreement if not accepted in writing by Owner and returned to Contractor within that time.

D. WORK PERFORMED BY OWNER OR OWNER'S SEPARATE CONTRACTORS

Any labor or materials provided by the Owner's separate Subcontractors while Contractor is still working on this project must be supervised by Contractor. Profit and overhead at the rate of __% will be charged on all labor and material furnished by Owner's separate Subcontractors while Contractor is still working on the project. Contractor has right to qualify and approve Owner's Subcontractors and require evidence of work experience, proper licensing, and insurance. If Owner wants to avoid paying Contractor's profit and overhead per this section, Owner must then bring in his separate Subcontractors only *before* or *after* Contractor has performed all of his work.

E. CHANGE ORDERS: CONCEALED CONDITIONS, ADDITIONAL WORK, AND CHANGES IN THE WORK

1. PEOPLE AUTHORIZED TO SIGN CHANGE ORDERS: The following people are authorized to sign Change Orders:

_____

_____

(Please fill in line(s) above at time of signing Agreement)

2. CONCEALED CONDITIONS: This Agreement is based solely on the observations Contractor was able to make with the structure in its current condition at the time this Agreement was bid. If additional Concealed Conditions are discovered once work has commenced which were *not* visible at the time this proposal was bid, Contractor will stop work and point out these unforeseen Concealed Conditions to Owner so that Owner and Contractor can execute a Change Order for any Additional Work.

3. CHANGES IN THE WORK: During the course of the project, Owner may order changes in the work (both additions and deletions). The cost of these changes will be determined by the Contractor and the cost of this Additional Work will be added to Contractor's profit and overhead at the rate of __% in order to arrive at the net amount of any additional Change Order work.

Contractor's profit and overhead, and any supervisory labor will not be credited back to Owner with any deductive Change Orders (work deleted from Agreement by Owner).

4. DEVIATION FROM SCOPE OF WORK IN CONTRACT DOCUMENTS: Any alteration or deviation from the Scope of Work referred to in the Contract Documents involving extra costs of materials or labor, including any overage on **ALLOWANCE** work, will be executed upon a written Change Order issued by Contractor and should be signed by Contractor and Owner prior to the commencement of any Additional Work by the Contractor. This Change Order will become an extra charge over and above the Lump Sum Price referred to at the beginning of this Agreement.

5. CHANGES REQUIRED BY PLAN CHECKERS OR FIELD INSPECTORS: Any increase in the Scope of Work set forth in the Contract Documents which is required by plan checkers or field inspectors with city or county building/planning departments will be treated as Additional Work to this Agreement for which the Contractor will issue a Change Order.

6. RATES CHARGED FOR ALLOWANCE ONLY AND TIME-AND-MATERIALS WORK: Journeyman Carpenter: $_____ per hour; Apprentice Carpenter: $_____ per hour; Laborer: $_____ per hour; Contractor: $_____ per hour; Subcontractor: Amount charged by Subcontractor. *Note:* Contractor will charge for profit and overhead at the rate of ___% on all work performed on a Time-and-Materials basis (on both materials and labor rates set forth in this paragraph) and on all costs that exceed specifically stated **ALLOWANCE** estimates in the Agreement.

F. PAYMENT SCHEDULE AND PAYMENT TERMS

1. PAYMENT SCHEDULE:
* First Payment: $1,000 or 10% of contract amount (whichever is less) due when contract is signed and returned to Contractor.
Contract Deposit Payment: $ _____

* Second Payment (Materials Deposits): Any materials deposits — required for such items as woodstoves, cabinetry, carpets and vinyl, granite, tile, and any and all special order items which require the payment of a materials deposit — must be paid within 3 days of submittal of invoice by Contractor. These items will not be ordered until the deposits set forth below are received by Contractor. Materials Deposits required on this project include the following:

_____
_____ $ _____
* Total Due for All Materials Deposits: _____ $ _____

* Third Payment: _____
_____ $ _____

* Fourth Payment: _____
_____ $ _____

* Fifth Payment: _____

_____ $ _____

* Sixth Payment: _____

_____ $ _____

* Final Payment: Due upon Substantial Completion of all work under this Agreement: $ _____

2. PAYMENT OF CHANGE ORDERS: Payment for each Change Order is due upon completion of Change Order work and submittal of invoice by Contractor for this work.

3. ADDITIONAL PAYMENTS FOR ALLOWANCE WORK AND RELATED CREDITS: Payment for work designated in the Agreement as **ALLOWANCE** work has been initially factored into the Lump Sum Price and Payment Schedule set forth in this Agreement. If the actual cost of the **ALLOWANCE** work *exceeds* the line item **ALLOWANCE** amount in the Agreement, the difference between the cost and the line item **ALLOWANCE** amount stated in the Agreement will be written up by Contractor as a Change Order subject to Contractor's profit and overhead at the rate of __%.

If the cost of the **ALLOWANCE** work is *less* than the **ALLOWANCE** line item amount listed in the Agreement, a credit will be issued to Owner after all billings related to this particular line item **ALLOWANCE** work have been received by Contractor. This credit will be applied toward the final payment owing under the Agreement. Contractor profit and overhead and any supervisory labor will *not* be credited back to Owner for **ALLOWANCE** work.

4. FINAL CONTRACT PAYMENT: The final contract payment is due and payable upon "Substantial Completion" (not Final Completion) of all work under contract. "Substantial Completion" is defined as being the point at which the Building/Work of Improvement is suitable for its intended use, or the issuance of an Occupancy Consent, or final building department approval from the city or county building department, whichever occurs first.

5. HOLD BACK FROM FINAL PAYMENT FOR PUNCH LIST WORK: At time of making the final contract payment, Owner may hold back 150% of the value of all Punch List work. Owner and Contractor will place a fair and reasonable value on each Punch List item at time of Punch list walk-through with Owner. Contractor and Owner will then execute the Punch List form. This 150% hold back for Punch List work assures Owner that all Punch List work will be completed by Contractor in a timely manner.

6. PAYMENT FOR COMPLETED PUNCH LIST WORK: Payment for completed Punch List items is due and payable upon submittal of invoice for any part of the completed Punch List work covered by Contractor's invoice.

7. INTEREST CHARGES: Interest in the amount of __% per month will be charged on all late payments under this Agreement. "Late Payments" are defined as any payment not received within __ days of receipt of invoice from Contractor.

G. MISCELLANEOUS CONDITIONS

1. MATCHING EXISTING FINISHES: Contractor will use best efforts to match existing finishes and materials. However, an exact match is not guaranteed by Contractor due to such factors as discoloration from aging, a difference in dye lots, and the difficulty of exactly matching certain finishes, colors, and planes.

Unless custom milling of materials is specifically called out in the plans, specifications, or Scope of Work description, any material not readily available at local lumberyards or suppliers is *not* included in this Agreement.

If Owner requires an exact match of materials or textures in a particular area, Owner must inform Contractor of this requirement in writing within seven (7) days of signing this Agreement. Contractor will then provide Owner with either a materials sample or a test patch prior to the commencement of work involving the matching of existing finishes.

Owner must then approve or disapprove of the suitability of the match within 24 hours. After that time, or after Contractor has provided Owner with two or more test patches that have been rejected by Owner, all further test patches, materials submittals, or any removal and replacement of materials already installed in accordance with the terms of this section will be performed strictly as Extra Work on a time-and-materials basis by Contractor.

2. CONFLICT OF DOCUMENTS: If any conflict should arise between the plans, specifications, addenda to plans, and this Agreement, then the terms and conditions of this Agreement shall be controlling and binding upon the parties to this Agreement.

3. INSTALLATION OF OWNER-SUPPLIED FIXTURES AND MATERIALS: Contractor can not warrant any Owner-supplied materials or fixtures (whether new or used). If Owner-supplied fixtures or materials fail due to a defect in the materials or fixtures themselves, Contractor will charge for all labor and materials required to repair or replace both the defective materials or fixtures, and any surrounding work that is damaged by these defective materials or fixtures.

4. CONTROL AND DIRECTION OF EMPLOYEES AND SUBCONTRACTORS: Contractor, or his appointed Supervisor, shall be the sole supervisor of Contractor's Employees and Subcontractors. Owner must not order or request Contractor's Employees or Subcontractors to make changes in the work. All changes in the work are to be first discussed with Contractor and then performed according to the Change Order process as set forth in this Agreement.

5. OWNER COORDINATION WITH CONTRACTOR: Owner agrees to promptly furnish Contractor with all details and decisions about unspecified construction finishes, and to consent to or deny changes in the Scope of Work that may arise so as not to delay the progress of the Work. Owner agrees to furnish Contractor with continual access to the job site.

H. TERMINATION OF AGREEMENT FOR CONVENIENCE BY OWNER DURING PRECONSTRUCTION DESIGN PHASE *ONLY*

Owner may terminate this entire Agreement at any time *only* during the Preconstruction Design Services phase of the work by giving Contractor written notice of the termination and providing payment to Contractor for all services rendered through the date of termination. Upon termination and payment for all work performed to date, all drawings, details, and estimates performed through the termination date by Contractor and his agents will be delivered to Owner subject to the limitations in Section II.F "Use of Plans and Specifications," above.

I. WORK STOPPAGE AND TERMINATION OF AGREEMENT FOR DEFAULT

Once construction has commenced, Contractor shall have the right to stop all work on the project and keep the job idle if payments are not made to Contractor in accordance with the Payment Schedule in this Agreement, or if Owner repeatedly fails or refuses to furnish Contractor with access to the job site and /or product selections or information necessary for the advancement of Contractor's work. Simultaneous with stopping work on the project, the Contractor must give Owner written notice of the nature of Owner's default and must also give the Owner a 14-day period in which to cure this default.

If work is stopped due to any of the above reasons (or for any other material breach of contract by Owner) for a period of 14 days, and the Owner has failed to take significant steps to cure his default, then Contractor may, without prejudicing any other remedies Contractor may have, give written notice of termination of the Agreement to Owner and demand payment for all completed work and materials ordered through the date of work stoppage, and any other loss sustained by Contractor, including Contractor's Profit and Overhead at the rate of ___% on the balance of the incomplete work under the Agreement. Thereafter, Contractor is relieved from all other contractual duties, including all Punch List and warranty work.

J. LIEN RELEASES

Upon request by Owner, Contractor and Subcontractors will issue appropriate lien releases prior to receiving final payment from Owner.

K. DISPUTE RESOLUTION AND ATTORNEY'S FEES

Any controversy or claim arising out of or related to this Agreement involving an amount of *less* than $5,000 (or the maximum limit of the court) must be heard in the Small Claims Division of the Municipal Court in the county where the Contractor's office is located. Any controversy or claim arising out of or related to this Agreement which is over the dollar limit of the Small Claims Court must be settled by binding arbitration administered by the American Arbitration Association in accordance with the Construction Industry Arbitration Rules. Judgment upon the award may be entered in any Court having jurisdiction thereof.

The prevailing party in any legal proceeding related to this Agreement shall be entitled to payment of reasonable attorney's fees, costs, and expenses.

L. INSURANCE

Owner shall pay for and maintain "Course of Construction" or "Builder's Risk" or any other insurance that provides the same type of coverage to the Contractor's work in progress during the course of the project. It is Owner's express responsibility to insure dwelling and all work in progress against all damage caused by fire and Acts of God such as earthquakes, floods, etc.

M. INDEMNIFICATION

If the Contractor who is a party to this Agreement does not build this project, then he is fully released, held harmless, and indemnified by Owner from liability and claims of every kind (including attorney's fees) related to the use of these drawings and specifications by any and all persons subsequently engaged by Owner or a future Owner of this property to design or build the project.

The Contractor named in this Agreement is hereby released, held harmless, and indemnified from all claims of every kind whatsoever (including attorney's fees) brought by any person or entity that result from alleged errors or omissions existing in the plans or specifications of Contractor *if* these plans are relied on in any way by anyone other than this Contractor and his designated agents in building or designing the project.

N. WARRANTY

Contractor provides a limited warranty on all Contractor- and Subcontractor-supplied labor and materials used in this project for a period of one year following substantial completion of all work.

No warranty is provided by Contractor on any materials furnished by the Owner for installation. No warranty is provided on any existing materials that are moved and/or reinstalled by the Contractor within the dwelling (including any warranty that existing/used materials will not be damaged during the removal and reinstallation process). One year after substantial completion of the project, the Owner's sole remedy (for materials and labor) on all materials that are covered by a manufacturer's warranty is strictly with the manufacturer, not with the Contractor.

Repair of the following items is specifically excluded from Contractor's warranty: Damages resulting from lack of Owner maintenance; damages resulting from Owner abuse or ordinary wear and tear; deviations that arise such as the minor cracking of concrete, stucco and plaster; minor stress fractures in drywall due to the curing of lumber; warping and deflection of wood; shrinking/cracking of grouts and caulking; fading of paints and finishes exposed to sunlight.

THE EXPRESS WARRANTIES CONTAINED HEREIN ARE IN LIEU OF ALL OTHER WARRANTIES, EXPRESS OR IMPLIED, INCLUDING ANY WARRANTIES OF MERCHANTABILITY, HABITABILITY, OR FITNESS FOR A PARTICULAR USE OR PURPOSE. THIS LIMITED WARRANTY EXCLUDES CONSEQUENTIAL AND INCIDENTAL DAMAGES AND LIMITS THE DURATION OF IMPLIED WARRANTIES TO THE FULLEST EXTENT PERMISSIBLE UNDER STATE AND FEDERAL LAW.

O. ENTIRE AGREEMENT, SEVERABILITY, AND MODIFICATION

This Agreement represents and contains the entire agreement between the parties. Prior discussions or verbal representations by Contractor or Owner that are not contained in this Agreement are *not* a part of this Agreement. In the event that any provision of this Agreement is at any time held by a Court to be invalid or unenforceable, the parties agree that all other provisions of this Agreement will remain in full force and effect. Any future modification of this Agreement must be made in writing and executed by Owner and Contractor in order to be valid and binding upon the parties.

P. ADDITIONAL LEGAL NOTICES REQUIRED BY STATE OR FEDERAL LAW

See page(s) attached: _____Yes _____No

Q. ADDITIONAL TERMS AND CONDITIONS

See page(s) attached: _____Yes _____No

I have read and understood, and I agree to, all the terms and conditions contained in the Agreement above.

Date: _____ _____
                                      CONTRACTOR'S SIGNATURE

Date: _____ _____
                                      OWNER'S SIGNATURE

Date: _____ _____
                                      OWNER'S SIGNATURE

3.2 ■ DESIGN-BUILD AGREEMENT

ANNOTATED

Charlie Contractor Construction, Inc.
123 Hammer Lane
Anywhere, USA 33333
Phone: (123) 456-7890
Fax: (123) 456-7899
Lic#: 11111

DATE: **May 22, 2001**

OWNER'S NAME: **Mr. & Mrs. Harry Homeowner**
ADDRESS: **333 Swift St.**
Anywhere, USA 33333

PROJECT ADDRESS: **same**

I. PARTIES
This contract (hereinafter referred to as "Agreement") is made and entered into on this __**22nd**__ day of __**May**__ , 20__**01**__ , by and between __**Harry and Helen Homeowner**__ , (hereinafter referred to as "Owner"); and __**Charlie Contractor Construction, Inc.**__ , (hereinafter referred to as "Contractor"). In consideration of the mutual promises contained herein, Contractor agrees to perform the following work:

> ■ See annotation, Form 1.3: Section I. Parties.

II. GENERAL SCOPE OF WORK DESCRIPTION
The Contractor agrees to perform the following work, in phases, as set forth below. The price to be paid for each phase of work is indicated below.

A. PRECONSTRUCTION DESIGN SERVICES PHASE
This phase includes design, drawings, meetings with Owner, and bidding.

Owner agrees to meet with Contractor from time to time in order to provide information to Contractor regarding design, cost, function and aesthetics of the project.

INITIAL
CC
HH

Contractor will develop plans and certain specifications which are generally suitable for construction by Contractor.

> ■ The clause above is a type of recital clause — a clause that establishes the general purpose of the contract, the duty of the Owner to make himself available to the Contractor for design meetings, and the expectation that the design/build Contractor will bid and hopefully perform the construction work.

B. ANTICIPATED SCOPE OF WORK

The scope of the plans and specifications to be provided by Contractor is strictly limited to the following (both parties initial the appropriate items):

HH *CC*   Conceptual Design Sketches
HH *CC*   Floor Plan
HH *CC*   Foundation Plan
HH *CC*   Framing Plan
HH *CC*   Exterior Elevations
_____ Interior Elevations (including elevations for kitchen and baths)
_____ Interior Finish Schedule For Doors, Hardware and Trim
HH *CC*   Other: **Landscaping and Irrigation plans** ____
_____ Other:_____
HH *CC*   Engineering Details and Outside Consultants Listed Below (Owner will directly pay for and enter into separate agreements with these outside consultants and services)

1. Energy Consultant for Title 24 Energy Compliance
2. Soils Engineer to review, approve, and stamp foundation details and to visit and test site, as required.

Owner expressly authorizes Contractor to rely on the accuracy of any existing reports (e.g., surveys, engineering reports, etc.) furnished to Contractor by Owner.

> ■ The clause above is very important because it establishes the scope of the design-build Contractor's design work. As with construction work, clearly establishing the scope of work is critical to avoiding future misunderstandings. Pay particular attention to this if the design work is being performed on a fixed price basis.
>
>   The statement indicating that the Contractor can rely on the Owner's existing surveys, engineering reports, etc. is important because if one of the reports later turns out to be inaccurate, the design-build Contractor stands a better chance of not being held responsible for subsequent problems.

INITIAL
CC
HH

Contractor will inform Owner of the effects of Owner-requested changes which force the cost of work to exceed the Lump Sum Contract Amount set forth in this Agreement.

> ■ The clause above places the burden on the design-build Contractor to inform the Owner (make sure it's in writing) of Owner-requested upgrades which increase the contract amount.

C. TIME TO COMPLETE INITIAL DESIGN WORK ABOVE

The date when plans will be substantially complete is estimated to be: __July 1__, 20_01_.

> ■ The clause above simply establishes a reasonable amount of time for you to complete the basic design work. Be sure to give yourself adequate time to perform the work. If the Scope of Work changes after this Agreement is signed and work has begun, you should change this date in your Change Order.

D. CONTRACTOR'S COMPENSATION AND BILLING

Owner agrees to pay Contractor in one of the methods set forth below:

1. LUMP SUM PRICE FOR THE DESIGN PHASE OF THE WORK: Owner agrees to pay the Lump Sum Price of $ _4,500_ (exclusive of the cost of outside consultants, engineers, permit fees, and any work not specifically included above which will be contracted for and paid for directly by Owner). However, if the scope of the design work is increased by Owner, or involves design work in addition to that specified in the design criteria set forth below, then Contractor will adjust the Lump Sum Price of this Agreement with a written Change Order.

2. HOURLY RATE: Owner agrees to pay Contractor at the hourly rate of $ _N/A_ per hour to perform the Preconstruction Services noted above.

Contractor will bill for his time on a periodic basis. Payment is due within _seven (7)_ days of receipt of Contractor's invoice.

> ■ Some design-build Contractors prefer to bill hourly for their design work, others prefer to charge a fixed price. Both methods have their pros and cons similar to those pointed out for construction agreements. When billing is hourly, many Owners will still want a not-to-exceed price for the design work.

E. FUTURE CONSTRUCTION WORK AND CREDIT FOR DESIGN WORK

Owner has indicated a willingness to work with Contractor during the construction phase of the project, but is under absolutely no obligation to do so.

INITIAL
CC
HH

However, if Owner selects Contractor to perform the construction phase of the work, Contractor agrees to credit back to Owner __50__ % of the total fees paid to Contractor for his time spent performing work under this preconstruction services agreement.

■ Whether you choose to credit back some, all, or none of the design fee should not affect your legal status or liability regarding your design work. It is strictly a matter of your company policy and typically reflects the amount of the design contract relative to the amount of the construction contract.

However, crediting back to the Owner all or part of the design fee can be a powerful incentive for the Owner to award the construction contract to you. Also, the positive relationship that hopefully has developed between you and the Owner during the design phase and your intimate knowledge of the construction aspects of the project — since you have done the design work — are additional factors that usually place the design-build Contractor in the lead position to perform the construction work.

F. USE OF PLANS AND SPECIFICATIONS

The plans and specifications are being developed for the sole use of the Contractor named in this Agreement. If the Contractor named in this Agreement is not awarded the Construction Agreement, Owner is *strictly prohibited* from using these plans and specifications for *construction*. However, Owner may furnish these plans and specifications to an architect or other competent design professional to be used strictly for conceptual design ideas, *not for construction*. Owner's new architect or design professional will then take complete responsibility for the accuracy and suitability of *all* plans.

■ The clause above is intended to limit the design-build Contractor's liability in the event he does not perform the construction work and the Owner's future Contractor improperly performs the work and then alleges faulty design work on the part of the original design-build Contractor. The clause states that the design-build Contractor's plans may *not* be used for construction by a future Contractor. This is a very important clause that may help to limit your liability if another Contractor performs work based on your plans and then alleges design errors.

At the same time, the clause above allows the Owner to use the design-build Contractor's plans as *conceptual design ideas only* so that the Owner can obtain at least a minimal benefit from these plans if he decides not to have the design-build Contractor perform the construction phase of the project.

Even if the Owner doesn't want the design-build Contractor to perform the work, he is free to use the Contractor's plans (after he pays for the preconstruction design work), subject to the plans being used strictly as *conceptual design ideas* when having *new* plans and specifications prepared.

INITIAL
CC
HH

> Then, if another contractor performs the work, all actual construction must necessarily be performed on the basis of the Owner's *new* plans, *not*, in theory, on the basis of the original design-build Contractor's plans.

G. OBTAINING BUILDING PERMIT

If construction of project is approved by Owner and the Contractor named in this Agreement is to perform the construction phase of the work, Contractor will assist Owner in obtaining building permit. Owner to directly pay the cost of all governmental permit fees.

> ■ This clause states the Contractor will assist the Owner in obtaining the building permit, but the Owner is to directly pay the cost of the permit. I prefer to have the Owner pay for all outside fees such as those for permits and outside consultants. If a problem develops between the Owner and the design-build Contractor, the Contractor will not have as much financial exposure if he has not used his own money to pay the Owner's permit fees, outside consultant's fees, etc.

H. CONSTRUCTION OF PROJECT AND DESIGN CRITERIA

Contractor agrees to furnish all labor, materials, and equipment to construct the project in accordance with the design criteria, plans, and specifications set forth in this Agreement.

1. LUMP SUM PRICE FOR THE CONSTRUCTION PHASE OF PROJECT: This Lump Sum Price is subject to being increased or decreased by written Change Order if the Scope of Work as set forth by the design criteria below or the plans and specifications in existence at the time this Agreement is entered into is increased or decreased by Owner or any governmental agency: $ __139,669__

> ■ The lump sum price can be increased or decreased by change order when the Contractor is expected to perform additional work. If the Owner increases the scope of the design criteria or increases the Scope of Work once construction has commenced, this will of course increase the cost of the construction work. For this reason, it is critical that the Contractor define the design criteria below as accurately as possible and then issue written change orders increasing the Lump Sum contract amount when the Owner deviates from the design criteria.

Allowances that are a part of the Lump Sum Price above are designated as follows:

Item	*Allowance Amount*
1. Plumbing fixtures:	$ 2,000
2. Electrical fixtures:	$ 1,500
3. Floor coverings:	$ 5,500

INITIAL
CC
HH

__4. Tile materials:__	$ __1,800__
__5. Cabinets:__	$ __9,500__
__6. Doors and windows:__	$ __13,000__

> ■ Just as with a fixed price contract, unless the Owner has specified the exact fixtures to be used and you have costed out these fixtures prior to entering into the design-build contract, you should list fixtures as allowance items. This will give you flexibility to increase the Lump Sum Price if the Owner decides to upgrade the quality of the fixtures you based your contract on.

2. DESIGN CRITERIA FOR ESTABLISHING LUMP SUM PRICE: Due to the fact the plans and specifications may not have been fully developed at the time this Agreement has been entered into, the basic design criteria which the Lump Sum Price for construction has been based on includes the following:

__Furnish materials and labor to construct an approximately 1,200-square-foot, two-story addition to Owner's residence. Price based on allowances set forth above. New siding, windows, doors, and trim to match existing as closely as possible. Price also based on painting entire exterior of house, but painting only new areas of interior of house. Price based on no work being performed inside any part of existing house except the remodeling of the existing kitchen and the upgrading of the electrical service panel and furnace. Price based on new dimensional shingle, composition roof on new and existing roof. None of the work in the standard exclusions or project-specific exclusions section of the agreement has been included in the lump sum price (see exclusions section below).__

(See Additional Page Attached: _____ Yes; __x__ No)

> ■ The section above describes the general scope of the anticipated project. This should generally include the size of the project, the quality of the finishes, and any areas of the existing residence that will be worked on in conjunction with the new addition. Be as complete in your description as possible. These are the details your price is based on.

III. GENERAL CONDITIONS FOR THE AGREEMENT ABOVE

A. EXCLUSIONS

This Agreement does *not* include *labor or materials* for the following work (unless Owner selects one of these items as an Additional Alternate):

INITIAL
CC
HH

■ See annotation, Form 1.3: Section III.A. Exclusions.

1. PROJECT SPECIFIC EXCLUSIONS:
New gutters on existing roof, exterior concrete flatwork, french drains, pine-vaulted ceilings.

2. STANDARD EXCLUSIONS: Unless specifically included in the "General Scope of Work" section above, this Agreement does *not* include *labor or materials* for the following work (any Exclusions in this paragraph which have been lined out and initialed by the parties do not apply to this Agreement): Removal and disposal of any materials containing asbestos (or any other hazardous material as defined by the EPA). Custom milling of any wood for use in project. Moving Owner's property around the site. Labor or materials required to repair or replace any Owner-supplied materials. Repair of concealed underground utilities not located on prints or physically staked out by Owner which are damaged during construction. Surveying that may be required to establish accurate property boundaries for setback purposes (fences and old stakes may not be located on actual property lines). Final construction cleaning (Contractor will leave site in "broom swept" condition). Landscaping and irrigation work of any kind. Temporary sanitation, power, or fencing. Removal of soils under house in order to obtain 18 inches (or code-required height) of clear space between bottom of joists and soil. Removal of filled ground or rock or any other materials not removable by ordinary hand tools (unless heavy equipment is specified in Scope of Work section above), correction of existing out-of-plumb or out-of-level conditions in existing structure. Correction of concealed substandard framing. Rerouting/removal of vents, pipes, ducts, structural members, wiring or conduits, steel mesh which may be discovered in the removal of walls or the cutting of openings in walls. Removal and replacement of existing rot or insect infestation. Failure of surrounding part of existing structure, despite Contractor's good faith efforts to minimize damage, such as plaster or drywall cracking and popped nails in adjacent rooms, or blockage of pipes or plumbing fixtures caused by loosened rust within pipes. Construction of continuously level foundation around structure (if lot is sloped more than 6 inches from front to back or side to side, Contractor will step the foundation in accordance with the slope of the lot). Exact matching of existing finishes. Public or private utility connection fees. Repair of damage to existing roads, sidewalks, and driveways that could occur when construction equipment and vehicles are being used in the normal course of construction.

■ See annotation, Form 1.3: Section III.A.2. Standard Exclusions.

B. DATE OF WORK COMMENCEMENT AND SUBSTANTIAL COMPLETION
Commence work: **August 1, 2001** . Construction time through substantial completion: Approximately **4** to **5** months, *not* including delays and adjustments for delays caused by: holidays, inclement weather, accidents, shortage of labor or material, additional time required for performance of Change Order work (as specified in each Change Order), delays caused by Owner, and other delays unavoidable or beyond the control of the Contractor.

INITIAL
CC
HH

> ■ See annotation, Form 1.3: Section III.B. Date of Work Commencement and Substantial Completion.

C. EXPIRATION OF THIS AGREEMENT

This Agreement will expire 30 days after the date at the top of page one of this Agreement if not accepted in writing by Owner and returned to Contractor within that time.

> ■ See annotation, Form 1.3: Section III.C. Expiration of this Agreement.

D. WORK PERFORMED BY OWNER OR OWNER'S SEPARATE CONTRACTORS

Any labor or materials provided by the Owner's separate Subcontractors while Contractor is still working on this project must be supervised by Contractor. Profit and overhead at the rate of __20__ % will be charged on all labor and material furnished by Owner's separate Subcontractors while Contractor is still working on the project. Contractor has right to qualify and approve Owner's Subcontractors and require evidence of work experience, proper licensing, and insurance. If Owner wants to avoid paying Contractor's profit and overhead per this section, Owner must then bring in his separate Subcontractors only *before* or *after* Contractor has performed all of his work.

> ■ See annotation, Form 1.3: Section II.D. Work Performed by Owner or Owner's Separate Contractors.

E. CHANGE ORDERS: CONCEALED CONDITIONS, ADDITIONAL WORK, AND CHANGES IN THE WORK

1. PEOPLE AUTHORIZED TO SIGN CHANGE ORDERS: The following people are authorized to sign Change Orders:

Harry Homeowner

Helen Homeowner

(Please fill in line(s) above at time of signing Agreement)

> ■ See annotation, Form 1.3: Section III.E.1. People Authorized to Sign Change Orders.

2. CONCEALED CONDITIONS: This Agreement is based solely on the observations Contractor was able to make with the structure in its current condition at the time this Agreement was bid. If additional Concealed Conditions are discovered once work has commenced which were *not* visible at the time this proposal was bid, Contractor will stop work and point out these unforeseen Concealed Conditions to Owner so that Owner and Contractor can execute a Change Order for any Additional Work.

INITIAL
CC
HH

■ See annotation, Form 1.3: Section III.E.2. Concealed Conditions.

3. CHANGES IN THE WORK: During the course of the project, Owner may order changes in the work (both additions and deletions). The cost of these changes will be determined by the Contractor and the cost of this Additional Work will be added to Contractor's profit and overhead at the rate of __20__ % in order to arrive at the net amount of any additional Change Order work.

Contractor's profit and overhead, and any supervisory labor will not be credited back to Owner with any deductive Change Orders (work deleted from Agreement by Owner).

■ See annotation, Form 1.3: Section III.E.3. Changes in the Work.

4. DEVIATION FROM SCOPE OF WORK IN CONTRACT DOCUMENTS: Any alteration or deviation from the Scope of Work referred to in the Contract Documents involving extra costs of materials or labor, including any overage on **ALLOWANCE** work, will be executed upon a written Change Order issued by Contractor and should be signed by Contractor and Owner prior to the commencement of any Additional Work by the Contractor. This Change Order will become an extra charge over and above the Lump Sum Price referred to at the beginning of this Agreement.

■ See annotation, Form 1.3: Section III.E.4. Deviation from Scope of Work in Contract Documents.

5. CHANGES REQUIRED BY PLAN CHECKERS OR FIELD INSPECTORS: Any increase in the Scope of Work set forth in the Contract Documents which is required by plan checkers or field inspectors with city or county building/planning departments will be treated as Additional Work to this Agreement for which the Contractor will issue a Change Order.

■ See annotation, Form 1.3: Section III.E.5. Changes Required by Plan Checkers or Field Inspectors.

6. RATES CHARGED FOR ALLOWANCE ONLY AND TIME-AND-MATERIALS WORK:
Journeyman Carpenter: $__26__ per hour; Apprentice Carpenter: $__21__ per hour; Laborer: $__17__ per hour; Contractor: $__35__ per hour; Subcontractor: Amount charged by Subcontractor. *Note:* Contractor will charge for profit and overhead at the rate of __20__ % on all work performed on a Time-and-Materials basis (on both materials and labor rates set forth in this paragraph) and on all costs that exceed specifically stated **ALLOWANCE** estimates in the Agreement.

INITIAL
CC
HH

■ See annotation, Form 1.3: Section III.E.6. Rates Charged for Allowance Only and Time-and-Materials Work.

F. PAYMENT SCHEDULE AND PAYMENT TERMS

1. PAYMENT SCHEDULE:

First Payment: $1,000 or 10% of contract amount (whichever is less) due when contract is signed and returned to Contractor.

Contract Deposit Payment: $ __1,000__

* Second Payment (Materials Deposits): Any materials deposits — required for such items as woodstoves, cabinetry, carpets and vinyl, granite, tile, and any and all special order items which require the payment of a materials deposit — must be paid within 3 days of submittal of invoice by Contractor. These items will not be ordered until the deposits set forth below are received by Contractor. Materials Deposits required on this project include the following:

Deposit for cabinets: $ __2,000__

* Total Due for All Materials Deposits: $ __2,000__

* Third Payment: **10% of contract amount due upon completion of demolition, foundation work, and delivery of initial framing materials to site:** $ 13,996

* Fourth Payment: **25% of contract amount due upon completion of rough framing:** $ 34,917

* Fifth Payment: **15% of contract amount due upon completion of roofing, installation of gutters and exterior sheet metal, and installation of all exterior doors and windows:** $ 20,950

* Sixth Payment: **15% of contract amount due upon completion of rough plumbing, rough electrical, insulation, and hanging of drywall (not taping or texturing):** $ 20,950

*Seventh Payment: **15% of contract amount due upon installation of lath, stucco scratch coat, stucco finish coat, taping and texturing of drywall, installation of interior doors, closet poles and shelves, and installation of heating system:** $ 20,950

*Eighth Payment: **15% of contract amount due upon completion of interior paint, exterior paint, installation of cabinets, installation of rough and finish electrical fixtures:** $ 20,950

INITIAL
CC
HH

* Final Payment: Due upon Substantial Completion of all work under this Agreement: $ __1,986__

■ See annotation, Form 1.3: Section III.F.1. Payment Schedule.

2. PAYMENT OF CHANGE ORDERS: Payment for each Change Order is due upon completion of Change Order work and submittal of invoice by Contractor for this work.

■ See annotation, Form 1.3: Section III.F.2. Payment of Change Orders.

3. ADDITIONAL PAYMENTS FOR ALLOWANCE WORK AND RELATED CREDITS: Payment for work designated in the Agreement as **ALLOWANCE** work has been initially factored into the Lump Sum Price and Payment Schedule set forth in this Agreement. If the actual cost of the **ALLOWANCE** work *exceeds* the line item **ALLOWANCE** amount in the Agreement, the difference between the cost and the line item **ALLOWANCE** amount stated in the Agreement will be written up by Contractor as a Change Order subject to Contractor's profit and overhead at the rate of __20__%.

If the cost of the **ALLOWANCE** work is *less* than the **ALLOWANCE** line item amount listed in the Agreement, a credit will be issued to Owner after all billings related to this particular line item **ALLOWANCE** work have been received by Contractor. This credit will be applied toward the final payment owing under the Agreement. Contractor profit and overhead and any supervisory labor will *not* be credited back to Owner for **ALLOWANCE** work.

■ See annotation, Form 1.3: Section III.F.3. Additional Payments for Allowance Work and Related Credits.

4. FINAL CONTRACT PAYMENT: The final contract payment is due and payable upon "Substantial Completion" (not Final Completion) of all work under contract. "Substantial Completion" is defined as being the point at which the Building/Work of Improvement is suitable for its intended use, or the issuance of an Occupancy Consent, or final building department approval from the city or county building department, whichever occurs first.

■ See annotation, Form 1.3: Section III.F.4. Final Contract Payment.

5. HOLD BACK FROM FINAL PAYMENT FOR PUNCH LIST WORK: At time of making the final contract payment, Owner may hold back 150% of the value of all Punch List work. Owner and Contractor will place a fair and reasonable value on each Punch List item at time of Punch list walk-through with Owner. Contractor and Owner will then execute the Punch List form. This 150% hold back for Punch List work assures Owner that all Punch List work will be completed by Contractor in a timely manner.

INITIAL
CC
HH

> ■ See annotation, Form 1.3: Section III.F.5. Hold Back From Final Payment for Punch List Work.

6. PAYMENT FOR COMPLETED PUNCH LIST WORK: Payment for completed Punch List items is due and payable upon submittal of invoice for any part of the completed Punch List work covered by Contractor's invoice.

> ■ See annotation, Form 1.3: Section III.F.6. Payment for Completed Punch List Work.

7. INTEREST CHARGES: Interest in the amount of __1.5__ % per month will be charged on all late payments under this Agreement. "Late Payments" are defined as any payment not received within __5__ days of receipt of invoice from Contractor.

> ■ See annotation, Form 1.3: Section III.F.7. Interest Charges.

G. MISCELLANEOUS CONDITIONS

1. MATCHING EXISTING FINISHES: Contractor will use best efforts to match existing finishes and materials. However, an exact match is not guaranteed by Contractor due to such factors as discoloration from aging, a difference in dye lots, and the difficulty of exactly matching certain finishes, colors, and planes.

Unless custom milling of materials is specifically called out in the plans, specifications, or Scope of Work description, any material not readily available at local lumberyards or suppliers is *not* included in this Agreement.

If Owner requires an exact match of materials or textures in a particular area, Owner must inform Contractor of this requirement in writing within seven (7) days of signing this Agreement. Contractor will then provide Owner with either a materials sample or a test patch prior to the commencement of work involving the matching of existing finishes.

Owner must then approve or disapprove of the suitability of the match within 24 hours. After that time, or after Contractor has provided Owner with two or more test patches that have been rejected by Owner, all further test patches, materials submittals, or any removal and replacement of materials already installed in accordance with the terms of this section will be performed strictly as Extra Work on a time-and-materials basis by Contractor.

> ■ See annotation, Form 1.3: Section III.G.1. Matching Existing Finishes.

INITIAL
CC
HH

2. CONFLICT OF DOCUMENTS: If any conflict should arise between the plans, specifications, addenda to plans, and this Agreement, then the terms and conditions of this Agreement shall be controlling and binding upon the parties to this Agreement.

> ■ See annotation, Form 1.3: Section III.G.2. Conflict of Documents.

3. INSTALLATION OF OWNER-SUPPLIED FIXTURES AND MATERIALS: Contractor can not warrant any Owner-supplied materials or fixtures (whether new or used). If Owner-supplied fixtures or materials fail due to a defect in the materials or fixtures themselves, Contractor will charge for all labor and materials required to repair or replace both the defective materials or fixtures, and any surrounding work that is damaged by these defective materials or fixtures.

> ■ See annotation, Form 1.3: Section III.G.3. Installation of Owner-Supplied Fixtures and Materials.

4. CONTROL AND DIRECTION OF EMPLOYEES AND SUBCONTRACTORS: Contractor, or his appointed Supervisor, shall be the sole supervisor of Contractor's Employees and Subcontractors. Owner must not order or request Contractor's Employees or Subcontractors to make changes in the work. All changes in the work are to be first discussed with Contractor and then performed according to the Change Order process as set forth in this Agreement.

> ■ See annotation, Form 1.3: Section III.G.4. Control and Direction of Employees and Subcontractors.

5. OWNER COORDINATION WITH CONTRACTOR: Owner agrees to promptly furnish Contractor with all details and decisions about unspecified construction finishes, and to consent to or deny changes in the Scope of Work that may arise so as not to delay the progress of the Work. Owner agrees to furnish Contractor with continual access to the job site.

> ■ See annotation, Form 1.3: Section III.G.5. Owner Coordination with Contractor.

H. TERMINATION OF AGREEMENT FOR CONVENIENCE BY OWNER DURING PRECONSTRUCTION DESIGN PHASE *ONLY*

Owner may terminate this entire Agreement at any time *only* during the Preconstruction Design Services phase of the work by giving Contractor written notice of the termination and providing payment to Contractor for all services rendered through the date of termination. Upon termination and payment for all work performed to date, all drawings, details, and estimates performed through the termination date by Contractor and his agents will be delivered to Owner subject to the limitations in Section II.F "Use of Plans and Specifications," above.

INITIAL
CC
HH

■ The termination clause makes it easy for the Owner to terminate the Contractor if he is not satisfied with the Contractor's performance. If the Owner and design-build Contractor don't have a good relationship during the design phase, it's better to make it easy to end the relationship early on. Written notice is required if the Owner wants to terminate the design phase of the work and payment for all work performed to date is required.

The Contractor may terminate the Agreement if the Owner fails to make payments in accordance with the payment schedule. The Contractor also should give written notice of the termination and the grounds for termination.

I. WORK STOPPAGE AND TERMINATION OF AGREEMENT FOR DEFAULT

Once construction has commenced, Contractor shall have the right to stop all work on the project and keep the job idle if payments are not made to Contractor in accordance with the Payment Schedule in this Agreement, or if Owner repeatedly fails or refuses to furnish Contractor with access to the job site and /or product selections or information necessary for the advancement of Contractor's work. Simultaneous with stopping work on the project, the Contractor must give Owner written notice of the nature of Owner's default and must also give the Owner a 14-day period in which to cure this default.

If work is stopped due to any of the above reasons (or for any other material breach of contract by Owner) for a period of 14 days, and the Owner has failed to take significant steps to cure his default, then Contractor may, without prejudicing any other remedies Contractor may have, give written notice of termination of the Agreement to Owner and demand payment for all completed work and materials ordered through the date of work stoppage, and any other loss sustained by Contractor, including Contractor's Profit and Overhead at the rate of __20__ % on the balance of the incomplete work under the Agreement. Thereafter, Contractor is relieved from all other contractual duties, including all Punch List and warranty work.

■ See annotation, Form 1.3: Section III.H. Work Stoppage and Termination of Agreement for Default.

J. LIEN RELEASES

Upon request by Owner, Contractor and Subcontractors will issue appropriate lien releases prior to receiving final payment from Owner.

■ See annotation, Form 1.3: Section III.I. Lien Releases.

K. DISPUTE RESOLUTION AND ATTORNEY'S FEES

Any controversy or claim arising out of or related to this Agreement involving an amount of *less* than $5,000 (or the maximum limit of the court) must be heard in the Small Claims Division of the Municipal Court in the county where the Contractor's office is located. Any controversy or claim arising out of or related to this Agreement which is over the dollar limit of the Small Claims Court

INITIAL
CC
HH

must be settled by binding arbitration administered by the American Arbitration Association in accordance with the Construction Industry Arbitration Rules. Judgment upon the award may be entered in any Court having jurisdiction thereof.

The prevailing party in any legal proceeding related to this Agreement shall be entitled to payment of reasonable attorney's fees, costs, and expenses.

> ■ See annotation, Form 1.3: Section III.J. Dispute Resolution and Attorney's Fees.

L. INSURANCE

Owner shall pay for and maintain "Course of Construction" or "Builder's Risk" or any other insurance that provides the same type of coverage to the Contractor's work in progress during the course of the project. It is Owner's express responsibility to insure dwelling and all work in progress against all damage caused by fire and Acts of God such as earthquakes, floods, etc.

> ■ See annotation, Form 1.3: Section III.K. Insurance.

M. INDEMNIFICATION

If the Contractor who is a party to this Agreement does not build this project, then he is fully released, held harmless, and indemnified by Owner from liability and claims of every kind (including attorney's fees) related to the use of these drawings and specifications by any and all persons subsequently engaged by Owner or a future Owner of this property to design or build the project.

The Contractor named in this Agreement is hereby released, held harmless, and indemnified from all claims of every kind whatsoever (including attorney's fees) brought by any person or entity that result from alleged errors or omissions existing in the plans or specifications of Contractor *if* these plans are relied on in any way by anyone other than this Contractor and his designated agents in building or designing the project.

> ■ The indemnification clause above is critical and should be used in conjunction with the "Use of Plans and Specifications" clause. The indemnification clause states that the Owner releases the design-build Contractor from liability that results from another contractor working off the design-build Contractor's plans. Although indemnification clauses can be overruled in court in some instances, this is a very important clause that will help protect the interests of the design-build Contractor.

N. WARRANTY

Contractor provides a limited warranty on all Contractor- and Subcontractor-supplied labor and materials used in this project for a period of one year following substantial completion of all work.

INITIAL
CC
HH

No warranty is provided by Contractor on any materials furnished by the Owner for installation. No warranty is provided on any existing materials that are moved and/or reinstalled by the Contractor within the dwelling (including any warranty that existing/used materials will not be damaged during the removal and reinstallation process). One year after substantial completion of the project, the Owner's sole remedy (for materials and labor) on all materials that are covered by a manufacturer's warranty is strictly with the manufacturer, not with the Contractor.

Repair of the following items is specifically excluded from Contractor's warranty: Damages resulting from lack of Owner maintenance; damages resulting from Owner abuse or ordinary wear and tear; deviations that arise such as the minor cracking of concrete, stucco and plaster; minor stress fractures in drywall due to the curing of lumber; warping and deflection of wood; shrinking/cracking of grouts and caulking; fading of paints and finishes exposed to sunlight.

THE EXPRESS WARRANTIES CONTAINED HEREIN ARE IN LIEU OF ALL OTHER WARRANTIES, EXPRESS OR IMPLIED, INCLUDING ANY WARRANTIES OF MERCHANTABILITY, HABITABILITY, OR FITNESS FOR A PARTICULAR USE OR PURPOSE. THIS LIMITED WARRANTY EXCLUDES CONSEQUENTIAL AND INCIDENTAL DAMAGES AND LIMITS THE DURATION OF IMPLIED WARRANTIES TO THE FULLEST EXTENT PERMISSIBLE UNDER STATE AND FEDERAL LAW.

> ■ See annotation, Form 1.3: Section III.L. Warranty.

O. ENTIRE AGREEMENT, SEVERABILITY, AND MODIFICATION

This Agreement represents and contains the entire agreement between the parties. Prior discussions or verbal representations by Contractor or Owner that are not contained in this Agreement are *not* a part of this Agreement. In the event that any provision of this Agreement is at any time held by a Court to be invalid or unenforceable, the parties agree that all other provisions of this Agreement will remain in full force and effect. Any future modification of this Agreement must be made in writing and executed by Owner and Contractor in order to be valid and binding upon the parties.

> ■ See annotation, Form 1.3: Section III.M. Entire Agreement, Severability, and Modification.

P. ADDITIONAL LEGAL NOTICES REQUIRED BY STATE OR FEDERAL LAW

See page(s) attached: __**X**__ Yes; _____ No

> ■ See annotation, Form 1.3: Section III.N. Additional Legal Notices Required by State or Federal Law.

INITIAL
CC
HH

Q. ADDITIONAL TERMS AND CONDITIONS

See page(s) attached: _____ Yes; __X__ No

> ■ See annotation, Form 1.3: Section III.O. Additional Terms and Conditions.

I have read and understood, and I agree to, all of the terms and conditions in the Agreement above.

> ■ You need a statement that indicates the parties have read and agree to the terms and conditions of the Agreement. Be sure to have the Agreement signed by the Owner prior to the time you commence work. Make sure you keep a signed copy of the Agreement in your records — an unsigned copy won't do you any good later on if you have a dispute. Finally, have the Owner initial every page of your Agreement, including any supplemental attachments, such as materials or Subcontractor bids.

Date: _5/22/01_   *CHARLIE CONTRACTOR, PRESIDENT*
                      CHARLIE CONTRACTOR, PRESIDENT
                      CHARLIE CONTRACTOR CONSTRUCTION, INC.

> ■ If your business is a corporation, be sure to sign the Agreement using your corporate title and place the word, "Inc.," after your company's name. If you fail to do this, you may have personal liability under the Agreement.
> I then attach the Notice of Cancellation Form, any other notice required by state law, and all other bids listed in the Scope of Work as "enclosures" to this Agreement. Two copies of the Agreement should be given to the Owners, the original for signing, and a copy for their records. Each page of the original should be stamped with the initial stamp, and initialed by both you and the Owners. When the Owners have initialed and signed the Agreement, they should return it to you.

Date: _5/22/01_   *Harry Homeowner*
                      HARRY HOMEOWNER

Date: _5/22/01_   *Helen Homeowner*
                      HELEN HOMEOWNER

CHAPTER 4

CHANGE ORDERS

4.1 ▪ CHANGE ORDER CONTINGENCY FUND

4.2 ▪ CHANGE ORDER FORM AND ACCOUNTING SUMMARY

*E*very experienced contractor knows that the owner who tells you on Monday to finish the extra work at all costs ("Just get it done, whatever it takes!") may not feel the same way about paying for the extra work the following Monday when he gets your invoice.

In my experience, the most common owner/contractor dispute is over change orders. Change order disputes often result in the contractor losing money, as well as the trust and confidence of the owner (and sometimes his subs). And, if there have been other areas of dispute with the owner, disagreeing about change orders often pushes the contractor closer to the brink of litigation.

The best way to avoid disputes over change orders is to establish a clear policy in your agreement on what amounts to "extra work," "additional work," or "change order" work, and how this additional work will be identified, carried out, and paid for when it is encountered during the project.

Your agreements should call for extra work to be written up as a change order. Stick to the policy of getting extra work approved in writing by the owner before you do the work. Occasionally, you'll do some work before the owner signs the change order — just to keep the job moving — but 95% of the time, it's possible to get that signature first. A signed change order will almost entirely eliminate arguments over the owner's obligation to pay for the work.

The construction agreements in this book contain clauses that define "additional work." When something comes up that is not included in the agreement's scope of work, you will be glad to have a contractual basis for identifying it as additional work. You can turn to a paragraph in the agreement that clearly establishes the item as additional work and feel confident about bringing the extra work to the owner's attention, following your change order procedure, and getting paid fair and square for the work.

Get It in Writing

Once in litigation or arbitration, owners almost always challenge claims of verbal change orders by the contractor. However, written change orders that are signed by the parties before the work is done are only very rarely raised as a claim against the contractor.

What does this tell you? Don't do change order work without first writing up the change order and getting the owner's signature. It's really very simple: *get it in writing in order to be paid and to avoid future disputes.*

I don't know a single residential contractor who hasn't "eaten" hundreds or even thousands of dollars of legitimate change order work at the end of a job just so he could get the owner to release his final check.

Why would a contractor do this? The main reason is that he never put these legitimate change orders in writing. Why weren't they in writing? Good question. Let's have a look at a few of my favorite excuses for donating "free" change order work worth hundreds or thousands of dollars. (I hate to admit that I've used a few of these myself, and always regretted it later.)

The first line is what the contractor was thinking when he decided not to put the change order in writing. The text in *italics* is what he was thinking after he discovered he would never be paid for the additional work.

■ "I'm just too busy right now to take the time to write this up..."

"Man, this lawsuit is sure taking a lot longer than it would have taken to write up that $7,000 in change orders the owner said he would pay. Being too busy to write up those change orders is like being too busy to change the oil in my truck. Boy, was I stupid."

■ "I'm a nice guy. And the owner seems like a fair kind of guy. He won't have a problem paying for legitimate work later. He said he trusted me and that is why he gave me the job. I really trust him too."

"Now I remember what Dad always said about "nice guys." Why didn't I listen when he told me that business is business and personal is personal, and never to cross the lines?

■ "I really don't want to rock the boat and give the owner a change order this early in the project (even though it is legitimate). After all, the owner said just a couple of weeks ago in our meeting that he expected there to be no change orders on this job."

"Wish I had rocked the boat now. I gave up two legitimate change orders early on, and then the owner would not agree to sign the later ones because he thought if I could afford to give away those first few, I could afford to give away the rest as well."

■ "No way. The owner is my (take your choice: good friend, neighbor, relative, business associate, friend of my best friend, doctor, lawyer). He would never refuse to pay for this legitimate additional work — we trust each other, and after all, that's why we're friends."

"That jerk! We used to be friends. I'll never do that again, and after all this we'll never be friends again either! How could I have so badly misjudged that guy?"

■ "Why, I've got plenty of money in the job to cover a few extras in the early phases. I'm not going to be both-

ered right now writing up these few change orders."

"Man, I blew it. I may have made 10% more on the foundation than I expected, but after losing 15% on the framing and finish work, I realize I underbid the job as a whole! I thought the owner would pay me for the things we both know I forgot in my bid, but he wouldn't. I sure wish I'd charged for those legitimate extras when I could have!"

■ "The owner is loaded. He knows he wants quality work, and he'll be willing to pay for it later. That's why he hired me in the first place."

"Wow, the owner really is loaded. He has his own lawyer on staff and says it won't cost him a penny to fight me on these $6,500 worth of change orders. How come the lawyer I talked to wanted $3,500 up front just to take the case? Just the legal fees will crush me on this thing: I don't have an attorney's fees clause in my agreement so I'll probably never be able to recoup the legal costs (which ought to just about equal what the owner owes me now). Guess I lose either way."

■ "I hate this paperwork stuff! I'm sure I'll be able to remember these 17 changes six months from now, and I'm sure the owner won't mind me showing these changes to him then."

"What a fool I was again. I can't figure out which of the receipts in this brown paper bag filing system of mine are for additional work and which are for work that was covered by the contract. I can't even begin to sort out how much labor I spent on the extra work. Oh well, to heck with it...the owner won't pay me for it now anyway. We're behind schedule and the owner is out of money and boiling mad that I never told him about the extra work until now. Besides, he's holding my final $7,500 check as a hostage. I gotta' have that final check to pay my bills. Maybe next job I'll get organized..., yeah, that's it...next job I'll get organized and make some real money!"

Any of these sound familiar? Do all of them sound a little familiar? Enough said. Put it in writing *before* you do the additional work unless you really are willing to happily forget about the money you are owed. One simple trick is to have some duplicate or triplicate carbon copies of the change order form that appears in Section 4.2.

Carry these with you in your truck or briefcase at all times. When you're on a job and need to write the change order right away, you'll have the forms with you and can do it right there. Then you can either get the owner's signature on the spot, or leave a couple of copies for him to sign and then pick them up before you start the extra work.

Tracking Change Orders

Having and using change orders won't do you much good if you don't have some organized system for identifying which change order is for what work, how many change orders have been generated on a certain job, and whether or not the owner has signed a given change order.

My method of tracking change orders is to have a change order subdirectory in the computer. When a change order comes up, I copy the blank change order to a new file with a few letters of the owner's name, the number of the change order, and the date of the change order. For example, the file name "Smith1.515" means it's a change order for the Smith job, it's change order #1, and the date is 5/15 or May 15.

Once you have made a copy of the blank change order or "master," fill out all the information on the new change order. Print it twice, stamp one copy "AFTER REVIEWING PLEASE SIGN THIS COPY AND RETURN TO CONTRACTOR," and give it to the owner for signature. Keep the second unstamped copy in your project file until you get the signed version back from the owner.

For change order #2 for Mr. Smith, follow the same procedure except, rather than copying a blank change order form, copy the file named "Smith1.515" and relabel the new file to reflect the number of the change order you are now writing up and the current date. This system has worked really well for me. It's very convenient to have all your change orders stored in the computer.

If you can't get a signature on the change order prior to starting the additional work, keep a phone log of the time and date of the conversation in which the owner authorized the additional work. Then, follow up the verbal approval with a written change order ASAP! With fax machines, there is no reason not to get a written change order signed prior to or within a day or two of beginning change order work.

What if I Need to "Prove" a Change Order?

In the absence of a written change order signed by both parties, a commonly accepted definition of extra work in many states requires that the contractor prove through clear and convincing evidence the following:

- The extra work was outside the scope of work in the original contract.
- The extra work was ordered by the owner.
- The owner agreed to pay for the additional work, either by words or conduct.
- The extra work was not a gift, or given "without charge" by the contractor.
- The extra work was not required as a result of the negligence of the contractor or caused by a default of one of the Subcontractors.

If proving these five points in order to establish the validity of extra work sounds complicated, you're right. It is extremely complicated in contrast to the simplicity of organizing your work in the field to incorporate written change orders for extra work prior to actually performing the work.

Why spend hours of time arguing over who will pay for change orders, accept unreasonable offsets in change order payments (or no money at all in some cases), or be forced into litigation or arbitration over disputed change orders, when using a simple change order procedure can prevent 98% of all of these problems from ever occurring?

Take the time to incorporate written change orders into your normal work routine and you will be rewarded with payment for extra work and better profit margins. You'll also have greater peace of mind as a result of not having to argue with the owner, after the fact, about getting paid for extra work.

Change Order Contingency Fund

I always provide the owner with a sample of my change order form at the time of contract signing. We review the change order form and I explain to him that some change orders are to be expected on practically every job.

I also give the owner a form letter, printed on company letterhead, that is really an information sheet explaining what leads to change orders and how they can be kept to a minimum. This letter recommends that the owner set aside some money, a "change order contingency fund," in order to budget for change orders. See Form 4.1 below.

Before typing in the customer's name, make your best effort to "guesstimate" the anticipated amount of change orders that may arise. Fill in the percentage of the total amount of the job that you anticipate these change orders will amount to. I often see 5% to 15% in change orders on a typical remodeling job. With a complicated job that has inadequate plans or the potential for considerable concealed conditions, this percentage can, of course, be much higher.

Then be sure to explain to the owner that your figure is only a guesstimate, but should point him in the right direction in budgeting for the overall "ballpark" project cost. Tell the owner that each change order form provides a detailed financial summary so that he will always know the adjusted gross contract amount at any given point in time during the construction process.

Most owners who are on a budget will appreciate your taking the time to educate them about change orders. You should explain the likelihood that some change orders will occur, how to help keep them to a minimum, and the need to budget for them with a contingency fund. If the owner is troubled by you raising the change order issue prior to beginning the job, he's probably a person with little construction experience or unrealistic expectations.

Explain that your intent is simply to inform him that certain unforeseeable expenses are typical on jobs like this and that your information will help him to better plan for all aspects of the project.

4.1 ▪ CHANGE ORDER CONTINGENCY FUND
CLIENT INFORMATION SHEET

CONTRACTOR'S NAME: _____

COMPANY NAME: _____

ADDRESS: _____

_____

PHONE: _____

FAX: _____

DATE: _____

Dear _____ :

Nobody likes change orders! For you, the owner, they present a sometimes unanticipated expense. For the contractor they tend to delay the project schedule and require a greater amount of administrative time in proportion to their cost. Nevertheless, we have found that even with the best of construction plans, some change orders are likely to arise.

So that you can better understand what leads to change orders, how to budget for them, and how to minimize their occurrence, we have included the information below and have enclosed a sample change order form so that you can ask us any questions you may have prior to the commencement of any work on your project.

Based on our experience with projects similar to this one, we anticipate that change orders may arise on this project as a result of both *owner-requested* upgrades of current plans (e.g., the owner decides to change metal windows to wood windows or to change vinyl flooring to tile flooring), and *contractor-requested* changes arising from either concealed conditions or actual site conditions that vary from those described in the construction drawings.

Therefore, we recommend that you initially budget approximately _____ % of the estimated project cost as a contingency fund for both owner-requested and contractor-requested change orders. This percentage is an estimate only. The actual total amount of all change orders for your project will vary based upon such currently unknown factors as:
• The number of owner-requested changes and upgrades requested during the course of the project.
• The possibility of encountering concealed conditions during the work.
• The degree of completeness and amount of detail shown in the construction drawings and the as-built drawings.

By obtaining a complete and detailed set of construction drawings, the owner can expect a smaller number of change orders during the course of the project. Detailed construction drawings will include such information as:
• accurate as-built drawings;
• all structural details and connections;
• electrical, plumbing, and mechanical plans (including the location of all phone and cable jacks and heat ducts);
• accurate door, window, and hardware schedules;
• finish plumbing and electrical fixture list;
• elevations showing all details in kitchen, bathrooms, and other rooms with cabinetry;
• inclusion of any exterior concrete flat work, decks, landings, irrigation and landscaping plans;
• finish floor schedule;
• appliance schedule, countertop schedule, etc.

This information should help you do a better job of budgeting for your project and keeping unanticipated expenses to a minimum.

Sincerely,

_____

CONTRACTOR'S NAME
COMPANY NAME

Keeping the owner (and yourself) informed and constantly updated on how the financial status of the project is affected by change orders is a critical part of the contractor's job. Nothing shakes the owner's confidence (or the contractor's bottom line) quite like not knowing just how much money the owner has paid toward contract payments and different change orders, and how the original contract amount has been affected by change orders. This includes both increases and deductive change orders (those that result in a credit to the owner).

By using an accounting summary on your change order form, you and the owner will always know how much work he has under contract and how much it has been adjusted up or down by change orders. And by starting out the accounting summary with change order #1, it will be very easy to keep track of the accounting aspects of your job as you issue future change orders. This will save you time, money, and aggravation in the long run.

Any contractor who has lost control of change order paperwork will recognize the value of tracking these forms and the corresponding accounting as the project moves forward, change order by change order. The owner will also appreciate being updated with this information on a regular basis.

The owner will have greater confidence in your ability to run the project when he sees how you keep track of basic accounting and how you properly and fairly execute change orders for additional work prior to doing that work.

Finally, because these change orders look professional, the typical owner will react more favorably when it is time to approve additional work. If the contractor's presentation looks professional and it is clear that the contractor is on top of things, I find that the owner's resistance to change orders decreases dramatically.

A Final Word
Remember, in order to make it easier to implement a good written change order system, you need to keep blank change order forms with you on the job site. Fill them out and submit them to the owner for signature whenever extra work arises that has to be done immediately. Either before or after a change order is signed, you can take it back to your office and fill in the "Accounting Summary" portion. And, again, send a copy of the signed change order with the proper accounting summary information to the owner for his records.

4.2 ■ CHANGE ORDER FORM AND ACCOUNTING SUMMARY

CONTRACTOR'S NAME: _____

ADDRESS: _____

_____

PHONE: _____

FAX: _____

LIC #: _____

DATE: _____

OWNER'S NAME: _____

ADDRESS: _____

_____

PROJECT ADDRESS: _____

_____

CONSTRUCTION CHANGE ORDER #_____

I. GENERAL SCOPE OF WORK DESCRIPTION

Pursuant to the Construction Agreement between Contractor and Owner dated _____ , 19_____ , Contractor agrees to perform the following additional work:

_____

_____

_____

LUMP SUM PRICE FOR ALL WORK ABOVE: $ _____

* Additional time needed to complete project as a result of this Change Order: _____ Days. (Add to completion date in Construction Agreement.)

II. ACCOUNTING SUMMARY

A. Original Contract Amount: $ _____

B. Net Change by all prior Change Orders: $ _____

C. Adjusted Gross Contract Amount prior to this Change Order: $ _____

D. Amount of this Change Order: $ _____

E. Adjusted Gross Contract Amount including this Change Order: $ _____

III. GENERAL CONDITIONS

A. PAYMENT

Payment for this Change Order is due upon completion of this Change Order work and submittal of invoice by Contractor.

B. INCORPORATION

This Change Order, by agreement of Owner and Contractor, is incorporated by reference into the Construction Agreement between Owner and Contractor. All terms and conditions in the "General Conditions" section of the Construction Agreement between Owner and Contractor apply to this Change Order.

I have read and understood the Change Order above, and I agree to all of its terms.

Date: _____   _____
                        OWNER'S SIGNATURE

Date: _____   _____
                        CONTRACTOR'S SIGNATURE

4.2 ■ Change Order Form And Accounting Summary

ANNOTATED

Charlie Contractor Construction, Inc.
123 Hammer Lane
Anywhere, USA 33333
Phone: (123) 456-7890
Fax: (123) 456-7899
Lic#: 11111

DATE: **June 16, 2001**

OWNER'S NAME: **Mr. & Mrs. Harry Homeowner**
ADDRESS: **234 Lumber Lane**
Anywhere, USA 33333

PROJECT ADDRESS: **same**

CONSTRUCTION CHANGE ORDER # 3

I. GENERAL SCOPE OF WORK DESCRIPTION

Pursuant to the Construction Agreement between Contractor and Owner dated **May 14**, 20 **01**, Contractor agrees to perform the following additional work:

> ■ Be sure to date and number all your Change Orders. You need some type of reference to the original or primary contract between the Owner and the Contractor such as the one above. The Scope of Work description should be fairly detailed so that there is no question about the exact nature of each item of additional work.

INITIAL
CC
HH

1. Provide excavation, framing, and materials for foundation and subframing of new redwood deck per plans by Art Architect, dated June 10, 2001.

2. Install *owner-supplied* 2x6 redwood decking after contractor sands, applies edge treatment, and acid washes this *owner-supplied redwood decking*.

3. Furnish materials and labor for stairs and stair pad per plans (no handrail or guardrail on deck or stairs; no deck sealer).

LUMP SUM PRICE FOR ALL WORK ABOVE: $ ___1,780___

> ■ I usually combine profit and overhead with the direct costs and show just one lump sum number (or total price) for each Change Order. If Change Orders have more than one item of extra work I show a total price for each line item.

* Additional time needed to complete project as a result of this Change Order: __7__ Days. (Add to completion date in Construction Agreement.)

> ■ It is absolutely critical that you add additional time for each Change Order so that your contract completion date is moved ahead. If you fail to add extra time for extra work, the owner may become upset and think that you are behind in your work when you are actually right on schedule. Adjusting contract completion dates with each Change Order will alleviate this problem.

II. ACCOUNTING SUMMARY

A. Original Contract Amount: ____(5/14/01)____ $ ___27,713___

B. Net Change by all prior Change Orders: $ ___9,864___

C. Adjusted Gross Contract Amount prior to this Change Order: $ ___37,577___

D. Amount of this Change Order: $ ___1,780___

E. Adjusted Gross Contract Amount including this Change Order: $ ___39,357___

III. GENERAL CONDITIONS

A. PAYMENT
Payment for this Change Order is due upon completion of this Change Order work and submittal of invoice by Contractor.

INITIAL
CC
HH

> ■ This clause makes it clear that payment is due whenever the work in the Change Order has been completed by the Contractor and an invoice for the completed work has been given to the Owner.

B. INCORPORATION

This Change Order, by agreement of Owner and Contractor, is incorporated by reference into the Construction Agreement between Owner and Contractor. All terms and conditions in the "General Conditions" section of the Construction Agreement between Owner and Contractor apply to this Change Order.

> ■ This clause indicates that all the terms and conditions affecting the Contractor and the Owner in the General Conditions section of your primary Construction Agreement with the Owner also govern this Change Order.

I have read and understood the Change Order above, and I agree to all of its terms.

Date: _____6/16/01_____   *CHARLIE CONTRACTOR, PRESIDENT*
                              CHARLIE CONTRACTOR, PRESIDENT
                              CHARLIE CONTRACTOR CONSTRUCTION, INC.

Date: _____6/16/01_____   *Harry Homeowner*
                              HARRY HOMEOWNER

CHAPTER 5

CONSTRUCTION SUBCONTRACTS

5.1 ▪ SHORT-FORM SUBCONTRACT AGREEMENT

5.2 ▪ LONG-FORM SUBCONTRACT AGREEMENT

5.3 ▪ SUBCONTRACTOR INFORMATION FORM

Construction subcontract agreements are often ignored by residential general contractors because of the close working relationships that are commonly established between select subcontractors and the general contractor.

While this close working relationship and established history of working together is extremely valuable and often makes work more predictable and enjoyable, contractors occasionally will work with new subs or will have a problem with an established sub over issues such as liability, scope of work, warranties, or payments.

If and when this happens, you will probably wish you had entered into a written subcontract with the subcontractor. A reasonable written subcontract can help you bring about a fair and fast solution to most problems.

For example, one general contractor I know obtained a bid for work from one of his regular subs. This subcontractor then furnished the general contractor with a typical bid form with a brief scope of work and a lump sum price (about $18,000) that was itemized into a few different categories.

When it was time to start the job, however, the sub was busy, so he subcontracted out the work to a friend of his (the lower-tier subcontractor) who was basically unknown to the general contractor. The lower-tier subcontractor performed most of the work. However, after tempers flared over some disagreement on the site, the lower-tier subcontractor refused to complete all the work, and the general contractor wouldn't pay for any of the work until the job was done right. To make matters worse, no change orders had been put in writing, and the lower-tier subcontractor had performed some work which had not been authorized by the general contractor.

Nobody could agree on the fair value of the work performed, nobody had a signed contract, so everyone brought in their lawyers. The general contractor was unhappy. The owner was unhappy when he received the mechanic's lien and withheld the final payment from the contractor. The subcontractor and lower-tier subcontractor were unhappy. Longstanding relationships were irreparably damaged. The situation was a mess.

While problems with construction are not uncommon, it is clear that the situation described above could have been prevented if the general contractor and the sub had signed a subcontract prior to commencing work. As with contractor-owner relations, it is easier to figure out who has deviated from the main road and what needs to be done in order to get back on the path when you have a road map. Having a road map, i.e., a good subcontract, is certainly no guarantee that problems will not arise, but it makes it a lot easier to resolve the problems if they do occur.

Accordingly, there are two different length subcontract agreements included in this chapter. As with the three fixed price agreements in Chapter One, decide which length subcontract agreement to use by assessing various factors: in this case, the quality of the subcontractor's work, the size of the job, and your previous relationship with the subcontractor. The longer the agreement, the more protection is provided to the general contractor.

Avoid relying on *verbal* subcontractor bids even if they are from your regular, well-known subs. Confusion over scope of work issues is too common and normally results in either the general contractor or the subcontractor paying for the subcontractor's oversight or lack of clarity in describing the work to be performed.

Even the short-form subcontract agreement requires the subcontractor to complete his work in conformance with a specified set of plans and/or specifications.

When the subcontractor knows he must sign a subcontract agreement, he may pay just a little more attention and spend a little more time thoroughly bidding the plans because he knows that if he misses details in the plans, he will be the one who is asked to perform and pay for the work that was inadvertently overlooked at bid time.

In any event, whether you are working with a subcontractor you have known for 10 years or are working with one for the first time, use some form of subcontract agreement that, at minimum, contains the following clauses:
• Proper licensing, bonding (if required in your state), and insurance.
• A clear scope of work to be performed by the subcontractor and time to complete the work.
• A clear lump-sum amount, payment schedule, and, if desired or required by your contract with the owner, retention.
• An indemnification clause and warranty terms.
• Any other miscellaneous clauses you feel are appropriate considering the scope of work and your prior history of working together.

Regarding insurance, you should require that every subcontractor who works for you carry worker's compensation and comprehensive general liability policies. At the minimum, your general liability coverage should be $500,000, although this amount can vary depending upon the requirements of your contract with the owner, or the advice of your attorney.

In fact, a contractor's comprehensive general liability policy usually requires that you work only with subcontractors who *also* carry comprehensive general liability insurance. Therefore, you should keep a file in your office with these certificates from your subs so they are easily accessible when you need them. Your insurance company doesn't want you working with subs who don't carry comprehensive liability insurance because it increases their exposure to claims. If you do work with a sub who doesn't carry comprehensive general liability insurance, you will probably be hit with an additional premium from *your* liability carrier when your annual audit is performed.

Before you let a sub start work on your project, make sure his insurance agent sends you an insurance certificate which names you, the general contractor, as an "additional insured" under his general liability policy. This may give you additional rights under the sub's policy in the event of a future claim.

The binder naming you as an additional insured must also contain a 10-day cancellation notice clause. This clause requires the sub's insurance agent to send you a written cancellation notice ten days prior to cancellation of the sub's general liability insurance. (If you receive one of these cancellation notices, you should contact your subcontractor to obtain a certificate of insurance from his *new* liability carrier.)

In general, it's safest to work with subcontractors who are large enough to carry a worker's compensation policy. However, if you work with a one-man-shop subcontractor, make it clear that you will *never* allow him to have an uninsured helper on your job site. Even a one-man sub may occasionally hire a laborer or helper, and if the sub carries no worker's comp coverage, that helper may make a claim against *your* worker's comp policy if he is injured on the site. This will cost you plenty. Obviously you can't always control who is working on your job site, so there is some inherent risk associated with a one-man-shop subcontractor (if he doesn't carry worker's compensation).

Some very small subcontractors work alone and never have employees. The sole owner of a business that has no employees is *not* required by law in most states to carry a worker's compensation policy to insure himself. The idea behind worker's comp insurance is to protect employees. The owner of the business is considered free to take his own risks or provide his own medical insurance that will cover any job-site injury to himself.

However, this exception to the general rule requiring worker's compensation insurance *extends only to properly and actively licensed subcontractors who operate with no employees.* It does not extend to unlicensed tradespersons who are working for you or the subcontractor on a "contract basis." Neither you nor your subcontractor should ever contractually agree to hire an unlicensed person as a "quasi-subcontractor" or "captive sub" who is responsible for his own worker's compensation insurance. It will never work. You or your sub should instead put this person on the payroll and make him a legal employee so you do not expose yourself to significant liability — not to mention problems with the IRS and state over tax withholdings.

In summary, try to work only with subcontractors who carry both liability and worker's compensation insurance. But if you do work with a sub who doesn't carry worker's comp because he has no employees, check with a construction attorney or worker's comp attorney in your state to verify that the sub is complying with the current laws regarding worker's compensation insurance. Further, be sure to include the clause in the subcontract that relates to "assumption of the duty to provide for safety procedures and abide by government regulations." Again, consult your attorney and your insurance agent to determine whether you and your subcontractors are adequately insured.

Finally, make sure you get the subcontractor's federal tax i.d. number or his social security number before allowing work to begin so that you have this information when it comes time to file the federal 1099 tax form at the end of the year. According to federal law, if you subcontract more than about $600 in a year to a given subcontractor, you are required by law to disclose the amount you paid the subcontractor during the prior year on this federal 1099 tax form.

This short-form subcontract can work well on both residential and light commercial jobs where you are working with subcontractors you have known and worked well with over an extended period of time. This short form covers many of the basic clauses I think should be included in any subcontract.

In the past, I have given my short-form subcontract agreement to regular subcontractors and told them to return this agreement to me as if it was their "bid." Before sending the blank agreement to the subcontractor, I fill in all portions except the payment schedule, lump-sum price, exclusions, and completion time. The subcontractor then fills in these portions, signs the subcontract, and returns it to me.

Be sure to review the subcontract for accuracy as soon as your receive it from the subcontractor. Assuming the bid deadline has not elapsed, you can get any questions addressed prior to signing a contract with the owner.

This way, I already have a subcontract for the work when I get the bid. If I get the job, all that's required is a quick review of the few areas the subcontractor has filled in, and my signature. This simplifies the paperwork procedure because the regular subcontractors are turning in their proposals on my subcontract form.

If you use this procedure, carefully review and clarify what the subcontractor has written on the form to make sure he has not excluded items you thought were within the scope of work, and to approve the other terms and conditions he may have added to the form.

5.1 ■ SHORT-FORM SUBCONTRACT AGREEMENT

CONTRACTOR'S NAME: _____

ADDRESS: _____

_____

PHONE: _____

FAX: _____

LIC #: _____

DATE: _____

OWNER'S NAME: _____

ADDRESS: _____

_____

PROJECT ADDRESS: _____

_____

CONSTRUCTION LENDER: _____

ADDRESS: _____

_____

I. PARTIES

This Subcontract (hereinafter referred to as "Agreement") is being entered into on the _____ day of _____, 19___, and is between _____ , (hereinafter referred to as "Contractor"); and _____ , (hereinafter referred to as "Subcontractor"). Subcontractor's license number is: _____ . Subcontractor warrants that he is properly licensed to perform the type of work described in this Agreement and that he is an *independent contractor* and *not* an agent or employee of the Contractor.

SUBCONTRACTOR'S ADDRESS: _____

_____

PHONE: _____

FAX: _____

Subcontractor's business is a: ___ Sole Proprietorship; ___ Partnership; ___ Corporation

Subcontractor's Federal Tax I.D.# or S.S.#: _____

In consideration of the mutual promises contained herein, the parties agree as follows:

II. GENERAL SCOPE OF WORK DESCRIPTION AND SUBCONTRACT AMOUNT

Subcontractor will furnish all labor, equipment, tools, materials, transportation, supervision, and all other items required for safe operations to complete the following work which will comply with the latest edition of

all applicable building codes and the Contract Documents referred to below (if conflict between plans, specifications, Subcontractor's proposal, or this Agreement arises, then this Agreement is controlling):

_____

_____

_____

(Additional Work Description page(s) attached: ____Yes; ____ No)

LUMP SUM SUBCONTRACT AMOUNT AND PAYMENT SCHEDULE: Contractor to pay Subcontractor the total Lump Sum Amount of: $_____

Receipt of payment from Owner is a condition precedent to paying Subcontractor under this Agreement.

Payments to be made to Subcontractor as follows:

_____

_____

_____

_____

III. GENERAL CONDITIONS FOR THE SUBCONTRACT AGREEMENT ABOVE

A. EXCLUSIONS FROM SUBCONTRACTOR'S SCOPE OF WORK
Labor and materials for the following work have *not* been included by Subcontractor:

_____

_____

_____

_____

_____

B. CONTRACT DOCUMENTS
Subcontractor will perform its work in accordance with all Contract Documents, which are identified as follows:

- This Construction Agreement

- Plans: _____

- Specifications: _____

- Addenda: _____

- Miscellaneous: _____

_____

Subcontractor warrants that he has been furnished all Contract Documents referred to above and has thoroughly familiarized himself with all Contract Documents and the existing site conditions.

C. WORK COMMENCEMENT AND COMPLETION TIME

Work shall commence on _____ and take approximately _____ calendar days to complete. **TIME IS OF THE ESSENCE** in all aspects of Subcontractor's performance.

D. INDEMNIFICATION

Subcontractor (and his agents) shall at all times indemnify, protect, defend, and hold harmless Contractor and Owner from all loss and damage, and against all lawsuits, arbitrations, mechanic's liens, legal actions of any kind whatsoever, attorney's fees, and any costs and expenses which are directly or indirectly caused or contributed to, or claimed to be caused or contributed to, by any act or omission, fault or negligence, whether passive or active, of Subcontractor or his agents or employees, in connection with or incidental to the work under this Agreement.

E. SUBCONTRACTOR'S INSURANCE

Before commencing work on the project, Subcontractor and its Subcontractors of every tier will supply to Contractor duly issued Certificates of Insurance, naming Contractor as an "additional insured," showing in force the following insurance for comprehensive general liability, automobile liability, and worker's compensation*:

• comprehensive general liability with limits of not less than $_____ per occurrence;

• automobile liability in comprehensive form with coverage for owned, hired, and non-owned automobiles;

• worker's compensation insurance in statutory form.

* All insurance binders must contain a clause indicating that certificate holders be given a minimum of 10 days written notice prior to cancellation of Subcontractor's insurance. Subcontractor must furnish the insurance binder referred to above as an express condition precedent to the Contractor's duty to make any progress payments to Subcontractor pursuant to this Agreement.

F. EXPRESS WARRANTY

At the request of Contractor, Subcontractor will promptly replace or repair any work, equipment, or materials that fail to function properly for a period of one year after completion of the project, at Subcontractor's own expense. Subcontractor will also repair any surrounding parts of the structure that are damaged due to any failure in Subcontractor's work during the warranty period.

G. LAWS, REGULATIONS, AND SAFETY

Subcontractor will comply with all statutes and regulations that establish safety requirements (including, but not limited to those of OSHA and any state agency regulating job-site safety). By signing this Agreement, Subcontractor knowingly and willingly accepts *full responsibility* for the safe operation of all of its activities and the protection of other persons and property during the course of this project.

H. DISPUTE RESOLUTION AND ATTORNEY'S FEES

Any controversy or claim arising out of or related to this Agreement involving an amount *less* than $5,000 (or the maximum limit of the court) must be heard in the Small Claims Division of the Municipal Court in the county where the Contractor's office is located. Any controversy or claim arising out of or related to this Agreement which is over the maximum dollar limit of the Small Claims Court must be settled by binding arbitration administered by the American Arbitration Association in accordance with the Construction

Industry Arbitration Rules. Judgment upon the award may be entered in any Court having jurisdiction thereof.

Subcontractor agrees to contractually make this AAA Arbitration Dispute Resolution Clause irrevocably bind and "flow down" to all lower-tier Subcontractors. This Agreement is not assignable.

The prevailing party in any legal proceeding related to this Agreement shall be entitled to payment of reasonable attorney's fees, costs, and expenses.

I. ENTIRE AGREEMENT, MODIFICATION, SEVERABILITY

This Subcontract Agreement represents the entire agreement of the parties. Prior discussions or verbal representations that are not contained in this Agreement are *not* a part of this Agreement. This Agreement cannot be modified by oral agreements but can only be modified by a written agreement which is signed by both parties.

J. ADDITIONAL LEGAL NOTICES REQUIRED BY STATE OR FEDERAL LAW

(See page attached: Yes____; No____)

K. ADDITIONAL TERMS AND CONDITIIONS

(See page attached: Yes____; No____)

I have read and understood, and I agree to, all of the terms and conditions contained in the Agreement above.

Date: _____   _____

                                CONTRACTOR

Date: _____   _____

                                SUBCONTRACTOR

5.1 ▪ SHORT-FORM SUBCONTRACT AGREEMENT

ANNOTATED

■ This is a typical short-form Subcontract Agreement for electrical work on a residential remodeling project. The form has been filled out with sample language and most of the clauses have annotations under them explaining the legal significance or practical importance of the contract language.

Compare the blank form with this annotated, filled-out form to become more familiar with how to use this Agreement.

Charlie Contractor Construction, Inc.
123 Hammer Lane
Anywhere, USA 33333
Phone: (123) 456-7890
Fax: (123) 456-7899
Lic#: 11111

DATE: **June 1, 2001**

OWNER'S NAME: **Mr. & Mrs. Harry Homeowner**
ADDRESS: **123 Project Place**
Anywhere, USA 33333

PROJECT ADDRESS: **123 Project Place**
Anywhere, USA 33333

CONSTRUCTION LENDER: **Lucky Lou's Construction Loans**
ADDRESS: **777 Lucky Lender Lane**
Anywhere, USA 33333

■ It's a good idea to list the name and address of the Construction Lender (required by law in certain states), if there is one on the project. This will give the Subcontractor all or nearly all the information he needs to send out any required Preliminary Notice Forms to the Owner.

INITIAL
CC
EE

I. PARTIES

This Subcontract (hereinafter referred to as "Agreement") is being entered into on the __**first**__ day of __**June**__ , 20__**01**__, and is between __**Charlie Contractor Construction, Inc.**__ (hereinafter referred to as "Contractor"); and __**Eddie's Electric**__ (hereinafter referred to as "Subcontractor"). Subcontractor's license number is: __**22222**__ . Subcontractor warrants that he is properly licensed to perform the type of work described in this Agreement and that he is an *independent contractor* and *not* an agent or employee of the Contractor.

> ■ Be sure your Subcontractor is properly licensed to perform the work in the Agreement.

SUBSONTRACTOR'S ADDRESS: __**123 Kilowatt Street**__
                                   __**Anywhere, USA 33333**__
               PHONE: __**(012) 111-2222**__
                  FAX: __**(012) 111-3333**__

Subcontractor's business is a: __**X**__ Sole Proprietorship; _____ Partnership; _____ Corporation

Subcontractor's Federal Tax I.D.# or S.S.#: __**77-01234**__

> ■ By listing address, fax, phone, type of business, and tax i.d. information all in one area, it will be easy to reference this information when you need it for 1099 tax information or for any other reason. If the Subcontractor is incorporated, you may not need to file a 1099 form. Verify this with your accountant.

In consideration of the mutual promises contained herein, the parties agree as follows:

> ■ The phrase, "In consideration of the mutual promises contained herein, the parties agree as follows:" is good boilerplate language that should remain in your Agreement. It means there is a *bargained-for exchange*, which is necessary for a contract to be valid. The Contractor will pay the Subcontractor the stated sum, and in exchange, the Subcontractor will complete the work described in the Subcontract Agreement.

II. GENERAL SCOPE OF WORK DESCRIPTION AND SUBCONTRACT AMOUNT

Subcontractor will furnish all labor, equipment, tools, materials, transportation, supervision, and all other items required for safe operations to complete the following work which will comply with the latest edition of all applicable building codes and the Contract Documents referred to below (if conflict between plans, specifications, Subcontractor's proposal, or this Agreement arises, then this Agreement is controlling):

1. __**Furnish labor and materials for all rough electrical work per plans.**__
2. __**Furnish labor to install all Owner-provided electrical fixtures**__
    __**and exhaust fans.**__

> INITIAL
> *CC*
> *EE*

3. Furnish labor and materials to install phone, computer, and t.v. cable wiring and receptacles per plans.

4. All receptacles, switches, and face plates to be white.

5. Hook up electrical appliances, as required.

(Additional Work Description page(s) attached: _____Yes; __x__ No)

> ■ Be sure to obtain a complete description of the work. Ambiguities in the Scope of Work cause most disputes. If the Subcontractor's work requires equipment such as scaffolding, be sure to find out whether it is included by the Subcontractor and whether or not it can be used by any other workers on the site. I expect the Subcontractor to know the building codes that relate to his trade and to perform his work in conformance with those building codes.

LUMP SUM SUBCONTRACT AMOUNT AND PAYMENT SCHEDULE: Contractor to pay Subcontractor the total Lump Sum Amount of: $ __3,800__ .

> ■ I rarely enter into time-and-materials agreements with Subcontractors (and I never do if I'm giving the Owner a fixed price agreement with a Lump Sum Amount). I almost always find that a fixed price contract or a time-and-materials contract *with a guaranteed not-to-exceed price* works out better than giving the Subcontractor a pure time-and-materials contract. One exception could be if the Subcontractor is very well known to you and the job is very small.

Receipt of payment from Owner is a condition precedent to paying Subcontractor under this Agreement.

> ■ The General Contractor may be paid on a draw system and it is often "understood" that the Subcontractor may need to wait a brief period for the Contractor to complete all the work specified for a given progress payment in order to be paid. Once the Owner has paid the progress payment incorporating the work of the Subcontractor, the Contractor should pay the Subcontractor within 10 days. State laws may govern this time period as well.
>
> The phrase above indicating that the Contractor's receipt of payment from the Owner is "a condition precedent to paying Subcontractor" protects the Contractor from always having to finance the progress payments to the Subcontractors.

INITIAL
CC
EE

Payments to be made to Subcontractor as follows:

1. 65% of contract amount due upon completion of all rough wiring: $ 2,470

2. Final contract amount due upon completion of all work under this

   Agreement: $ 1,330

> ■ The payment schedule is an important part of every subcontract. Be sure you do not pay for work prior to its completion. Where there is no with-held retention, leave a large enough final payment to cover Punch List work specified for a given progress payment.

III. GENERAL CONDITIONS FOR THE SUBCONTRACT AGREEMENT ABOVE

A. EXCLUSIONS FROM SUBCONTRACTOR'S SCOPE OF WORK

Labor and materials for the following work have *not* been included by Subcontractor:

1. Any finish light fixtures or exhaust fans.

2. Permits or permit fees.

> ■ Make sure you understand what is *not* being included in the Subcontractor's bid. That way, you can either exclude those items from your Agreement with the Owner or have the Subcontractor give you a price for the excluded work so you can include it in your bid.

B. CONTRACT DOCUMENTS

Subcontractor will perform its work in accordance with all Contract Documents, which are identified as follows:

   • This Construction Agreement

   • Plans: **By Art Architect, 5 pages dated April 1, 2001**

   • Specifications: **See Plans referred to above**

   • Addenda: **#1, by Art Architect, 2 pages dated May 13, 2001**

   • Miscellaneous: **N/A**

Subcontractor warrants that he has been furnished all Contract Documents referred to above and has thoroughly familiarized himself with all Contract Documents and the existing site conditions.

> ■ It is very important to accurately list all the plans, addenda, and any other contract documents that provide details about the Scope of Work to be performed by the Subcontractor. The Agreement refers to the "Contract Documents" in several locations and you need to make sure that this section of the Agreement is properly and fully filled out.

INITIAL
CC
EE

C. WORK COMMENCEMENT AND COMPLETION TIME

Work shall commence on __Negotiable__ and take approximately __7__ calendar days to complete __(rough work only. Finish to take approximately 2 days to set)__. TIME IS OF THE ESSENCE in all aspects of Subcontractor's performance.

> ■ With the uncertainties of exactly when a Subcontractor can start his phase of the work, it can be difficult to pin down exact dates to start and finish work. In some cases, you can break the Subcontractor's work into distinct phases (such as rough and finish work) and state approximately how long each phase of the work will take.
>
> The clause, "TIME IS OF THE ESSENCE," is typical legal language that means it really *does* matter how long it takes to complete the work, and, if the Subcontractor's performance is substantially behind schedule, the Owner (and you, the General Contractor) may suffer damages and have a legal action against the Subcontractor as a result of his untimely performance.
>
> Whether you see this clause directed at you in a prime contract with the Owner or whether you direct this clause toward your Subcontractor, it simply means that the stated time to complete the work should be taken seriously because damages may be suffered if the work is delayed without a legal excuse or time extension (such as clear Owner-caused delays, Change Orders, etc.).

D. INDEMNIFICATION

Subcontractor (and his agents) shall at all times indemnify, protect, defend, and hold harmless Contractor and Owner from all loss and damage, and against all lawsuits, arbitrations, mechanic's liens, legal actions of any kind whatsoever, attorney's fees, and any costs and expenses which are directly or indirectly caused or contributed to, or claimed to be caused or contributed to, by any act or omission, fault or negligence, whether passive or active, of Subcontractor or his agents or employees, in connection with or incidental to the work under this Agreement.

> ■ It is very important to have an indemnification clause in your Subcontract Agreement with every Subcontractor. This language requires the Subcontractor to "...indemnify, protect, defend, and hold harmless" the Contractor from the negligent acts and omissions of the Subcontractor. Most jobs will never have an event that triggers the need for this clause.
>
> However, you'll appreciate this clause the one time when a major problem occurs and someone sues you for a negligent act or omission by the Subcontractor or one of his employees or agents. Remember that "General Contractor" means general liability.
>
> If a party sues you for the actual or even alleged negligent acts or omissions of your Subcontractors (e.g., personal injury, property damage, Me-

INITIAL
CC
EE

chanic's Liens that were improperly filed, etc.), this clause contractually requires the Subcontractor to "...indemnify, protect, defend, and hold [you] harmless" from that party. In practice, this puts you in a strong position to sue the Subcontractor for any losses, should that be necessary. For example, let's say your painter is behind schedule on painting the exterior of your project and doesn't show up prior to your crew's installation of the tile roof. When the painter's van finally arrives, it is his helper that jumps out with the airless paint sprayer, because his boss has the flu.

The helper is a bit inexperienced and doesn't realize that the soft breeze that's blowing is just strong enough to blow the elastomeric paint (which coincidentally is just a few shades lighter than the color of the roof) all over the new $18,000 tile roof.

By the time the painter's helper notices the overspray on the roof, a small, angry crowd has gathered at the foot of his ladder to demand that his insurance company repaint their cars — which also have been misted by the overspray in their nearby parking spaces. About this same time, the Owner's wife shows up (who has great eyesight) and starts ranting and raving about lawsuits and "never working in this town again"!

In this case, you, the Contractor, will be responsible to the Owner for the negligent acts of your painting Subcontractor. Your first step is to tell your Subcontractor to submit this claim to his comprehensive general liability insurance carrier. (You may also want to notify your own carrier about the property damage in case you need to submit a future claim.)

If you're lucky, your painter has an "overspray" endorsement on his policy that covers the repair of items that are inadvertently oversprayed. If he does, the painter's insurance company should take care of all the damage done. If the painter's insurance company *doesn't* pay out, you may need to submit the claim to your insurance company.

If the insurance companies don't settle the matter under the policy coverage available and litigation begins between the damaged parties, the negligent painter, and you, the General Contractor, will need to rely on the indemnification clause in your subcontract with the painter in order to have the court or arbitrator find the painting Subcontractor responsible for all the damages, and award you attorney's fees in defending yourself in the lawsuit that was brought by the damaged parties.

If there is no insurance coverage available, the painter had better have enough assets to satisfy the judgment against him (or you) and pay for your attorney's fees. The indemnification clause gives you the contractual ground to demand this from the Subcontractor.

This example also illustrates the importance of working with Subcontractors who are properly insured and financially stable. If the Subcontractor has no insurance coverage, has insufficient assets to cover the judgment, or files for bankruptcy, the Contractor will likely be ordered to

INITIAL
CC
EE

pay any shortfall that the Subcontractor cannot pay.

Although indemnification clauses can be overruled in court in some instances, this is a very important clause that helps protect the interests of the Contractor.

E. SUBCONTRACTOR'S INSURANCE

Before commencing work on the project, Subcontractor and its Subcontractors of every tier will supply to Contractor duly issued Certificates of Insurance, naming Contractor as an "additional insured," showing in force the following insurance for comprehensive general liability, automobile liability, and worker's compensation*:

• comprehensive general liability with limits of not less than $ __500,000__ per occurrence;

• automobile liability in comprehensive form with coverage for owned, hired, and non-owned automobiles;

• worker's compensation insurance in statutory form.

* All insurance binders must contain a clause indicating that certificate holders be given a minimum of 10 days written notice prior to cancellation of Subcontractor's insurance. Subcontractor must furnish the insurance binder referred to above as an express condition precedent to the Contractor's duty to make any progress payments to Subcontractor pursuant to this Agreement.

■ There are numerous reasons why you need a clause like the one above. First, Owners tend to go after any "deep pocket" that is available to recover a real or alleged loss. If your Subcontractor negligently causes damage to the Owner but carries no insurance (and has few assets), the Owner will find the insured General Contractor a welcome target, since he is ordinarily liable for the negligent acts and omissions of his Subcontractors. However, if the Subcontractor carries insurance, the Subcontractor's insurance company may step in and take care of the loss without the need for involving the General Contractor.

Second, every state in the U.S. requires employers to carry worker's compensation insurance that covers job-related accidents. If a Subcontractor has anyone working on the site who is not an owner of the business, you must make sure he holds a worker's compensation policy. If he doesn't, you become liable for any claims filed by his employees, and you may also be assessed an additional premium when your books are audited by your worker's comp insurer.

Third, the General Contractor should carry comprehensive general liability insurance. Assuming he does, most policies require that the General Contractor work only with Subcontractors who also carry comprehensive general liability insurance. If the Subcontractor does not carry comprehensive general liability insurance, the General Contractor will ordinarily be assessed an additional insurance premium when his books are audited by his insurance company.

Require your Subcontractors to carry this insurance and you won't have

INITIAL
CC
EE

many of the problems described above. Like all insurance, you buy it (or require it) and hope you'll never need it. However, the first time you have an accident that triggers the need for this insurance, you'll feel extremely fortunate that you required your Subcontractor to carry it.

Be sure to have your company named as an "additional insured" under the Subcontractor's policy with cancellation notice rights. As a named "additional insured," you may have rights under the policy you would not otherwise have. You need cancellation notice rights so that you are automatically notified if a Subcontractor cancels his insurance (this happens all the time). If it is canceled, you can then require the Subcontractor to give you a new insurance certificate.

I state in the Subcontract that furnishing the insurance binder is a condition precedent to the Contractor's duty to pay the Subcontractor progress payments. This emphasizes how important this insurance requirement is to the Contractor. Let your Subcontractors know about this insurance requirement before they bid for you so that you don't waste everyone's time having uninsured Subcontractors bid your work.

Finally, the amount of coverage and the number of additional riders you should buy (and require your subs to carry) in order to properly protect your business is a matter to discuss with your attorney and insurance agent. I recommend requiring a rider that covers "automobile liability in comprehensive form with coverage for owned, hired, and non-owned automobiles." For the sake of example only, a $500,000 limit was required in this sample Subcontract.

This ensures that your Subcontractors (and their employees) have liability insurance that covers them while they are driving company vehicles (or rented vehicles) for work on your project. This binder is not an expensive add-on to most liability policies.

F. EXPRESS WARRANTY

At the request of Contractor, Subcontractor will promptly replace or repair any work, equipment, or materials that fail to function properly for a period of one year after completion of the project, at Subcontractor's own expense. Subcontractor will also repair any surrounding parts of the structure that are damaged due to any failure in Subcontractor's work during the warranty period.

■ Be sure to state your expectations about how long the Subcontractor will provide an express warranty for his work. Refer to your Agreement with the Owner and make sure your Subcontractor's warranty is at least as long and as comprehensive as your warranty with the Owner.

I also indicate that any surrounding parts of the structure that are damaged due to a failure in the Subcontractor's work will be repaired by Subcontractor.

For instance, if the plumber's spa leaks after four months and ruins the ceiling, carpet, and piano in the first-floor living room, the plumber will be

INITIAL
CC
EE

> expected to repair the leak, repair the ceiling and paint, and repair or re-place the carpet and the piano. This type of consequential damage can be far more costly to repair than the actual defect itself. In this example, the Subcontractor's comprehensive general liability insurance would likely cover much of the consequential damage — which highlights the importance of both the Subcontractor's warranty and insurance coverage.

G. LAWS, REGULATIONS, AND SAFETY

Subcontractor will comply with all statutes and regulations that establish safety requirements (including, but not limited to those of OSHA and any state agency regulating job-site safety). By signing this Agreement, Subcontractor knowingly and willingly accepts *full responsibility* for the safe operation of all of its activities and the protection of other persons and property during the course of this project.

> ■ This clause indicates that the Subcontractor clearly assumes full respon-sibility for the safe operation of all its activities and the protection of other persons and property during the course of the project. Once again, you will seldom have to rely on this clause in a courtroom setting.
>
> However, let's say the Subcontractor digs a trench which later collapses and injures one of the Owner's kids. You will be in a better position legally if it is clearly understood that the Subcontractor was contractually respon-sible for monitoring and assuring the safety of all its activities. However, re-gardless of a clause like this, the General Contractor can not delegate all of his duty to monitor and supervise the job to Subcontractors. You must still vigilantly monitor the overall safety of the job site.
>
> In this situation, the General Contractor may also be sued (and should always be alert to job-site safety issues). However, a clause like this one (along with the indemnification clause above) will help to shift the liability away from the General Contractor and toward the primarily responsible party — the Subcontractor. As you can see, many of these "boilerplate" legal clauses (e.g., indemnification, insurance, warranty, safety, etc.) work together to protect the General Contractor in the event of serious personal injury or property damage.

H. DISPUTE RESOLUTION AND ATTORNEY'S FEES

Any controversy or claim arising out of or related to this Agreement involving an amount *less* than $5,000 (or the maximum limit of the court) must be heard in the Small Claims Division of the Municipal Court in the county where the Contractor's office is located. Any controversy or claim arising out of or related to this Agreement which is over the maximum dollar limit of the Small Claims Court must be settled by binding arbitration administered by the American Arbitration Association in accordance with the Construction Industry Arbitration Rules. Judgment upon the award may be entered in any Court having jurisdiction thereof.

Subcontractor agrees to contractually make this AAA Arbitration Dispute Resolution Clause irrevocably bind and "flow down" to all lower-tier Subcontractors. This Agreement is not assignable.

INITIAL
CC
EE

■ The dispute resolution clause in your Agreement with the Subcontractor should call for the same type of resolution as that in your Agreement with the Owner. If a complicated dispute arises which involves the Owner, Contractor, and the Subcontractor, it will be cheaper and faster to resolve this dispute with all three parties in the same dispute resolution forum (such as binding arbitration). Consult your attorney about the best dispute resolution clause for your business.

The prevailing party in any legal proceeding related to this Agreement shall be entitled to payment of reasonable attorney's fees, costs, and expenses.

■ Whether a dispute will be resolved in the normal court system or through binding arbitration will be determined by your Agreement. For small disputes, I prefer a Small Claims Court where you can ordinarily get a court date within a few months, where the maximum dollar limit is ordinarily between $2,500 and $5,000, and where you can't take a lawyer. It's fast (usually less than a 20-minute hearing in my area), easy, and cheap. Verify maximum dollar limits with your local court or attorney.

Very limited appeals, if any, are available (depending upon whether you are the plaintiff or defendant). You can't take a lawyer into a Small Claims Court proceeding in many jurisdictions, but prior to the hearing you can consult a lawyer familiar with construction law to help you identify the legal issues and prepare an oral outline to follow. He can also help you come up with a written statement of your position so you can present your case well and thereby increase your odds of receiving a judgment in your favor.

For disputes over the jurisdictional limits of the local Small Claims Court, I prefer binding arbitration through either privately selected arbitrators or through the American Arbitration Association (AAA). AAA arbitration is ordinarily much faster than the court system — usually several months from application to completion of hearing. It is perhaps a bit cheaper than the court system (cheaper if the court system involves appeals), and it allows you the option of bringing an attorney to the arbitration hearing.

In addition, the rules of evidence are much more relaxed in arbitration. The hearing is less formal than the court system hearing and a big advantage is that you may end up with an arbitrator who is familiar with the construction business.

One *disadvantage* to arbitration is that some arbitrators have a tendency to "split the difference" if they don't feel the case is clearly in favor of one side or the other. This makes it important to get legal representation (or at least advice) *prior* to going into an arbitration hearing. The failure to get some legal advice has made many a contractor unhappy when they open up the letter that contains the decision against them! Once again, an ounce of prevention...

INITIAL
CC
EE

> Attorney's fees are ordinarily awarded only if they have been agreed to in the Construction Agreement, although the AAA arbitrator has the authority to award attorney's fees and costs. An attorney's-fees clause is a very good idea if you do good work and don't think you will be in the wrong if a dispute arises. Being able to tell the other side, "You'll not only have to pay what you owe me, but also my attorney's fees," can be a helpful negotiating lever when a dispute arises.

I. ENTIRE AGREEMENT, MODIFICATION, SEVERABILITY

This Subcontract Agreement represents the entire agreement of the parties. Prior discussions or verbal representations that are not contained in this Agreement are *not* a part of this Agreement. This Agreement cannot be modified by oral agreements but can only be modified by a written agreement which is signed by both parties.

> ■ The clause above is necessary because it states that the agreement of the parties is limited to what is actually in the written contract. Pre-contract signing or verbal representations from the Contractor to the Subcontractor or from the Subcontractor to the Contractor that are not included in the Agreement are not legal and binding parts of the Agreement.
>
> This clause reduces the possibility of the Subcontractor coming back to you and saying, "You said you'd pay me an extra $500 if I got the job done by Sunday. It's Sunday: give me my money." In this type of situation the Contractor can point to this clause and say that the contract contained their entire agreement and did not include what the Subcontractor is now demanding.

J. ADDITIONAL LEGAL NOTICES REQUIRED BY STATE OR FEDERAL LAW

(See page attached: Yes **x** ; No _____)

> ■ If your state requires any legal notices to be included in your subcontracts, include them here or attach them to your Agreement. Most states don't have very many requirements for agreements between a Contractor and Subcontractor, but consult your own attorney to be sure.

K. ADDITIONAL TERMS AND CONDITIONS

(See page attached: Yes _____ ; No **x**)

> ■ The clause above has been placed in the Agreement to remind you that you are acting as your own attorney in using these agreements and to remind you that the forms as presented are a starting point and not necessarily entirely suited in their present form to your purpose without some additional contract language or modifications.

INITIAL
CC
EE

This clause specifically indicates that you may need to add additional clauses to the Agreement based on the unique needs of your business, the disposition of the Owner, or the particular job that is the subject of the contract. The additional contract language you need may be beyond the scope of any form agreement in this book.

You may also decide to delete certain clauses from the Agreement depending upon the same factors mentioned above. If you work with a word processor, modifications to the Agreement will be fast and simple to make. Simply add (or delete) any clauses necessary to your agreement and then delete the last clause, "ADDITIONAL TERMS AND CONDITIONS." Remember to consult an attorney familiar with construction before making significant changes to any agreement.

I have read and understood, and I agree to, all of the terms and conditions contained in the Agreement above.

Date: _6/01/01_   *CHARLIE CONTRACTOR, PRESIDENT*
_____
CHARLIE CONTRACTOR, PRESIDENT
CHARLIE CONTRACTOR CONSTRUCTION, INC.

Date: _6/01/01_   *Eddie's Electric*
_____
EDDIE'S ELECTRIC

■ Be sure you have an Agreement in your files which is signed by the Subcontractor and contains his initials on each page.

The long-form subcontract can be used on both residential and light-commercial jobs. It contains several expanded and additional clauses that give added protection to the general contractor. It is well suited for jobs with more complicated types of subtrade work or on jobs where you don't have a good, established working relationship with the subcontractor who will be performing the work.

The main changes in the long-form subcontract are as follows:

The payment schedule (II.A) has additional language regarding retention and lien releases. It requires the subcontractor to provide progress payment lien releases before progress payments are made. It also gives the contractor the right to withhold 5% or 10% from all progress payments until the owner has paid the contractor for this work and the sub has provided the necessary lien releases.

A clause addressing the intent of the contract documents has been added to the section on Contract Documents (III.B). This helps guard against being billed for unreasonable extras — such as copper tubing and fittings for a plumber.

The Changes in the Work clause (III.C) is an important addition that guards against the owner attempting to deal directly with your subcontractor regarding changes.

The Back Charges clause (III.E) gives you leverage to get a sub to correct substandard work or repair any damage he may have caused to other work. This can be particularly useful if you're working with a new sub or a large sub with many crews.

The Subcontractor's Insurance clause (III.G) is expanded to give the general contractor further protection against claims arising from negligent or uninsured subs. This is very important since, in many cases, the general contractor will be held partly or fully liable for any claims against the subcontractor which are not covered by the sub's insurance.

The clause about Cleanup (III.H) has been added for obvious reasons: a messy job site is unsafe and makes working less efficient.

The Laws, Regulations, and Safety clause (III.J) has been expanded to cover legal compliance more broadly, since the authorities may come after the general if a subcontractor fails to meet certain legal obligations.

Finally, the Subcontractor's Default clause (III.K) gives the contractor the right to terminate the sub's contract if he does not perform, and the right to collect any difference owed back to the contractor. The Attorney's Fees clause puts extra teeth in the contractor's ability to collect.

For some jobs, you may wish to add some of these clauses, but not all — or only the specific language that is relevant to a given job. Keep in mind that there is no absolute rule to help you decide which specific language to use on a given job. Generally, the more costly and complicated the subcontract, and the less well-established your relationship with the sub, the longer the agreement you will want to use.

5.2 ■ LONG-FORM SUBCONTRACT AGREEMENT

CONTRACTOR'S NAME: _____

ADDRESS: _____

_____

PHONE: _____

FAX: _____

LIC #: _____

DATE: _____

OWNER'S NAME: _____

ADDRESS: _____

_____

PROJECT ADDRESS: _____

_____

CONSTRUCTION LENDER: _____

ADDRESS: _____

_____

I. PARTIES

This Subcontract (hereinafter referred to as "Agreement") is being entered into on the _____ day of _____,
19___, and is between _____, (hereinafter referred to as "Contractor"); and
_____, (hereinafter referred to as "Subcontractor"). Subcontractor's license number is:
_____. By signing this Agreement, Subcontractor warrants that he is fully experienced, properly
licensed, and insured to perform the type of work described in this Agreement, and that he is an *independent
contractor* and *not* an agent or employee of the Contractor.

SUBCONTRACTOR'S ADDRESS: _____

_____

PHONE: _____

FAX: _____

Subcontractor's business is a: ___ Sole Proprietorship; ___ Partnership; ___ Corporation

Subcontractor's Federal Tax I.D.# or S.S.#: _____

In consideration of the mutual promises contained herein, the parties agree as follows:

II. GENERAL SCOPE OF WORK DESCRIPTION AND SUBCONTRACT AMOUNT

Subcontractor will furnish all labor, equipment, tools, materials, transportation, supervision, and all other items required for safe operations to complete the following work which will comply with the latest edition of all applicable building codes and the Contract Documents referred to below (if conflict between plans, specifications, Subcontractor's proposal, or this Agreement arises, then this Agreement is controlling):

_____

_____

_____

_____

_____

(Additional Work Description page(s) attached: _____Yes; _____ No)

LUMP SUM SUBCONTRACT AMOUNT AND PAYMENT SCHEDULE: Contractor to pay Subcontractor the total Lump Sum Amount of: $_____

Payments to Subcontractor to follow payment schedule below. At Contractor's discretion, a 5% or 10% retention may be withheld from all progress payments to Subcontractor. Subcontractor to be paid progress payments after General Contractor has received payment from Owner for project work that has been successfully completed by Subcontractor.

Receipt of payment from Owner is a condition precedent to paying Subcontractor under this Agreement. Progress payment lien releases to be furnished by Subcontractor upon request by Contractor.

Payments to be made to Subcontractor as follows:

_____

_____

_____

_____

_____

Retention of: _____ 5% _____ 10%
_____ will _____ will not
be withheld from all progress payments to Subcontractors.

Withheld retention to be paid to Subcontractor *only when all of the following conditions have been satisfied:* completion of all work by Subcontractor; 35 days has elapsed since filing a Notice of Completion by Owner (or Final Completion of the Project); Subcontractor has provided Contractor with Unconditional Lien Releases for all work and materials suppliers paid to date by Contractor and a Conditional Lien Release for the final retention payment. Subcontractor has furnished all warranty information and operation manuals to Contractor.

III. GENERAL CONDITIONS FOR THE SUBCONTRACT AGREEMENT ABOVE

A. EXCLUSIONS FROM SUBCONTRACTOR'S SCOPE OF WORK
Labor and materials for the following work have *not* been included by Subcontractor:

_____

_____

_____

_____

_____

B. CONTRACT DOCUMENTS
Subcontractor will perform its work in accordance with all Contract Documents, which are identified as follows:
- This Construction Agreement
- Plans: _____
- Specifications: _____
- Addenda: _____
- Miscellaneous: _____

_____

_____

Subcontractor warrants that he has been furnished all Contract Documents referred to above and has thoroughly familiarized himself with all Contract Documents and the existing site conditions.

The intent of the Contract Documents and this Agreement is to obtain a *complete and professional job.* Subcontractor agrees that the Scope of Work covered by this Agreement includes all labor and materials that are both specified and reasonably implied by the Contract Documents.

C. WORK COMMENCEMENT AND COMPLETION TIME
Work shall commence on _____ and take approximately _____ calendar days to complete.
TIME IS OF THE ESSENCE in all aspects of Subcontractor's performance.

D. CHANGES IN THE WORK
Only the Contractor shall have the right to order changes in the scope of Subcontractor's work (both additions and deletions). These changes shall be made in writing and signed by both Subcontractor and Contractor prior to commencement of any Change Order work by Subcontractor.

E. BACK CHARGES AND PROTECTION OF THE WORK
Contractor has the right to deduct from progress payments due to Subcontractor the cost of repairing damage caused by Subcontractor or the cost of repairing/replacing Subcontractor's defective work if Subcontractor fails to take significant steps toward correcting this damage or non-conforming or defective work within 7 days after receiving written notice from Contractor. Subcontractor agrees to be responsible for protecting all of its work in progress.

F. INDEMNIFICATION

All work performed by Subcontractor pursuant to this Agreement shall be done at the sole risk of the Subcontractor. Subcontractor (and his agents) shall at all times indemnify, protect, defend, and hold harmless Contractor and Owner from all loss and damage, and against all lawsuits, arbitrations, mechanic's liens, legal actions, legal or administrative proceedings, claims, debts, demands, awards, fines, judgments, damages, interest, attorney's fees, and any costs and expenses which are directly or indirectly caused or contributed to, or claimed to be caused or contributed to by any act or omission, fault or negligence, whether passive or active, of Subcontractor or his agents or employees, in connection with or incidental to the work under this Agreement.

G. SUBCONTRACTOR'S INSURANCE

Before commencing work on the project, Subcontractor and its Subcontractors of every tier will supply to Contractor duly issued Certificates of Insurance, naming Contractor as an "additional insured," showing in force the following insurance for comprehensive general liability, automobile liability, and worker's compensation*:

• comprehensive general liability with limits of not less than $_____ per occurrence;
• automobile liability in comprehensive form with coverage for owned, hired, and non-owned automobiles;
• worker's compensation insurance in statutory form.

* All insurance binders must contain a clause indicating that certificate holders be given a minimum of 10 days written notice prior to cancellation of Subcontractor's insurance. Subcontractor must furnish the insurance binder referred to above as an express condition precedent to the Contractor's duty to make any progress payments to Subcontractor pursuant to this Agreement.

Subcontractor's insurance shall be the primary insurance and neither Contractor's nor Owner's insurance shall be called on to contribute to a loss caused in whole or part by the negligence of Subcontractor.

Any Subcontractor who does not carry worker's compensation insurance coverage to protect himself *personally* from work-related injuries hereby fully releases, holds harmless, and indemnifies Contractor from any injuries that may occur to the Subcontractor himself during the course of this project. In no way does this provision affect the *absolute duty* of every Subcontractor to provide worker's compensation insurance coverage to each and every one of his employees according to the provisions of this Agreement and all applicable state and federal laws.

H. CLEANUP

Subcontractor will continuously clean up its work areas, and keep them in a safe, sanitary condition, and remove all of its debris on a periodic basis. Contractor may back charge Subcontractor at the rate of $35 per hour if cleanup and debris removal is not performed on a regular basis by Subcontractor.

I. EXPRESS WARRANTY

At the request of Contractor, Subcontractor will promptly replace or repair any work, equipment, or materials that fail to function properly for a period of one year after completion of the project, at Subcontractor's own expense. Subcontractor will also repair any surrounding parts of the structure that are damaged due to any failure in Subcontractor's work during the warranty period.

J. LAWS, REGULATIONS, AND SAFETY

Subcontractor and its employees and representatives shall at all times comply with *all* applicable laws, ordinances, rules and regulations, whether federal, state, or municipal, particularly those relating to wages, hours, working conditions, safe operations, all applicable union contributions, and the payment of all taxes.

Subcontractor will comply with all statutes and regulations that establish safety requirements (including, but not limited to those of OSHA and any state agency regulating job-site safety). By signing this Agreement, Subcontractor knowingly and willingly accepts *full responsibility* for the safe operation of all of its activities and the protection of other persons and property during the course of this project.

K. SUBCONTRACTOR DEFAULT

If Subcontractor fails to diligently complete work under this Agreement or fails in any way to perform in accordance with all the terms and conditions of this Agreement, then Contractor may, without prejudicing any other rights he may have, give a 72-hour Notice to Subcontractor to cure his default. If Subcontractor does not take significant steps to cure his default within 72 hours of receiving notice, then Contractor may terminate this Agreement for cause by giving Subcontractor written notice of termination of this Agreement.

Contractor will then have no duty to pay Subcontractor any remaining funds due until the project has been completed. If the cost to complete Subcontractor's work and the amount of funds paid to Subcontractor to date exceeds the contract amount of this Agreement, Subcontractor will then be responsible for immediately paying this difference to Contractor. Subcontractor is responsible for paying all of Contractor's attorney's fees and court costs in connection with the enforcement of this clause.

L. ASSIGNMENT

Any assignment of any part of this contract is prohibited and void without the prior written consent of Contractor.

M. DISPUTE RESOLUTION AND ATTORNEY'S FEES

Any controversy or claim arising out of or related to this Agreement involving an amount *less* than $5,000 (or the maximum limit of the court) must be heard in the Small Claims Division of the Municipal Court in the county where the Contractor's office is located. Any controversy or claim arising out of or related to this Agreement which is over the maximum dollar limit of the Small Claims Court must be settled by binding arbitration administered by the American Arbitration Association in accordance with the Construction Industry Arbitration Rules. Judgment upon the award may be entered in any Court having jurisdiction thereof.

Subcontractor agrees to contractually make this AAA Arbitration Dispute Resolution Clause irrevocably bind and "flow down" to all lower-tier Subcontractors. This Agreement is not assignable.

The prevailing party in any legal proceeding related to this Agreement shall be entitled to payment of reasonable attorney's fees, costs, and expenses.

N. ENTIRE AGREEMENT, MODIFICATION, SEVERABILITY

This Subcontract Agreement represents the entire agreement of the parties. This Agreement cannot be

modified by oral agreements but can only be modified by a written agreement which is signed by both parties. Should any portion of this Agreement be deemed unenforceable or invalid by law, same shall not invalidate this Agreement but the remaining portion of this Agreement shall remain in full force and effect.

O. ADDITIONAL LEGAL NOTICES REQUIRED BY STATE OR FEDERAL LAW
See page(s) attached: Yes ___; No ___

P. ADDITIONAL TERMS AND CONDITIONS
See page(s) attached: Yes ___; No ___

I have read and understood, and I agree to, all of the terms and conditions contained in the Agreement above.

Date: _____   _____
                         CONTRACTOR

Date: _____   _____
                         SUBCONTRACTOR

5.2 ∎ Long-Form Subcontract Agreement

ANNOTATED

∎ This is a typical long-form Subcontract Agreement for foundation and framing work on a residential remodeling project. The form has been filled out with sample language and most of the clauses have annotations under them explaining the legal significance or practical importance of the contract language.

Compare the blank form with this annotated, filled-out form to become more familiar with how to use this Agreement.

Charlie Contractor Construction, Inc.
123 Hammer Lane

Anywhere, USA 33333
Phone: (123) 456-7890
Fax: (123) 456-7899
Lic#: 11111

DATE: **June 1, 2001**

OWNER'S NAME: **Mr. & Mrs. Harry Homeowner**
ADDRESS: **123 Project Place**
Anywhere, USA 33333

PROJECT ADDRESS: **123 Project Place**
Anywhere, USA 33333

CONSTRUCTION LENDER: **Lucky Lou's Construction Loans**
ADDRESS: **777 Lucky Lender Lane**
Anywhere, USA 33333

∎ It's a good idea to list the name and address of the Construction Lender (required by law in certain states), if there is one on the project. This will give the Subcontractor all or nearly all the information he needs to send out any required Preliminary Notice Forms to the Owner.

I. PARTIES
This Subcontract (hereinafter referred to as "Agreement") is being entered into on the   **first**

INITIAL
CC
77

day of **June**, 20**01**, and is between **Charlie Contractor Construction, Inc.** (hereinafter referred to as "Contractor"); and **Frank's Framing** (hereinafter referred to as "Subcontractor"). Subcontractor's license number is: **22222**. Subcontractor warrants that he is properly licensed to perform the type of work described in this Agreement and that he is an *independent contractor* and *not* an agent or employee of the Contractor.

■ Be sure your Subcontractor is properly licensed to perform the work in the Agreement.

SUBCONTRACTOR'S ADDRESS: **123 Wood Lane**

Anywhere, USA 33333

PHONE: **(011) 222-3333**

FAX: **(011) 222-3334**

Subcontractor's business is a: **XX** Sole Proprietorship; ____ Partnership; ____ Corporation

Subcontractor's Federal Tax I.D.# or S.S.#: **88-01000**

In consideration of the mutual promises contained herein, the parties agree as follows:

■ See annotation, Form 5.1: Section I. Parties to the Agreement.

II. GENERAL SCOPE OF WORK DESCRIPTION AND SUBCONTRACT AMOUNT

Subcontractor will furnish all labor, equipment, tools, materials, transportation, supervision, and all other items required for safe operations to complete the following work which will comply with the latest edition of all applicable building codes and the Contract Documents referred to below (if conflict between plans, specifications, Subcontractor's proposal, or this Agreement arises, then this Agreement is controlling):

1. **Furnish labor and materials for all foundation work, including all backfill, and rough and finish grading within six feet of foundation. Remove all dirt spoils from site. Install all wet-set foundation hardware and holddowns.**

2. **Furnish all labor and materials to rough frame residence. Includes installation of all exterior doors and windows, all exterior door and window trim, vapor barrier, and all wood siding.**

3. **Furnish and install all blocking and cutouts required by Subcontractors on project.**

(Additional Work Description page(s) attached: _____ Yes; **X** No)

INITIAL
CC
FF

■ See annotation, Form 1.3: Section II. General Scope of Work Description and Subcontract Amount.

LUMP SUM SUBCONTRACT AMOUNT AND PAYMENT SCHEDULE: Contractor to pay Subcontractor the total Lump Sum Amount of: $ _57,200_ .

Payments to Subcontractor to follow payment schedule below. At Contractor's discretion, a 5% or 10% retention may be withheld from all progress payments to Subcontractor. Subcontractor to be paid progress payments after General Contractor has received payment from Owner for project work that has been successfully completed by Subcontractor.

■ The clause above gives the Contractor the option of withholding a 5% or 10% retention from payments due to the Subcontractor under the Subcontract Agreement. If the Owner is withholding retention from your payments (which hopefully is not the case very often), you'd better be withholding the same retention from payments to your subs.

Or, if you have reason to question a subcontractor's willingness to promptly take care of Punch List work, having a retention can be a carrot that will sometimes make certain subs work harder at the end of a job. Practices may vary regarding the withholding of retention depending upon your local area. Make sure you are familiar with local law and custom in your area.

Receipt of payment from Owner is a condition precedent to paying Subcontractor under this Agreement. Progress payment lien releases to be furnished by Subcontractor upon request by Contractor.

■ This clause states that the Subcontractor must furnish appropriate mechanic's lien releases upon request of the Contractor. If the Owner requires you to furnish these releases, you will also need to obtain the same type of releases from your subcontractors.

Many courts interpret the phrase "Receipt of payment from Owner is a condition precedent to paying Subcontractor..." as determining the time when the payment will be made to the Subcontractor (not *if* payment will be made to the Subcontractor). In other words, by relying on this clause, the sub gets paid shortly after the Contractor is paid for the sub's work— usually within 10 days, or less. If you can't afford to finance the payments to your subs, this clause can do a lot to improve your cash flow.

However, if the Owner refuses to pay you for work performed and his refusal has nothing to do with a defect in the Subcontractor's work, this

INITIAL
_CC_
_77_

clause will probably not deter the Subcontractor from bringing a breach of contract action against you for failing to pay him (even though the Owner refused to pay you for the sub's work). After all, you signed the contract with the Subcontractor, the Owner didn't.

Payments to be made to Subcontractor as follows:

1. Upon completion of foundation: $ 9,200

2. Upon completion of all underfloor framing and installation of subfloor and delivery of rough framing lumber to site: $ 8,000

3. Upon completion of first and second-floor framing and installation of roof sheathing: $ 20,000

4. Upon installation of exterior doors and windows, delivery of siding material to site, and installation of shear plywood on all exterior walls: $ 10,000

5. Balance due upon completion of all work: $ 10,000

Retention of: __N/A__ 5% _____ 10%
_____ will _____ will not
be withheld from all progress payments to Subcontractors.

Withheld retention to be paid to Subcontractor *only when all of the following conditions have been satisfied:* completion of all work by Subcontractor; 35 days has elapsed since filing a Notice of Completion by Owner (or Final Completion of the Project); Subcontractor has provided Contractor with Unconditional Lien Releases for all work and materials suppliers paid to date by Contractor and a Conditional Lien Release for the final retention payment. Subcontractor has furnished all warranty information and operation manuals to Contractor.

■ The clause above simply states when the Subcontractor is entitled to be paid the withheld retention. As a condition precedent to payment, the Subcontractor must furnish lien releases, warranty information, and any operation manuals, and wait 35 days from the date of the filing of Notice of Completion (or substantial completion of the project).

In some states, the Subcontractor only has lien rights for the 30-day period following the Owner's recording of the Notice of Completion. By waiting 35 days after recording of the Notice of Completion to make the final payment, the Owner and the Contractor can be assured that the Subcontractor will have no further legal right to file a lien against the project.

In addition, by waiting the 35 day period, any lien filed by the Subcontractor during the 30 days following the recording of the Notice of Completion should be known to the Owner and the Contractor, and any final

INITIAL
CC
77

> payment can be reduced accordingly. Once conditional and unconditional lien releases have been obtained from the Subcontractor covering the entire project (and the Punch List work is 100% complete), many Contractors will be comfortable issuing the final check to the Subcontractor if they have already been given their final check from the Owner.

III. GENERAL CONDITIONS FOR THE SUBCONTRACT AGREEMENT ABOVE

A. EXCLUSIONS FROM SUBCONTRACTOR'S SCOPE OF WORK

Labor and materials for the following work have *not* been included by Subcontractor:

1. Furnishing exterior doors and windows.

> Make sure you understand what is *not* being included in the Subcontractor's bid. That way, you can either exclude those items from your Agreement with the Owner or have the Subcontractor give you a price for the excluded work so you can include it in your bid.

B. CONTRACT DOCUMENTS

Subcontractor will perform its work in accordance with all Contract Documents, which are identified as follows:

- This Construction Agreement
- Plans: **By Art Architect, 5 pages dated April 1, 2001**
- Specifications: **See Plans referred to above**
- Addenda: **#1, by Art Architect, 2 pages dated May 13, 2001**
- Miscellaneous: **N/A**

Subcontractor warrants that he has been furnished all Contract Documents referred to above and has thoroughly familiarized himself with all Contract Documents and the existing site conditions.

The intent of the Contract Documents and this Agreement is to obtain a *complete and professional job.* Subcontractor agrees that the Scope of Work covered by this Agreement includes all labor and materials that are both specified and reasonably implied by the Contract Documents.

> ■ The language in this paragraph states that the subcontractor must perform any work that is "reasonably implied" by the contract documents (plans, specifications, addenda, etc.) as well as the work that is clearly specified. This can prove useful, since no contract documents are 100% descriptive.
>
> For instance, even though the specification for the painting on a large remodel may not call out caulking or protecting the landscaping prior to

INITIAL
CC
77

> painting, caulking in certain locations and protecting the landscaping from paint should be considered reasonably implied by the Contract Documents in order to obtain a "complete and professional" job and will need to be done by the painter.

C. WORK COMMENCEMENT AND COMPLETION TIME

Work shall commence on __July 1, 2001__ and take approximately __70__ calendar days to complete. **TIME IS OF THE ESSENCE** in all aspects of Subcontractor's performance.

D. CHANGES IN THE WORK

Only the Contractor shall have the right to order changes in the scope of Subcontractor's work (both additions and deletions). These changes shall be made in writing and signed by both Subcontractor and Contractor prior to commencement of any Change Order work by Subcontractor.

> ■ This clause makes it clear that if a Subcontractor wants to be paid for any additional work, he'd better get a written Change Order for that work. This clause can prevent a lot of problems with an Owner who doesn't want to pay for additional work that he might have directly asked the sub to perform.
>
> When extra subtrade work arises, first get a Change Order from the Subcontractor, then incorporate the Sub's Scope of Work and pricing into a Change Order which you issue to the Owner. Once the Owner signs your Change Order for the Sub's work, you can then authorize the Sub to perform the extra work.

E. BACK CHARGES AND PROTECTION OF THE WORK

Contractor has the right to deduct from progress payments due to Subcontractor the cost of repairing damage caused by Subcontractor or the cost of repairing/replacing Subcontractor's defective work if Subcontractor fails to take significant steps toward correcting this damage or non-conforming or defective work within three days after receiving written notice from Contractor. Subcontractor agrees to be responsible for protecting all of its work in progress.

> ■ This clause requires the Subcontractor to protect all of his work in progress. This could mean taping off tile that has just been set with warning tape or tarping a roof that has just been torn off on a remodel project. This clause also gives the Contractor a contractual right to back charge the Sub (or withhold money from payments due him) if the Sub damages work on the site or performs his work in a defective or substandard manner.
>
> This clause requires the Contractor to give the Sub written notice of defective work or damage to the property and gives the Sub three days in which to correct the damage. It's only fair to give the Subcontractor notice

INITIAL
CC
77

and an opportunity to correct substandard work or his property damage. However, offsetting payments to the Subcontractor or back charging the Sub can provide a powerful incentive to the Sub to perform properly and correct known defects.

F. INDEMNIFICATION

All work performed by Subcontractor pursuant to this Agreement shall be done at the sole risk of the Subcontractor. Subcontractor (and his agents) shall at all times indemnify, protect, defend, and hold harmless Contractor and Owner from all loss and damage, and against all lawsuits, arbitrations, mechanic's liens, legal actions, legal or administrative proceedings, claims, debts, demands, awards, fines, judgments, damages, interest, attorney's fees, and any costs and expenses which are directly or indirectly caused or contributed to, or claimed to be caused or contributed to by any act or omission, fault or negligence, whether passive or active, of Subcontractor or his agents or employees, in connection with or incidental to the work under this Agreement.

■ See annotation, Form 5.1: Section III.D. Indemnification.

G. SUBCONTRACTOR'S INSURANCE

Before commencing work on the project, Subcontractor and its Subcontractors of every tier will supply to Contractor duly issued Certificates of Insurance, naming Contractor as an "additional insured," showing in force the following insurance for comprehensive general liability, automobile liability, and worker's compensation*:

• comprehensive general liability with limits of not less than __$500,000__ per occurrence;
• automobile liability in comprehensive form with coverage for owned, hired, and non-owned automobiles;
• worker's compensation insurance in statutory form.

* All insurance binders must contain a clause indicating that certificate holders be given a minimum of 10 days written notice prior to cancellation of Subcontractor's insurance. Subcontractor must furnish the insurance binder referred to above as an express condition precedent to the Contractor's duty to make any progress payments to Subcontractor pursuant to this Agreement.

■ See annotation, Form 5.1: Section III.E. Subcontractor's Insurance.

Subcontractor's insurance shall be the primary insurance and neither Contractor's nor Owner's insurance shall be called on to contribute to a loss caused in whole or part by the negligence of Subcontractor.

■ The clause above indicates that the Subcontractor's insurance shall be the primary insurance in the event of a loss caused in whole or in part by the negligence of the Subcontractor. Sometimes more than one party may have insurance coverage for a particular loss.
   Naturally, no party wants a claim against their insurance. This clause

INITIAL
CC
77

> makes it clear that in the event a loss is caused by the Subcontractor, the Sub's insurance will be the policy against which the claim is made, not the Owner's or Contractor's insurance.

Any Subcontractor who does not carry worker's compensation insurance coverage to protect himself *personally* from work-related injuries hereby fully releases, holds harmless, and indemnifies Contractor from any injuries that may occur to the Subcontractor himself during the course of this project. In no way does this provision affect the *absolute duty* of every Subcontractor to provide worker's compensation insurance coverage to each and every one of his employees according to the provisions of this Agreement and all applicable state and federal laws.

> ■ The law generally places a high degree of responsibility for safe working practices on the Subcontractor when he is performing his own work.
>
> However, in some situations, if the Subcontractor himself is injured on your job and does not carry health insurance to pay for his injuries, he may try to sue you for failing to provide a safe work environment. This clause will certainly not provide absolute protection from that type of suit, but it may help somewhat as long as you were not negligent in your duty to monitor the safety aspects of the site.

H. CLEANUP

Subcontractor will continuously clean up its work areas, and keep them in a safe, sanitary condition, and remove all of its debris on a periodic basis. Contractor may back charge Subcontractor at the rate of $35 per hour if cleanup and debris removal is not performed on a regular basis by Subcontractor.

> ■ This clause addresses one of the more troublesome aspects of many projects — whether the Sub or the General Contractor cleans up the site and removes the accumulated debris. This clause squarely places this responsibility on the Sub and allows the Contractor to back charge the Sub for failing to clean up his work area and remove his debris.
>
> Some Contractors typically make arrangements to place a dumpster on site or locate a central debris pile which the Subs can use and they will haul away. If you do this, simply make a note of that in your subcontract in the Scope of Work section so there is no confusion about responsibilities.

I. EXPRESS WARRANTY

At the request of Contractor, Subcontractor will promptly replace or repair any work, equipment, or materials that fail to function properly for a period of one year after completion of the project, at Subcontractor's own expense. Subcontractor will also repair any surrounding parts of the structure that are damaged due to any failure in Subcontractor's work during the warranty period.

INITIAL
CC
77

■ Be sure to state your expectations about how long the Subcontractor will provide an express warranty for his work. Refer to your Agreement with the Owner and make sure your Subcontractor's warranty is at least as long and as comprehensive as your warranty with the Owner.

     I also indicate that any surrounding parts of the structure that are damaged due to a failure in the Subcontractor's work will be repaired by Subcontractor.

     For instance, if the plumber's spa leaks after four months and ruins the ceiling, carpet, and piano in the first-floor living room, the plumber will be expected to repair the leak, repair the ceiling and paint, and repair or replace the carpet and the piano. This type of *consequential damage* can be far more costly to repair than the actual defect itself. In this example, the Subcontractor's comprehensive general liability insurance would likely cover much of the consequential damage — which highlights the importance of both the Subcontractor's warranty and insurance coverage.

J. LAWS, REGULATIONS, AND SAFETY

Subcontractor and its employees and representatives shall at all times comply with *all* applicable laws, ordinances, rules and regulations, whether federal, state, or municipal, particularly those relating to wages, hours, working conditions, safe operations, all applicable union contributions, and the payment of all taxes.

■ There are already laws in place that require the Sub to do these things, but his failure to do so may complicate your project, so it's a good idea to make these contractual obligations. In some instances, the authorities may come after the General Contractor if the Subcontractor fails to meet certain legal obligations.

Subcontractor will comply with all statutes and regulations that establish safety requirements (including, but not limited to those of OSHA and any state agency regulating job-site safety). By signing this Agreement, Subcontractor knowingly and willingly accepts *full responsibility* for the safe operation of all of its activities and the protection of other persons and property during the course of this project.

K. SUBCONTRACTOR DEFAULT

If Subcontractor fails to diligently complete work under this Agreement or fails in any way to perform in accordance with all the terms and conditions of this Agreement, then Contractor may, without prejudicing any other rights he may have, give a 72-hour Notice to Subcontractor to cure his default. If Subcontractor does not take significant steps to cure his default within 72 hours of receiving notice, then Contractor may terminate this Agreement for cause by giving Subcontractor written notice of termination of this Agreement.

Contractor will then have no duty to pay Subcontractor any remaining funds due until the project has been completed. If the cost to complete Subcontractor's work and the amount of funds paid to Subcontractor to date exceeds the contract amount of this Agreement, Subcontractor will then be

INITIAL
CC
77

responsible for immediately paying this difference to Contractor. Subcontractor is responsible for paying all of Contractor's attorney's fees and court costs in connection with the enforcement of this clause.

> ■ This clause sets forth a very important procedure for terminating a Sub-contractor who is not performing as expected on your project. You must be careful that the Sub's failure to perform amounts to a *material breach* of your contract with him. See discussion of material vs. minor breaches in Chapter 11 for important considerations prior to terminating an agreement.
>
> Be sure to carefully follow this procedure and consult an attorney if this situation arises and you are not 100% certain that you have a strong legal basis for terminating the subcontract. If you terminate a subcontract without legal grounds, you may end up paying the Sub his lost profit and overhead on the work he didn't get a chance to perform (or other damages) due to your improper termination of the agreement.
>
> Nevertheless, when the Sub has materially breached the terms of the subcontract, it will be a great advantage to you to have a specific procedure like this to follow to properly terminate the subcontract so that you can bring in a new sub to complete the work.

L. ASSIGNMENT

Any assignment of any part of this contract is prohibited and void without the prior written consent of Contractor.

M. DISPUTE RESOLUTION AND ATTORNEY'S FEES

Any controversy or claim arising out of or related to this Agreement involving an amount *less* than $5,000 (or the maximum limit of the court) must be heard in the Small Claims Division of the Municipal Court in the county where the Contractor's office is located. Any controversy or claim arising out of or related to this Agreement which is over the maximum dollar limit of the Small Claims Court must be settled by binding arbitration administered by the American Arbitration Association in accordance with the Construction Industry Arbitration Rules. Judgment upon the award may be entered in any Court having jurisdiction thereof.

Subcontractor agrees to contractually make this AAA Arbitration Dispute Resolution Clause irrevocably bind and "flow down" to all lower-tier Subcontractors. This Agreement is not assignable.

> ■ See annotation, Form 5.1: Section III.M. Dispute Resolution and Attorney's Fees.

The prevailing party in any legal proceeding related to this Agreement shall be entitled to payment of reasonable attorney's fees, costs, and expenses.

INITIAL
CC
77

N. ENTIRE AGREEMENT, MODIFICATION, SEVERABILITY

This Subcontract Agreement represents the entire agreement of the parties. This Agreement cannot be modified by oral agreements but can only be modified by a written agreement which is signed by both parties. Should any portion of this Agreement be deemed unenforceable or invalid by law, same shall not invalidate this Agreement but the remaining portion of this Agreement shall remain in full force and effect.

> ■ See annotation, Form 5.1: Section III.N. Entire Agreement, Modification, Severability.

O. ADDITIONAL LEGAL NOTICES REQUIRED BY STATE OR FEDERAL LAW

(See page attached: Yes_____; No___**X**___)

> ■ If your state requires any legal notices to be included in your subcontracts, include them here or attach them to your Agreement. Most states don't have very many requirements for agreements between a Contractor and Subcontractor, but consult your own attorney to be sure.

P. ADDITIONAL TERMS AND CONDITIONS

(See page attached: Yes_____; No___**X**___)

> ■ See annotation, Form 5.1: Section III.P. Additional Terms and Conditions.

I have read and understood, and I agree to, all of the terms and conditions contained in the Agreement above.

Date: _____6/01/01_____　　　*CHARLIE CONTRACTOR, PRESIDENT*
　　　　　　　　　　　　　　CHARLIE CONTRACTOR, PRESIDENT
　　　　　　　　　　　　　　CHARLIE CONTRACTOR CONSTRUCTION, INC.

Date: _____6/01/01_____　　　*Frank's Framing*
　　　　　　　　　　　　　　FRANK'S FRAMING

> ■ Be sure you have an Agreement in your files which is signed by the Subcontractor and contains his initials on each page.

It's a good idea to send a Subcontractor Information Form to new subcontractors when you give them plans to bid. You can also send it to your regular subcontractors to get their tax identification numbers, and to inform them about your current insurance requirements.

Review your insurance requirements with a good construction attorney and an insurance agent familiar with the construction industry, and fill in the amount required by your company for subcontractors' comprehensive general liability insurance (Section I, item B, of the form). Make sure the dollar amount you write in on this form matches the amount of insurance you require the sub to carry in your subcontract agreement and the amount the owner requires you to carry.

With new subs, the purpose of the Subcontractor Information Form is to "prequalify" them by having them provide evidence of proper licensing and insurance. This form also requires the subcontractor to provide his federal tax i.d. number or social security number (you'll need this to properly file your tax return) and his type of business (e.g., sole proprietorship, corporation, partnership, etc.).

With your regular subcontractors, it's still a good idea to send this form out on an annual basis to help you stay on top of any changes in their businesses that could have an impact on your working relationship. Your insurance carrier will probably want to audit the insurance certificates of all your subcontractors so it is important to have this information readily available when the auditor comes calling. To make things easy, I typically keep the subcontractors' insurance certificates and licensing information in a single file.

If this information is not readily available, you will probably be hit with an additional insurance premium for each subcontractor you worked with for whom you didn't have an insurance certificate on file. It's well worth it to take the time to carefully review these insurance audits. In my experience, they are often inaccurate and revised only when the contractor digs into the extra charges and provides additional documentation to refute them.

These surprise premiums, if not successfully challenged with documentation, can cost you hundreds or thousands of dollars a year.

This subcontractor information form doesn't request customer references from the subcontractor. If you have never worked with the sub and have never heard of him before, you may want to ask around and get some references from other contractors who have worked with him in the recent past.

If your state has a contractor's state license department, you may want to contact this agency and find out the nature and number of any complaints against the subcontractor. You also should verify that a license has been issued to the business you are contracting with. Do not work with a subcontractor who is "borrowing" the license of another.

Finally, in order to reduce the possibility of change order disputes, the subcontractor information form states your company's policy of requiring all change orders to be in writing prior to beginning any change order work by the subcontractor. This form also states that the subcontractor's payment requests must be backed up by written invoices, and it informs the subcontractor that he may need to furnish mechanic's lien releases if requested by the contractor.

5.3 ■ SUBCONTRACTOR INFORMATION FORM

CONTRACTOR'S NAME: _____

ADDRESS: _____

_____

PHONE: _____

FAX: _____

LIC #: _____

DATE: _____

SUBCONTRACTOR'S NAME: _____

ADDRESS: _____

_____

PHONE: _____

FAX: _____

Please review the information below and provide the information requested prior to commencing any work for Contractor.

I. GENERAL SUBCONTRACTOR REQUIREMENTS

A. CONSTRUCTION SUBCONTRACT AGREEMENT

Carefully review all contract terms and conditions in the Contractor's Subcontract Agreement prior to signing it. Do not commence work without a signed Subcontract Agreement issued by the Contractor.

B. INSURANCE AND LICENSING

Carefully review all insurance and licensing requirements in the Subcontract Agreement. Subcontractor is strictly prohibited from working on any project for Contractor at any period of time when either Subcontractor's contractor's license or insurance (worker's compensation in statutory form and comprehensive general liability insurance in the amount of $ _____) is not current, active, and in conformance with the requirements set forth in the Subcontract Agreement.

Send Contractor a photocopy of Subcontractor's current contractor's license prior to commencing work for Contractor.

C. CHANGE ORDERS

Do not proceed with any Change Order work without first receiving written approval from Contractor.

II. PAYMENT PROCEDURE

A. INVOICING

Submit an invoice for all payments requested. With each invoice, clearly identify the project and the phase of

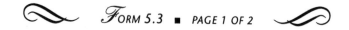

work being invoiced. Invoicing must follow the payment schedule and provisions set forth in the payment schedule portion of the Subcontract Agreement.

B. MECHANIC'S LIEN RELEASES

If requested by Contractor, furnish appropriate Mechanic's Lien Releases (for Subcontractor and Subcontractor's suppliers) with invoices. Contractor reserves the right to issue payment in the form of joint checks made out to Subcontractor and materials suppliers.

C. WITHHOLDING OF PAYMENT

No payments of any kind will be issued to Subcontractor if insurance certificates, as specified in the Subcontract Agreement, are not filed in Contractor's office *prior* to the time of Subcontractor's payment request.

III. GENERAL INFORMATION AND TAX IDENTIFICATION NUMBER

In order to facilitate the federal government's requirement to file Form 1099 reporting statements annually, Subcontractor *must* provide the following information to Contractor *prior* to commencing any work for Contractor:

A. GENERAL SUBCONTRACTOR INFORMATION

BUSINESS NAME: _____

NAME OF OWNER(S): _____

ADDRESS: _____

PHONE: _____

FAX: _____

LICENSE #: _____

LICENSE CLASSIFICATION: _____

YEAR THIS BUSINESS WAS STARTED: _____

B. BUSINESS FORM

( ) Individual or Sole Proprietorship

( ) Partnership

( ) Corporation

( ) Other:

C. FEDERAL TAX I.D. NUMBER OR SOCIAL SECURITY NUMBER

Federal Tax I.D.#: _____

Social Security #: _____

Thank you for taking the time to review this form and return it to us prior to commencing any work. We look forward to working with you on this project.

Sincerely,

_____

CONTRACTOR'S NAME

COMPANY NAME

CHAPTER 6

INVOICING FOR YOUR WORK

6.1 ▪ LONG-FORM CONSTRUCTION INVOICE

6.2 ▪ SHORT-FORM CONSTRUCTION INVOICE

There's an old saying, "We pay off of invoices and work off of contracts!" This is a good rule, no matter which end of the money you're on. Perform your work in accordance with a signed contract and when you're ready to be paid for all or a portion of your work, give the owner an invoice. It's a simple procedure and it keeps everyone's accounting in order.

As a general rule, invoicing forms and procedures are not as important on small jobs. However, on medium-sized and larger jobs with multiple draws and numerous change orders, invoicing forms and procedures become indispensable.

Invoicing and Payment Schedules

The payment schedule set forth in your construction agreement is what structures your invoicing, and your invoicing is what structures the timing of your payments from the owner. Because most contractors can't afford to personally finance a large job, they need to count on a reliable and predictable incoming cash flow from the owner in order to get the job done. The best way to ensure this cash flow is to issue invoices in a timely fashion which coincides with the payment schedule and contract payment terms.

Contractors should tailor the payment schedules in their contracts based upon the requirements of each job, and take frequent draws that are roughly equivalent to the value of labor and materials installed at the job site. Then, follow the contract payment schedule when issuing invoices.

Some contractors are nearly devastated by sloppy payment schedules that leave them trying to finance big portions of the job out of funds they don't have. This leads to credit problems and hurts the contractor's relationship with employees and subs who rely on him for prompt payments.

Do *not* leave a large payment owing at the end of the job. Ideally, your final payment should not exceed 20% of the contract amount on small and medium-sized jobs and 5% to 10% of the contract amount on larger jobs. I always try to make the final contract payment equal to, or less than, the dollar limit of the local Small Claims Court. That way, if an owner decides to unjustly withhold my final payment for some reason, I know that I can turn to the Small Claims Court and have a chance of collecting my money without the need for a costly lawsuit.

Another mistake contractors sometimes make is drafting payment schedules and invoices for work that has not yet been completed. If you do this, you will probably make the owner nervous, and get a false sense of having earned money that you actually owe to your employees, subs, materials suppliers, and the government. This can also get you into a lot of trouble.

In certain states, it is law that the contractor's initial contract payment/deposit for home improvement contracts must not exceed 10% of the contract amount or $1,000, whichever is the smaller amount. In some states it is a violation of the contractor's state license law to invoice for more work than is actually completed.

Sometimes I forget just how much a well-organized invoicing system simplifies the tracking and collection of payments on larger jobs. Recently, I administered the contract on a large job for a contractor who let the owner make payments at any time for any reason. If the owner wanted to issue a check for any random amount, the contractor took it in and told me about it much later, if at all.

The amount of the owner's checks never matched the amount of the invoices we wrote up. Consequently, the invoice accounting summary was worthless, the change order accounting summary was worthless, and the contractor spent hours and hours at the end of the job trying to figure out what the owner had paid for and what was still owing. This experience reminded me how much time is saved and frustration avoided by following the invoicing and accounting summary procedure described below.

Be fair with your payment schedule and invoicing, but be prompt about getting your invoices to the owner and getting paid. And keep your invoices in sequence with the payment schedule in your contract with the owner. The outcome will be well worth your organizational efforts.

Two Types of Invoices

Some contractors draft payment schedules for fixed price agreements so that they are paid on a weekly or bimonthly basis for the percentage of work they have completed during the prior week or two. This *percentage-of-work-completed invoicing* method works well on time-and-materials jobs and with fixed price contracts as long as there are no disagreements about the amount or value of work actually completed by the contractor. However, this invoicing method can backfire on a contractor at the end of a job if he has invoiced for 90% of the job but only completed 75% of the work.

Other contractors draft payment schedules that trigger a specified percentage of the contract amount to become due immediately upon completion of certain phases of the work. This *specified-draw invoicing* procedure is typically used on residential projects in my part of the country.

I prefer the specified-draw method because, prior to beginning any work, everyone knows and contractually agrees that payments will be due when fixed amounts of work are completed. As work is completed in stages, there are fewer opportunities for disagreements over exactly how much money is due in each payment.

Construction lenders also tend to use a specified-draw system like this for residential construction (e.g., 15% of contract amount due upon completion of excavation for foundation, setting of forms and steel, pouring of concrete and stripping of forms; 15% of contract amount due upon completion of subfloor and all underfloor mechanical and plumbing work; 25% of contract amount due upon completion of above-floor rough framing, installation of roof and gutters, and setting of exterior doors and windows; etc.).

When drafting your own payment schedule, remember that frequent smaller payments are always better than a couple of bigger payments. If an owner is going to have financial trouble on a job and not be able to pay you, the sooner you find out about this, the better. For instance, if the owner has only $12,000 to pay toward your last $22,000 invoice, better that you discover this before you complete the last $10,000 worth of work. The smaller the payments, the less likely the owner will be to let you perform work that he can't pay for in a timely manner.

Providing the owner with a tight, concise, and fair payment schedule and then following this payment schedule with prompt invoicing and collection of draws is a key factor in a contractor's survival in the construction business.

The Accounting Summary

Compared to drafting contracts, invoicing is rather routine. Anyone can scratch out an invoice for a small job. For small jobs with one or two invoices, almost any invoice form in the world will do. However, on jobs with numerous payments and change orders, it's best to provide an accounting summary at the bottom of every invoice given to the owner.

This accounting summary serves as an aid to both the owner and the contractor. On a bigger job with lots of change orders and lots of contract draws, the accounting summary portion of your invoice can save you literally hours of time reviewing the job file in an attempt to piece together exactly how much the owner paid and whether the disbursements were for contract payments, change orders (if so, which ones?), or materials deposits, etc. It can even keep you from losing money through simple accounting errors.

In addition, an owner always has more confidence in a contractor when it is evident that the contractor has complete understanding and control of where the owner's money is being applied throughout the project. On a larger job, timely invoicing and use of the accounting summary increases the owner's trust in your ability to run the job professionally. In this respect, the invoice accounting summary is also a good marketing tool.

The invoice accounting summary is simple to use and quite similar to the change order accounting summary. You start off by including it on invoice #1 and it simply builds itself with each successive invoice. The accounting summary is made up of three sections which correspond to the three sections of the invoice itself: one section for itemizing contract payments, one section for change order payments, and a third for the total amount paid in both of the other categories. Separating out the change order payments and the contract payments makes reading the accounting summary (and the invoice itself) easier because these two categories (and their subtotals) will always match these same categories and "balance out" in the change order accounting summary (see Form 4.2).

Long-Form Construction Invoice 6.1

Getting invoices out on time can become a problem when you're running multiple jobs. One helpful way to remind yourself or your office when to invoice is to copy the payment schedule from the contract and give it to your lead carpenter, or, if that's you, put it in a visible place in your office as a reminder of when contract payments are due.

Then, a couple of days prior to the payment due date, the lead person on the job should inform the person who creates the invoice to draft it and have it ready to give to the owner right on time. Sometimes the invoice is mailed out (when the owner doesn't live at the site or is out of the area), and sometimes the invoice is simply left on site in an envelope for the owner to pick up or come home to.

If you have trouble remembering which invoices have been paid and which ones are outstanding, you may want to place a copy of each invoice in a "receivables" file. Then, when each check comes in, match it up with the corresponding invoice, stamp it "PAID," noting the date the check was received, make a copy of the check, and staple it to the paid invoice. Then file the paid invoice and the copy of the check in the job file for that project.

An invaluable step in this process is the photocopying of the check. If you copy each check you receive from the owner and put it in the job file, you will have immediate access to the date, amount, and check number of all payments. (If you don't have a copy machine, your fax machine may double as a lightweight copier and be used to copy checks.) I can't tell you how many times I have put my mind at ease because I was able to find a copy of the owner's check in my file and didn't have to go to him and ask, "Just how much have you paid me so far"?

In addition, if you keep a simple ledger sheet of the outstanding unpaid invoices and log the payments and new invoices as they come in, you will always know the total amount of money (receivables) owed to your company at any given time. Knowing the amount of your receivables can also help you to budget for payroll, tax payments, etc.

Getting Paid

If the owner hasn't offered a check within a couple of days of receiving the invoice, I'll sometimes be direct and ask if I can pick it up (rather than having it sent in the mail). Use discretion here. It's good to run a tight ship, but I don't like to appear anxious.

If the owner hasn't paid the invoice within a week or so, I might send or drop off another invoice in conjunction with a phone call in which I mention the need to pay suppliers and subcontractors on time. This can sometimes help the owner to realize that most of his money is flowing down the line to others and needs to continue to flow in a timely manner so that his project can continue.

If the check is approximately two weeks late, I figure there may be a problem and I will schedule a meeting with the owner. Then I can determine the source of the problem and what to do about it. Again, use your discretion. Every job is different, every owner is different, and some small payments are less significant than large ones which are crucial to your cash flow and your ability to keep your subs working and your suppliers paid up.

Paying Subcontractors and Suppliers

In some states, if a contractor has been paid a disbursement by the owner which is intended to cover labor and materials supplied by subcontractors and materials suppliers, he must pass on that portion of the disbursement within 10 days of receiving the owner's payment, unless otherwise agreed to in writing.

If the contractor does not pay the subcontractors and materials suppliers within 10 days of receipt of funds, he may be assessed penalties such as 2% interest per month on the unpaid debt and end up paying for attorney's fees in any action brought by a sub or materials supplier for payment.

(Be sure to investigate all payment discount options if you pay your suppliers prior to the tenth of the month. Some suppliers will give good, reliable contractors an extra 2% discount over and above the usual contractor discount if you approach them about this.)

Change Orders and Invoicing

As a general rule, avoid taking in money that is not a payment toward either the contract payment schedule or a change order (e.g., the owner asks you to buy that extra woodstove for the family room and have it installed at your cost, and wants to simply pay you the actual cost of the woodstove and installation once you receive the bill).

If you do any additional work on behalf of the owner in the form of extra labor or materials, write it up as a change order, and add your profit and overhead to the additional work. Then invoice for change orders as the extra work is completed. If you procrastinate in invoicing for change order work, you're much more likely to have trouble collecting.

Simply passing on the direct, actual cost of materials or labor as a favor to the owner under a fixed price contract will not fit into either the category of contract payments or change order payments, nor will it enhance your profits on the job.

Also, if you can afford to pass on the direct costs once, some owners will assume you can afford to do it every time and will expect you to pass every bit of additional work along to them at your direct cost without ever charging any profit and overhead. Once this happens, you lose control of the job and will feel like you're working for free.

If for some reason you decide to donate your profit and overhead on a change order or a small piece of work to the owner, be sure you let him know that this is indeed a donation. Let him know that you are breaking from your company's policy this one time as a gesture of appreciation or as a special favor, and that you can't repeat the favor every time additional work arises (or is requested by the owner).

Again, have a policy, and be extremely careful about breaking from this policy. However, if you deviate from this policy in a very rare circumstance, let the owner know what you are doing and why so that he appreciates your gesture.

6.1 ▪ Long-Form Construction Invoice

CONTRACTOR'S NAME: _____

ADDRESS: _____

_____

PHONE: _____

FAX: _____

LIC #: _____

DATE: _____

OWNER'S NAME: _____

ADDRESS: _____

_____

PROJECT ADDRESS: _____

_____

INVOICE

Dear _____ ,

This is an invoice for work completed per your request at the Project Address referred to above per our Construction Agreement dated _____, 19___.

A. CONTRACT PAYMENTS DUE NOW

1. _____ : $_____

2. _____ : $_____

3. _____ : $_____

* Subtotal of Contract Payments Due Now: $_____

B. CHANGE ORDER PAYMENTS DUE NOW

1. _____ : $_____

2. _____ : $_____

3. _____ : $_____

* Subtotal of Change Order Payments Due Now: $_____

C. TOTAL OF CHANGE ORDER & CONTRACT PAYMENTS DUE NOW: $_____

ACCOUNTING SUMMARY

A. CONTRACT PAYMENTS

After receipt of the payment above, the Owner will have made the following payments to Contractor toward the Contract Payment Schedule set forth in the Construction Agreement dated _____, 19__:

1. Initial payment : $_____

2. _____ : $_____

3. _____ : $_____

4. _____ : $_____

5. _____ : $_____

6. _____ : $_____

* Total Paid To Date Toward Contract Payments: $_____

B. CHANGE ORDER PAYMENTS

After receipt of the payment above, the Owner will have made the following payments to Contractor toward Change Orders related to the Construction Agreement dated _____, 19___:

1. _____ : $_____

2. _____ : $_____

3. _____ : $_____

4. _____ : $_____

5. _____ : $_____

6. _____ : $_____

* Total Paid To Date Toward Change Order Payments: $_____

C. TOTAL AMOUNT PAID TOWARD CHANGE ORDERS & CONTRACT PAYMENTS

1. Contract Payments Paid To Date: $_____

2. Change Orders Paid To Date: $_____

* Total Of All Funds Paid To Date By Owner: $_____

Thank you for giving us the opportunity to perform this work for you.

Very Truly Yours,

_____
CONTRACTOR
COMPANY NAME

6.1 ■ Long-Form Construction Invoice

ANNOTATED

Charlie Contractor Construction, Inc.
123 Hammer Lane
Anywhere, USA 33333
Phone: (123) 456-7890
Fax: (123) 456-7899
Lic#: 11111

DATE: __December 2, 2001__

OWNER'S NAME: __Oscar Owner__
ADDRESS: __444 Suburban St.__
__Anywhere, USA 33333__

PROJECT ADDRESS: __same__

INVOICE

Dear __Oscar__ ,

This is an invoice for work completed per your request at the Project Address referred to above per our Construction Agreement dated __October 10__ , 20 __01__ .

> ■ It's a good idea to refer to the Construction Agreement which is the basis for the invoice. Do *not* invoice for work that has not been completed.

A. CONTRACT PAYMENTS DUE NOW

1. __Sixth contract payment due__ : $ __7,283.80__
2. __Seventh contract payment due__ : $ __3,641.90__

* Subtotal of Contract Payments Due Now: $ __10,925.70__

> ■ It will simplify accounting tremendously if you always invoice for contract payments (per the payment schedule in your Construction Agreement) in one section of the invoice and change order payments in the following section of the invoice.

> Every Contractor I know has at one time or another become hopelessly confused and embarrassed on a large job when asked by the Owner, "Just how much have I paid you toward the original contract versus payments I've made toward the change orders"? Think of contract payments and change order payments as apples and oranges. You can bill for them on the same invoice, but always list them and track them separately on each invoice.
>
> I've literally spent days on some legal cases just trying to decipher how much the Owner actually paid the Contractor toward change orders and contract payments because the Contractor didn't have a clue. If you follow the invoice and accounting summary set forth below, you'll never have a problem with this.

B. CHANGE ORDER PAYMENTS DUE NOW

* Subtotal of Change Order Payments Due Now: $ ____0____

C. TOTAL OF CHANGE ORDER & CONTRACT PAYMENTS DUE NOW: $ **10,925.70**

> ■ On a job with more than a couple of payments, I'll always use an accounting summary like the one below to help me keep track of payments that have been made by the Owner. It is very simple to build this accounting summary with each additional invoice you issue.
>
> This may seem unimportant at first, however it can be a tremendous benefit in the latter stages of a long job where you may have six to eight contract payments and five or more change orders. The Owner is always reassured to see that you have accurately tracked and allocated the flow of his payments as the project has progressed.

ACCOUNTING SUMMARY

A. CONTRACT PAYMENTS

After receipt of the payment above, the Owner will have made the following payments to Contractor toward the Contract Payment Schedule set forth in the Construction Agreement dated **October 10** , 20**01** :

1. Initial Payment:		$ **1,000.00**
2. **Second Contract Payment**	:	$ **10,925.70**
3. **Third Contract (Partial Payment)**	:	$ **17,350.00**
4. **Balance of Third Contract Payment**	:	$ **4,501.40**
5. **Fourth Contract Payment**	:	$ **10,700.70**
6. **Fifth Contract Payment**	:	$ **10,925.70**
7. **Sixth Contract Payment**	:	$ **7,283.80**

8. __Seventh Contract Payment__ _____ : $ __3,641.90__

* **Total Paid To Date Toward Contract Payments:** $ __66,329.20__

B. CHANGE ORDER PAYMENTS

After receipt of the payment above, the Owner will have made the following payments to Contractor toward Change Orders related to the Construction Agreement dated __October 10__ , 20__01__ :

1. __Change Order #1__ _____ : $ __1,382.00__

2. __Change Order #2__ _____ : $ __2,030.00__

3. __Change Order #3__ _____ : $ __462.00__

4. __Change Order #4__ _____ : $ __450.00__

5. __Change Order #5__ _____ : $ __388.00__

* **Total Paid To Date Toward Change Order Payments:** $ __4,712.00__

C. TOTAL AMOUNT PAID TOWARD CHANGE ORDERS & CONTRACT PAYMENTS

1. Contract Payments To Date: $ __66,329.20__

2. Change Order Payments To Date: $ __4,712.00__

* **Total Of All Funds Paid To Date By Owner:**   **$ 71,041.20**

> ■ Prior to issuing the final invoice on your project, compare the amount of money received toward change orders and contract payments in the accounting summary above to the corresponding line items and amounts in the last change order you issued. These amounts should always match and the total of these two amounts will always equal the adjusted contract amount on your last change order.
>
> If these line items all match, you'll know that you have properly tracked every penny on the project and broken down the funds paid by the Owner between the contract payments and the change order payments. If these amounts do not match, review your accounting to determine what you've mixed up and who owes whom. This is a great way to verify that you have taken in all the money you were supposed to take in toward the contract payment schedule and the change orders.

Thank you for giving Charlie Contractor Construction, Inc. the opportunity to perform this work for you.

Very Truly Yours,

CHARLIE CONTRACTOR, PRESIDENT _____

CHARLIE CONTRACTOR, PRESIDENT
CHARLIE CONTRACTOR CONSTRUCTION, INC.

The following is a sample short-form invoice that is useful for small jobs with one or two payments where no accounting summary is needed. Even though this is a short-form invoice, it is still helpful to refer to the contract and project address that you are invoicing for, and separate out contract payments from change order payments. An organized, professional looking short-form invoice like this one will make a better impression on the owner than a lunch sack or piece of note paper with a few numbers scratched on the back.

6.2 ■ SHORT-FORM CONSTRUCTION INVOICE

CONTRACTOR'S NAME: _____

ADDRESS: _____

_____

PHONE: _____

FAX: _____

LIC #: _____

DATE: _____

OWNER'S NAME: _____

ADDRESS: _____

_____

PROJECT ADDRESS: _____

INVOICE

Dear _____,

This is an invoice for work completed per your request at the Project Address referred to above per our Construction Agreement dated _____, 19____.

A. CONTRACT PAYMENTS DUE NOW

1. _____: $_____
2. _____: $_____
3. _____: $_____
* Subtotal of Contract Payments Due Now: $_____

B. CHANGE ORDER PAYMENTS DUE NOW

1. _____: $_____
2. _____: $_____
3. _____: $_____
* Subtotal of Change Order Payments Due Now: $_____

C. TOTAL OF CHANGE ORDER & CONTRACT PAYMENTS DUE NOW: $_____

Thank you for giving us the opportunity to perform this work for you.

Very Truly Yours,

_____

CONTRACTOR
COMPANY NAME

6.2 ■ Short-Form Construction Invoice

ANNOTATED

Charlie Contractor Construction, Inc.
123 Hammer Lane
Anywhere, USA 33333
Phone: (123) 456-7890
Fax: (123) 456-7899
Lic#: 11111

DATE: **December 2, 2001**

OWNER'S NAME: **Oscar Owner**
ADDRESS: **444 Suburban St.**
Anywhere, USA 33333

PROJECT ADDRESS: **same**

INVOICE

Dear **Oscar**,

This is an invoice for work completed per your request at the Project Address referred to above per our Construction Agreement dated **October 10**, 20 **01**.

> ■ It's a good idea to refer to the Construction Agreement which is the basis for the invoice. Do *not* invoice for work that has not been completed.

A. CONTRACT PAYMENTS DUE NOW

1. **Final Contract Payment Due Now** : $ **2,300.00**

* Subtotal of Contract Payments Due Now: $ **2,300.00**

> ■ It will simplify accounting tremendously if you always invoice for contract payments (per your contract payment schedule in your construction agreement) in one section of the invoice and change order payments in the following section of the invoice.

B. CHANGE ORDER PAYMENTS DUE NOW

1. <u>**Change Order #1 Due Now**</u>   : $ <u>**345.00**</u>

* Subtotal of Change Order Payments Due Now: $ <u>**345.00**</u>

C. TOTAL OF CHANGE ORDER & CONTRACT PAYMENTS DUE NOW: $ <u>**2,645.00**</u>

Thank you for giving Charlie Contractor Construction, Inc. the opportunity to perform this work for you.

Very Truly Yours,

CHARLIE CONTRACTOR, PRESIDENT

CHARLIE CONTRACTOR, PRESIDENT
CHARLIE CONTRACTOR CONSTRUCTION, INC.

CHAPTER 7

MARKETING AND CUSTOMER RELATIONS

Contractors use a wide range of methods to market their companies and attract new business. Some contractors make fancy brochures, highly visible job-site signs and quarterly newsletters, and display photos of their work at local home shows. Others play golf and join all the "animal" clubs (e.g., Elks, Lions, etc.) in an effort to get the word out that they are civic minded *and* available to service all your construction needs. Some depend on Yellow Pages and newspaper advertising that costs thousands of dollars each year. Still others make it a point never to advertise — letting you know that they work only with referral customers (all contractors should be so lucky!).

The purpose of this chapter is not to survey these and other marketing methods — plenty of other books do that well. Rather, this chapter briefly looks at a few strategies for good communication which can have a positive impact on a contractor's legal and personal relationships with owners.

Whenever you conduct business in a positive way with current or prospective customers, you are doing a good job of marketing your business because you are building a referral base of satisfied customers who will recommend you to others.

If, on the other hand, your style of doing business leaves owners upset and has you always on guard about disputes or litigation, then your methods of operating and communicating need to be reviewed. Also, when there is constant tension between the owner and contractor, any advertising you do will be much less effective due to the negative comments being passed around town about your business and style of operating.

As they say in another business, "Your reputation precedes you."

Good Marketing Is Good Communication

Why focus on communication? First, because marketing is a form of communication and most contractors are "marketing" more by how they operate than through the dollars they spend on advertising. In this sense, a contractor can often improve his marketing without spending a penny on traditional advertising.

Second, many, if not most, legal disputes I've seen could have been avoided by better communication both in the contract and between the owner and contractor in other interactions. Remember, the dynamic of the construction process is new to the owner. The contractor is the seasoned professional in this area and should therefore be the one to set the tone and style of communication.

In my experience, good communicators in this business have a better shot at being successful and are less likely to end up with unhappy customers and legal disputes. Whether they know it or not, good communicators are doing good marketing, and operating in a manner that helps prevent legal problems.

Value Vs. Price

Owners who aren't just looking for the lowest bid *are* looking for something called value. Your style of conducting business and your ability to communicate, both in person and in writing, may convince this type of owner that, while your price may be a bit higher, you will deliver more value for the money spent and therefore should get the job.

Higher-end customers expect higher-end, professional service from their contractor. Many of these owners are in the position to be able to hire you because they have been successful in their own businesses; they know the difference between a well-run business and a seat-of-the-pants operation.

Therefore, your business style needs to communicate value, competence, professionalism, and integrity at every stage of your dealings with the owner. If you succeed in doing this, you won't get every job you bid, but you'll be in the running for the types of jobs you'd rather be working on.

Returning Phone Calls

Almost too basic to mention, but too common to overlook, is that some contractors don't return phone calls promptly. Most owners find this frustrating. The owner reasons, "If the contractor doesn't have time to return my calls, how will he have time to perform my work? If he doesn't care enough to make the call, will he care enough to do a good job"?

Make yourself accessible, at least during specified business hours. With the advent of pagers and answering machines with remote access, there is no reason to be inaccessible to your current and future customers.

Prequalifying the Job

It doesn't make financial sense to bid every job that comes your way. In fact, you'll go broke if you spend all your time bidding. The alternative is to prequalify job prospects prior to deciding to invest your time in bidding the job. How do you determine which jobs are worth bidding and which are likely a waste of time for

your company? Consider the following factors:

- Is the type of work required a specialty of yours and an area where you have been competitive and profitable in the past? If you aren't fully experienced in the type of work the owner needs, you probably will be less competitive.

- How many contractors will bid on the job? If it's more than three, the odds are stacked against you.

- How much time will it take to prepare the bid? The more time required, the more important it is to determine in advance whether bidding is a good gamble or a bad investment.

- How complete are the plans and specifications? How serious does the owner appear to be about proceeding in the near future? Is a permit available or have the plans been submitted? An owner who already has invested the time and money to obtain a complete set of plans and specifications is more likely to go through with his project in the near future. You can waste a lot of time with "tire kickers" who primarily want some free time from you to discuss projects that may never be built.

- Do the owners seem like people you can get along with personally? Is the architect someone you can work with comfortably? Every once in a while, personalities grate on each other. If you notice this tension at the start, bidding the job may be a mistake.

- How busy are you at the time the project needs to be bid and completed? If you are too busy to make a thorough and timely bid, and are too busy to perform the work with your existing crew when the owner is ready to start, you may want to pass on bidding the work.

- Which other contractors are bidding the project? Ask the owner. Get to know how your frequent competitors typically bid against you. If you typically bid higher than the two other bidders and you have plenty of work, you may decide not to invest the time required to bid this project.

If you decide the job is *not* worth bidding based on the considerations above, let the owner know how much you appreciated the opportunity to discuss the project and that you'd like to consider other jobs for him in the future. If you take the time to send a letter to this effect, the owner is apt to appreciate your efforts even though you aren't bidding the job.

Once you have decided to bid the job, there are many ways to sway the owner toward working with you. During the bidding phase, it is especially important to demonstrate to the owner that you are an experienced, conscientious contractor who is qualified to bid and perform the project.

Be careful, however, that your advertising, brochures, informational letters, and even conversations, don't make promises you can't keep, since you may be creating implied warranties. For instance, stating that your company is the best and most experienced company and that your roofs "never leak" or your concrete "never cracks" almost certainly will create false impressions and express warranties which you shouldn't make if there is even a remote possibility that you can't live up to them.

Being somewhat creative in your advertising is okay, but if you bluster too much about how great you are and how totally satisfied you will make the owner, you may be going beyond the bounds of ordinary advertising and creating unintentional express warranties for your company. If you think you may be on the border, consult an attorney. (See Chapter 10 for more on warranties.)

If your bid is too high, chances are you won't get the job. On the other hand, your bid on a job may be right in line with the others, in which case the contractor who appears most professional, experienced, and in tune with the owner's concerns may end up getting the job — even if he is not the low bidder. With this in mind, I convey my professionalism to prospective customers by providing the following information during the bidding stage:

General contractor reference list. Providing a list of previous satisfied customers to the owner either during or just after your initial contact will help convince him that you're both experienced and good at what you do. Provide a brief description of the type of work performed on each job. You may also want to highlight the projects that were similar to the project you are being asked to bid.

Subcontractor reference list. This list can be useful on larger projects if a lot of the work will be performed by subcontractors. Because the success of the project relies so much on the quality of the subs, it can be helpful to let the owner know that you have developed working relationships with respected subcontractors in the area, many of whom may be working on his project.

Evidence of contractor's insurance. It is also helpful to give the owner a copy of your insurance binder showing that you have adequate coverage. This binder typically includes worker's compensation insurance and comprehensive general liability insurance. However, during the bidding phase, do *not* make the owner an "additional insured" under your policy since there may be extra costs involved which you probably don't want to incur.

Providing the owner with proof of insurance may motivate him to ask the other bidders if they also carry this insurance. If one or more does not, you may find that your bid is viewed more favorably, and uninsured bidders may even be eliminated from the bid list.

Also, federal law requires all employers to carry worker's compensation insurance for their employees. Showing the owner that you carry this type of insurance demonstrates that you are aware of the labor laws applicable to contractors and that you will not put yourself, the owner (and his property), or your employees at risk in the event an employee is injured on the owner's job.

Contractor's license. If you give the owner a copy of your contractor's license there will be no doubt in his mind that he is working with a registered contractor.

Again, providing the owner with this information may motivate him to confirm that the other bidders are also licensed. You may even want to confirm that the owner is obtaining bids only from licensed contractors. (If the owner is soliciting bids from unlicensed contractors, it's a waste of time to bid the job since licensed contractors can rarely compete price-wise with unlicensed contractors.)

Photographs of prior projects. Another way to impress potential customers is to maintain and show a professional-looking portfolio of prior projects. Many potential customers are favorably impressed and even excited to spend a few minutes looking through pictures of projects you've done in the past.

After I have bid the job and have been selected for the work — or have a strong feeling that I will be — I provide the owner with one or more of the following forms to review before he signs the contract:

• Preconstruction Conference Form (Form 7.3)
• Ups and Downs of Remodeling: Customer Information Sheet (Form 7.4)
• Change Order Contingency Fund: Customer Information Sheet (Form 4.1)
• Sample Construction Schedule (below)
• Sample Bar Chart Construction Schedule (below)

These forms are not contractual. Rather they are strictly informational and meant to communicate the contractor's expectations about how the project will proceed. This is an ideal time to begin to temper any unrealistic expectations the owner may have about how the work will proceed — a critical factor in preventing future misunderstandings and disputes.

If you are working for a husband and wife, it is helpful to have a contract meeting (or preconstruction conference) with *both* of them to review these forms along with the unsigned contract. Usually it is a waste of time to meet with either the husband or the wife individually since most decisions are made jointly.

If, for some reason, there is no contract meeting or preconstruction conference, I send these forms to the owners when I send them their contract for signature.

Change Order Contingency Fund: Customer Information Sheet

It is important to educate owners in this early stage about change orders. The construction contract indicates that under normal circumstances change orders will be issued in writing and should be signed by the owner before change order work is started. But the topic merits further explanation in the preconstruction conference. I use a letter I call the "Change Order Contingency Fund: Customer Information Sheet" (Form 4.1) to describe how change orders arise and how to keep them to a minimum.

It's simply being honest to tell the owner that a few change orders are very common on most jobs and that it only makes sense to have a certain amount of money budgeted for them prior to beginning the project. I typically recommend that the owner budget between five and ten percent of the initial project cost for possible change orders. With more complicated projects, or projects that have the potential for many concealed conditions or have very incomplete finish specifications, this recommended percentage should be higher.

Ironically, many owners will appreciate your telling them they will probably need to budget more money for change orders. They also appreciate your telling them what they can do to reduce the number of potential change orders over the course of the project. See Chapter 4, Section 4.1 for a full discussion of change orders and the change order contingency fund.

Coming In On Time: The Construction Schedule

A lot of tension can arise if the owner expects that the job will take three months to complete but you know that four or five months is more reasonable.

Providing the owner with even the most basic construction schedule along with the contract is a good way to put him or her at ease while better organizing your own job performance. A simple written outline of the commencement and completion dates of most phases of the work, like the one that follows, can work well for most customers. Make sure to put in realistic and achievable time periods for work completion. Do *not* underestimate the time it takes to perform the work since this schedule will create definite expectations on the part of the owner. Ordinarily, it will also create a legal contract obligation for the contractor.

When you provide a schedule, you guard against the owner later claiming that you are significantly behind schedule when that is not the case. On the other hand, giving this schedule to the owner can work against you if you fall significantly behind the projected completion date. So be sure to document additional work days for completion with your change orders (a place for this is included in the sample change order form in Chapter 4, Section 4.2).

Of course, creating and using this schedule also helps you to better organize your subcontractors and the ordering of materials with long lead times. There are many good computerized scheduling programs available for contractors today. The first sample provided is very basic and requires no specialized software. The bar chart also does not require software, but is usually easier to generate on a computer.

■ **Sample Construction Schedule** ■

Charlie Contractor
123 Hammer Way
Anytown, USA 33333
(123) 456-7890
Lic#: 11111

TENTATIVE CONSTRUCTION SCHEDULE
PROJECT: __Harry Homeowner residence__

Note: This schedule only approximates the commencement and completion dates of various phases of the work. Some variation in these dates is common and will likely occur. The substantial and final completion dates are subject to change based on factors such as additional time required for change order work, inclement weather, and any other delays beyond the control of Contractor.

I. Phase I (Work outside existing kitchen):
  1. *April 21, 2001:* Commence work; demolition.

  2. *April 22 - 25:* Frame subfloor and under-floor plumbing and mechanical.

  3. *April 26 - May 3:* Framing of rear addition.

  4. *May 1 - 2:* Owners clear out rear pantry, clear all items off countertops in entire kitchen in preparation for demolition of north wall of house.

II. Phase II (Work inside existing kitchen):
  1. *May 3 - 7:* Complete framing of addition; install sheet metal on roof; run plumbing vents through roof; inspect roof plywood; install roof; install skylights; demolish much of north wall of house; install headers; frame some new walls in kitchen area *(no penetrations will be made into dining room area during this week)*. Remove doors and windows from north side of house. Take to door company for rehanging for reuse in new walls. Begin electrical work toward end of this week. *Note: Plywood will be placed over door and window openings on north wall after doors and windows have been removed. Also, work done on May 3 and May 4 will result in extremely dusty conditions inside existing kitchen.*

  2. *May 10 - 14:* Paper exterior plywood sheathing; install exterior doors and windows; complete rough electrical work; inspect all rough framing; complete interior rough mechanical and plumbing. Cut openings in wall into dining room. *Owner to confirm design and layout of custom painted tiles with custom tile painter.*

(continued next page)

3. *May 17 - 21:* Inspect all rough plumbing, mechanical, electrical; insulate; inspect insulation; hang drywall; complete siding work. Begin exterior painting. Drywall nailing inspection.

4. *May 24 - 28:* Begin finishing of drywall during this week. Complete exterior paint work.

5. *May 31 - June 4:* Complete finishing of drywall; begin installation of interior doors, casings, and moldings. Install hardwood floors. Begin interior paint. *Owner to select door hardware and order.*

6. *June 7 - 11:* Continue interior paint; set all cabinetry; begin painting cabinets; install plywood tops on cabinets. *Owner to schedule appliance delivery for week of June 21 - June 25.*

7. *June 14 - 18:* Install all tile and granite work; finish paint work; begin sanding of floors.

8. *June 21 - 25:* Complete floor finishing; begin installation of finish electrical and finish plumbing. Install appliances. Weatherstrip doors. Install door hardware.

9. *June 28 - July 2:* Complete installation of finish plumbing and finish electrical; begin punchlist work.

10. *July 2, 2001:* **SUBSTANTIAL COMPLETION.**

11. *July 5 - 14:* Complete all punchlist work.

12. *July 14, 2001:* **FINAL COMPLETION.**

■ Sample Bar Chart Construction Schedule ■

Charlie Contractor
123 Hammer Way
Anytown, USA 33333
(123) 456-7890
Lic#: 11111

TENTATIVE CONSTRUCTION SCHEDULE
PROJECT: **Harry Homeowner residence**

Note: This schedule only approximates the commencement and completion dates of various phases of the work. Some variation in these dates is common and will likely occur. The substantial and final completion dates are subject to change based on factors such as additional time required for change order work, inclement weather, and any other delays beyond the control of Contractor.

Work Description	May				June				July				August				September				October			
1. Demolition:	>	>																						
2. Excavation:		>	>																					
3. Foundation:			>	>																				
4. Framing:					>	>	>	>																
5. Gutters & Roofing:									>	>														
6. Rough Plumbing:									>	>	>													
7. Rough Electrical:											>	>												
8. Rough Mechanical:											>	>												
9. Insulation:												>	>											
10. Drywall:													>	>	>									
11. Finish Carpentry:															>	>								
12. Paint:																	>	>						
13. Countertops:																		>	>					
14. Finish Plumbing & Electrical Fixtures:																				>				
15. Finish Flooring:																					>	>		
16. Punchlist:																						>	>	

You may want to use a preconstruction conference form like the following one on your medium-sized and larger projects. I typically present the form to the owners at the meeting where they will be signing the contract — the preconstruction conference — and have them review and sign this form at the same time.

The purpose of this form is to assure that important aspects of the project and the accompanying paperwork have been reviewed by the owner and the contractor prior to the signing of the construction agreement. In addition, I usually add clauses that establish some ground rules for job-site procedures and logistical details, such as safety on the site, work hours, use of a job-site sign, and access to a phone and a bathroom. Some of these clauses may seem nit-picking, but they cover issues that come up on almost every job site and which, if not addressed *before* work begins, can lead to tension and misunderstandings between the contractor and the owner and his family, especially when doing residential remodeling and repair work.

You can modify this form — based on your own experiences — by simply deleting the clauses that don't apply on your individual jobs. You can also add new clauses to address specific situations not included here.

7.3 ■ PRECONSTRUCTION CONFERENCE FORM

CONTRACTOR'S NAME: _____

ADDRESS: _____

_____

PHONE: _____

FAX: _____

LIC #: _____

DATE: _____

OWNER'S NAME: _____

ADDRESS: _____

_____

PROJECT ADDRESS: _____

_____

I. CONSTRUCTION AGREEMENT AND NOTICES

It is acknowledged that Contractor has furnished Owner with the following documents and requests that the Owner carefully review all these documents prior to entering into an Agreement with Contractor:

A. Detailed Construction Agreement.

B. Other: _____

_____

II. JOB-SITE PROCEDURES

No matter how much the Owner and the Contractor prepare for it, remodeling is disruptive in certain ways. However, to help us minimize this disruption, please read our company policies on each item and list the information asked for in the spaces provided. Thank you for helping us carefully consider many of the logistical details of the job site prior to the commencement of our work.

A. CHILDREN AND PET SAFETY

Please keep children and pets away from work areas at all times, including construction trenches and the debris pile.

We do _____ do not _____ have a dog or cat that must be kept in a specific location. Instructions concerning dog or cat:

_____

_____

B. OWNER AND VISITOR SAFETY

Please do not enter work area without proper protective footwear. Due to safety concerns, do not bring friends or neighbors onto the site until the project has been completed unless you are willing to be entirely responsible for their safety. Steps may have been removed, trenches may be open, and other temporary conditions may present safety hazards to the Owner, Owner's family, and any guests.

C. SANITATION AND PHONE USE

It is understood that unless portable sanitation and a job-site phone have been specifically included in the bid amount, Contractor's employees will use the Owner's bathroom and telephone. Any long-distance calls made by Contractor (which are not job-related) will be promptly paid by Contractor upon submission of phone bill to Contractor.

Contractor's employees and Subcontractors may use the bathroom located:

_____

Contractor's employees and Subcontractors may use the telephone located:

_____

D. SECURITY

If a key is to be hidden on site, Owner wants key to be hidden in the following location:

_____

Key is to be issued to Contractor to keep during the project: _____ Yes _____ No

E. NORMAL WORK HOURS

We normally work during the winter from _____ A.M. to _____ P.M. During the summer we normally work from _____ A.M. to _____ P.M. We rarely have crews working on weekends. We want our crews to rest on the weekends and be fresh for a productive work week on Monday morning. The hours above may vary somewhat depending on the need for materials pickups, scheduling of Subcontractors, etc. If you have any special requirements, please let us know by stating them below.

How early may we start work? _____ A.M.

By what hour should we be gone? _____ P.M.

F. PARKING

Parking arrangements for Contractor's, employees', and Subcontractor's vehicles are as follows:

_____

_____

G. LOCATION OF CONTRACTOR'S TOOLS AND EQUIPMENT

The best on-site location for the Contractor's tools and equipment is: _____

_____

H. CONTRACTOR'S SIGN
It is understood that Contractor may place a sign at the front of the property for advertising and to help delivery trucks, Subcontractors, and others to easily locate the property.

I. DUST IN THE WORK AREA/DEBRIS PILE OR DUMPSTER
1. Remodeling will be dusty at times. Please cover any and all items in your house (e.g., computer equipment, stereo and TV equipment, musical instruments, bookshelves, China cabinets, food, etc.) that are dust-sensitive prior to the commencement of any work by Contractor. Be sure to have the work area cleared of all personal property prior to commencement of work by Contractor.
2. The best place for a debris pile or dumpster is: _____
_____

J. INSPECTIONS
Please inspect our work on a regular basis and let the Contractor know about any changes you would like as early in each phase of the work as possible.

K. OWNER-SELECTED MATERIALS
All Owner selections (which are not already specified on plans) regarding finish plumbing and electrical fixtures, cabinetry, medicine chests, towel bars, floor and wall coverings, tile and countertop materials, paint and stain colors, appliances, drywall and stucco texture, door hardware, etc., should be given to Contractor within _____ weeks of commencement of work by Contractor in order to avoid schedule delays.

L. MISCELLANEOUS INFORMATION AND INSTRUCTIONS FOR CONTRACTOR
_____
_____
_____

Information in this Preconstruction Conference Form has been reviewed, acknowledged, and agreed to by Owner.

Date: _____   _____
                              OWNER'S SIGNATURE

Ups and Downs of Remodeling

A simple information letter that describes the common high and low points that occur during the course of a residential remodeling project can help owners better prepare for and cope with what lies ahead.

By informing the owner about these things in advance, it's less of a surprise when the framing goes fast and is exciting, but the rough plumbing, electrical, mechanical, drywall, painting, finish woodworking, flooring, and countertop stages seem endless.

This letter also tells the owner that his normal lifestyle will be disrupted to some degree as a result of disorder, dust, temporary loss of the driveway or utilities, and other inconveniences created by every remodeling project. Realistically preparing the owner up front for these typical problems may help later when tensions rise over inconveniences that invariably occur despite your best efforts.

7.4 ■ UPS AND DOWNS OF REMODELING: CUSTOMER INFORMATION SHEET

CONTRACTOR'S NAME: _____

COMPANY NAME: _____

ADDRESS: _____

_____

PHONE: _____

FAX: _____

LIC #: _____

DATE: _____

Dear _____ :

Remodeling projects seem to have some common high and low points for many owners. If you are a seasoned remodeling veteran, you may already know what to expect and may not need to read this letter. If you have *not* lived through a remodeling project before, however, you are probably approaching your project with some degree of both anxiety and excitement. The purpose of this letter is not to cast a shadow of doubt on your project before it starts, but rather to simply point out some of these high and low points that naturally occur in the course of most remodeling projects. This way, you will a have a realistic idea of what to expect.

First, *every remodeling job creates some degree of disorder, dust, uncertainty, and inconvenience.* People will be tearing apart your house and putting it back together again — many of them people you have never met. We understand that this can be an unsettling experience and, accordingly, we expect our crews and subcontractors to respect the fact that this is your *home* they are working in and not a vacant warehouse. We want to know immediately if you have any problems in this area so that we can take immediate steps to correct them.

Second, *remodeling proceeds in stages.* One of the most difficult stages is working through the plan and permit process. Hopefully, by the time you receive this letter, you are finished with this process and are well into the stage of having your ideas turned into working construction drawings.

Another difficult early stage is the demolition phase. You will see new faces in and around your home on a regular basis. The insides of your house will be exposed. Electrical, heating, or plumbing services may be intermittently interrupted during this time. Dust, dirt, debris piles, and dumpsters will be visible in and around the work area. All of this can be rather stressful.

However, don't worry — the demolition phase goes quickly, and will be cleaned up just as rapidly. Once the framing nears completion, people usually start to see the light at the end of the tunnel.

After the framing is completed and the plumbing, electrical, and mechanical work are under way, the project can appear to slow down because the progress is not as dramatic and visible as in the framing phase. However, a lot of detail work is done at this time. For instance, there are many required inspections by building officials. It's critical that the work be done thoroughly at this point prior to insulating and closing the walls with drywall.

Next comes the drywall stage which most people are excited about. When the walls are covered with drywall, suddenly the rooms take on their true proportions and people start to imagine what it will be like to move back in.

Unfortunately, the final phase of all the work, after the drywall, can seem to take a long time. The finish work — grading and exterior concrete flat work; interior and exterior painting; installation of all interior doors and finish woodwork; installation of cabinets, tile, and floor coverings; installation of finish plumbing and electrical fixtures; installation of shelving, closet poles, mirrors, glass shower doors, hardware, appliances, etc. — requires a fair amount of time and the efforts of many different subcontractors.

Nevertheless, thanks to a well-planned and coordinated scheduling effort during this phase, the day arrives when your project is completed. Finally, your house is once again your private residence, free of the constant construction activity that has transformed your ideas and plans into the new spaces that we hope you will enjoy.

Please let us know if you have any questions about this process or any suggestions about how we can minimize the disruption to your daily routine. Thank you.

Sincerely,

_____

CONTRACTOR'S NAME
COMPANY NAME

Once you have bid the job, been selected as the contractor, obtained a signed contract, and furnished the owner with all the legal notices required by state and federal law, as well as the informational forms you think appropriate, you've effectively laid the groundwork for a successful job — and the first nail hasn't even been driven.

If you've done a good job with all of the above, you'll find most jobs will go more smoothly because you've given the owner and yourself a good "road map" to follow. The owner has been educated and has effectively agreed to your style of conducting business. You've also tempered some of the owner's unrealistic expectations. Many disputes will be avoided because of the groundwork you have laid up to this point in time.

After the work begins, however, there are still numerous steps you should take — in addition to performing the work on time and in a workmanlike manner — to maintain good customer relations.

Some of the following ideas are intended to improve communications and avoid misunderstandings, and some of them are meant to clarify the changing legal obligations of both parties. Keeping up with customer relations as the job proceeds can get you more referrals, help avoid legal disputes, and improve your bottom line.

Take Photos Before Work Starts

Before starting work on a project, photograph any existing damage to surrounding parts of the structure or site that you could be held responsible for later (make sure you date the photos). Then review these areas with the owner. If you are nervous about the owner or want to be extremely cautious, document the existing damage in a quick letter to the owner.

Examples include damaged roadways, driveways, or sidewalks, cracked tiles, scratched floors, stained carpets, broken windows, sticking doors, etc. Remodeling is tough enough without being held responsible for damage that occurred before you arrived on the site. If you run into this problem, a dated picture is truly worth a thousand words — maybe even a thousand dollars.

Make and Return Phone Calls

Keep in mind that the owner's expectations change as the job progresses. If you and your crew are going to be away from the job for a half day or more — for any reason — call the owner and let him know. Many owners don't understand why you aren't working during downtime and assume the worst. A simple phone call can alleviate their concerns and help maintain the professional image you have fostered up to this point. And, as mentioned earlier, return phone calls promptly.

Regular Meetings and Contact With Owner

Meet with the owner on a regular basis and communicate often about the job. Regular punch list walk-throughs with the owner are an effective way to communicate your concern over quality and can lead to a much smaller punch list at the end of the job.

Refer to the Contract

If you sense a dispute arising, you can "de-personalize" the situation by referring to your contract to see if the issue in question is covered by a section of the contract (many potential disputes are covered by various clauses in the contracts in this book). If it is, you have brought the disagreement back to the business level by simply referring back to the contract for instructions on how the owner and contractor are to proceed.

If you are uncertain about the possible legal ramifications of what you want to do, place a quick call to your construction attorney to get a fresh perspective and some advice.

An Orderly and Courteous Job Site

Post a job-site sign and keep the job site picked up. Make sure your subcontractors and employees clean up after breaks and lunch, and after performing their work. Construction is stressful enough for owners as it is; if the site is disorganized and debris-filled every time they come by, their confidence in you is bound to go way down. If the owner is living on the job site during the time of construction, these concerns are especially worth your attention.

Insist that your employees and subs show common courtesies to the owner and his family and are respectful of the owner's property. For instance, keep the volume of any job-site music at a respectable level and warn your employees against using offensive or inappropriate language. Demand courteous driving habits around the site. Inform your employees and your subs of any policies your customers have about inside smoking.

These items may seem petty, but to some owners they make the difference between a contractor they will want to rehire and one they never want to see again. A little common sense and common courtesy in these areas can go a long way toward avoiding petty disputes. And the last thing you need is a conflict between the owner and one of your crew or subs.

Put Important Communications In Writing

Keep up with the following written communications to the owner and architect during the course of the project. The importance of putting certain communications in writing can't be overstated.

Change orders. Obtain all change orders in writing. If you can't get the change order in writing prior to beginning the work, get a verbal approval and make a note of this conversation. Then, quickly follow up with the written change order. If the owner delays returning the signed change order after having verbally approved it, send the owner a letter detailing how you were verbally instructed to perform the change order work without a written change order and that now you must have a signed copy for your files.

Changes in specs or finishes. If a specification is changed by the owner *which has no impact on the cost of construction* (e.g., a change in type of cabinets or paint called out in plans) send a letter to the owner or issue a change order with no change in the contract price, detailing this change. Disputes arise sometimes over changes made by the owner in the specifications or finishes. Avoid this confusion and these disputes by putting deviations from specifications and finishes in writing — even if they do not affect the contract price.

Architect and engineer directives or approvals. If the architect or engineer verbally tells you that a particular deviation from the plans or specifications is acceptable, make a note of that conversation and send a quick memo confirming your understanding of the situation to the architect or engineer and send a copy to the owner.

If you have misunderstood something, or the architect or engineer later has a memory lapse, you'll be glad you documented the conversation prior to implementing the change in the scope of work. You'll also be happy you sent the owner a copy so that he can object to the change if he needs to.

Potentially faulty design. If the plans, specifications, or instructions from the owner call for you to build something which you think is substandard in some respect, send the owner and architect a letter indicating that you will follow the plans and specifications if instructed to do so, but noting that you recommend the owner *not* employ the specified details due to a likelihood of premature failure.

Then, after sending this letter, if the owner still refuses to change the specification, it is unlikely you will be found negligent for constructing or failing to warn the owner of a bad design. If the owner heeds your warning, *be sure the architect does the redesign. Don't do it yourself or you may well be liable for any failures if the newly specified work fails in any way!*

Warn the owner of details that don't make sense, but be aware that if you deviate from the plans without written approval you may well be found liable for any failure that results from your deviation.

Finally, if you are certain that a detail violates a section of your local code, bring this to the attention of the owner and architect and insist that the owner have the architect provide a new detail that conforms to the local governing code.

Changes Requiring Building Department Approval

If the owner requests a change after work has begun which you think might require additional building department approval (and new architectural drawings) or engineering, talk to the inspector and your architect or engineer prior to giving the owner a change order or telling him the requested change is "no problem."

Changes in Plans Called For by Plan Checkers

Pay close attention to any changes made to the plans by the plan checkers at the building department. Be sure that you pick up these changes by carefully reviewing the Job Copy of the plans issued by the building department. Charge for the changes if they impact the price. If you overlook these changes and the building inspector forgets to bring them to your attention, you may be held responsible for the higher cost of retroactively installing these changes out of their normal sequence.

Punch List Work

Don't take forever to complete the punch list. Delaying the punch list work will only aggravate any tensions that have developed between you and the owner over the course of the job. It's also at this time that the owner is showing off his new project to friends and being asked, "How was the contractor"? Even though it can be difficult, putting in some extra effort at this point will always pay off!

Owner's Possessions

First, don't borrow the owner's tools! This inevitably leads to problems. Remember, Murphy's First Law of Construction states: "Any tool that breaks on the job will always be the owner's tool in the hands of your laborer."

You may want to mark your common tools with a band of bright paint so they're less easily confused with the owner's tools on the job site.

Second, don't agree to move or store the owner's possessions! I always make this an excluded item in the contract. Similar to the owner's tools, Murphy's Second Law of Construction states: "Any piece of furniture that can break, will break in the hands of your laborers."

If the owner's property *must* be moved, don't do it — delay starting the job if necessary until the owner moves his own property.

If you absolutely *can't* delay the job, write up a change order for moving the property and put in a clause about not being responsible for any damage that occurs. Have the owner sign the change order, thereby releasing you from liability for damage done to his property while it is being moved.

Third, insist on a work area that is free of the owner's furniture and possessions. You can't properly perform your work if the owner's belongings are constantly in the way. Also, we all know what condition the owner's exposed property will be in after six months of remodeling: not a pretty sight. Better to insist that the owner find temporary storage in an area that is not in your way and not subject to the dust and debris of the construction area. You set the ground rules. In many areas, the owner can rent a lockable steel storage container for $100 to $200 per month.

Job Safety, Security, and Utilities

Monitor the safety and security of the site on a daily basis in order to prevent accidents and show that you are conscientious about the inherent risks involved in construction. Keep a written log of your safety meetings. Make sure doors and windows are locked when you leave. For remodels, make sure utility disconnections are kept to an absolute minimum and notice is given to the owner prior to shut-off.

The Warranty Phase

If all has gone well up to this point, the project has been completed well and on time, the contractor has been paid in full for all contract work and all change orders, the punch list is complete, the subcontractors and materials suppliers are all paid, the property is free of liens, and the owner and contractor are appreciative of what each has done for the other.

The owner who is satisfied at this point is one of your best sources of future work and referrals. Your good work and professional handling of the paperwork and communication aspects of the job will have paid off in numerous ways and hasn't added much, if anything, to your job costs or overhead expenses.

Customer-Satisfaction Questionnaires

Some contractors send a standard customer-satisfaction questionnaire to the owner around the time the punch list work is completed. By evaluating the responses on this questionnaire, they are able to improve their customer relations and marketing policies.

Warranty and Maintenance Information

Another good practice used by conscientious contractors at this phase of the project is to give the owner an information packet with the appliance and fixture warranty information and a description of common areas of owner maintenance.

By informing the owner that wood shrinks, grout can crack at the tub to wall connection, filters need replacing, and gutters and drains need cleaning, etc., the owner may be less surprised and upset when minor maintenance items like this occur. Some contractors make it a point to leave touch-up paints and grout so the owners can perform this minor maintenance work themselves.

Job Photo Album

Finally, owners often appreciate receiving ten or twenty photographs of the various stages of their project. Because I photograph jobs for myself, it's easy to have duplicate prints made and then present this small album of photographs when the work is complete. Giving this photo album to the owner along with the information on maintenance and product warranties makes a nice closing presentation for your project.

Following Up on Warranty Complaints

Even though the contractor has been paid in full at this point in time, he still has a legal obligation to promptly investigate and perhaps repair any work that fails or any defects that arise during the express warranty period and also during any implied warranty period that is required by law in your state.

In fact, some contractors make it a point to call the owner six months and twelve months after the project is complete to see if any warranty work has arisen and to answer any questions the owner may have. While some contractors would view this as knowingly and willfully sticking their head back in the lion's mouth, taking this approach is sure to leave a lasting positive impression in the mind of the owner which will pay off in the long run with increased referrals.

Consult your local construction attorney to determine how long you may be liable for both *latent* and *patent* defects under your state law. Depending upon the type of failure, the contractor's responsibility may well extend past the one-year express warranty period specified in the contract.

Consumer protection laws are on the rise in many states. Nonresponsiveness to a legitimate complaint can wind up costing you plenty in terms of loss of reputation and the dollars you may have to reimburse the owners — especially if they have hired another contractor to correct a defect for which you are legally responsible.

Better Communication: The Payoff

Marketing is communication and, whether they know it or not, most contractors are marketing more by how they operate day to day than through the dollars they spend on advertising. By implementing some of the ideas in this chapter, the contractor can improve his marketing and expand his customer base without spending a penny more on traditional forms of advertising.

In my experience, nine out of ten owners really appreciate the contractor's efforts to better communicate expectations and maintain positive customer relations. Also, a contractor's business will run much more efficiently when he adds the few steps required to communicate effectively with the owner before, during, and after the project.

Again, if the owner is less surprised about certain events because you have prepared him up front and tempered his expectations, your relationship will be much less strained. Consequently, the potential for disputes and litigation decreases, while the potential for a successful project (with both profits and peace of mind intact) will be significantly increased.

Not all of these forms will be appropriate to every builder or every job. Customize and selectively use the forms to suit your business.

Gauge your use of the forms in this chapter upon both the size of the job and your past history of working with the owner. Generally, the smaller the job, the fewer the informational forms you need to give the owner. The larger the job and the less you know the owner, the more important these forms can be.

Regardless of the size of the job, the importance of showing common courtesies and communicating well with the owner both verbally and in writing can't be overestimated.

CHAPTER 8

ESTIMATE AND BID PREPARATION

8.1 ▪ **S**UBCONTRACTOR **A**ND **S**UPPLIER **B**IDDING **C**HECKLIST

8.2 ▪ **E**STIMATING **W**ORKSHEET

8.3 ▪ **P**RELIMINARY **C**OST **E**STIMATE

*I*t is assumed that the contractor reading this chapter has good knowledge of the trades, knows the abilities of his crew, and knows the approximate labor hours required to perform the various tasks being estimated. He also has a good "ballpark" knowledge of what similar local contractors charge for similar tasks.

It is also assumed that the contractor has a pool of time-tested, regular subcontractors (hopefully two good ones in each subtrade) who are quality-minded, competitively priced, and eager to bid his projects. And it is likewise assumed that the contractor has a good knowledge of which materials suppliers provide the best prices and the best service. All of these skills, relationships, and characteristics are crucial to successful bidding.

The forms in this chapter can help improve your estimating, primarily by helping you to remember to include everything necessary in your bid. They can also help by reminding you to clarify with your subcontractors and suppliers numerous small items which can and often do fall through the cracks, and which usually leave the contractor to absorb the costs of poor communication.

This chapter contains a bidding checklist and an estimating worksheet. These forms are included for the benefit of those contractors who have not yet developed their own forms for regularly organizing their estimat-ing and bidding systems. They will be especially helpful for organizing bidding on medium-sized and larger projects.

There are many different systems for organizing bidding. Some contractors find that computerized bidding programs work to their advantage. On the other hand, many residential contractors still prefer to rely on their memory, stick-by-stick estimates, and a few pieces of scratch paper in assembling bids.

This latter approach, however, can lead to disputes and costly misunderstandings due to the delicate balance of interests and relationships between the contractor, owner, subs, suppliers, and architect. It is easy to make false assumptions about what is included in a quick quote from a subcontractor or supplier.

While your years of experience may help you remember most items required when you compile a bid, there's nothing wrong with assisting your memory with a couple of organized forms and notes.

Charging for Estimates

Should the contractor charge for an estimate? The answer depends upon each given situation. Generally, I think most contractors are usually not in a position to charge for estimates. When a contractor does charge for an estimate, he should have an agreement specifying *how much* he will charge. This type of agreement often contains a provision stipulating that the contractor will credit back the amount charged for the estimate if he is later awarded the job. This often seems fair.

In many areas, competition is just too fierce for contractors to even think of approaching an owner with the concept of charging for an estimate. If you can't charge for an estimate, you have to develop your abilities to "prequalify" job prospects so you don't waste your time bidding the wrong jobs. (See Chapter 7, Section 7.1 "The Bidding Phase.")

If you are faced with bidding a project that has involved recent legal trouble with a prior contractor, be extremely cautious. You may even want to charge for this type of bid. You could find yourself being subpoenaed to court or an arbitration hearing to testify about your bid. Accordingly, have an agreement stating whether or not you'll be paid for this time, and, if so, how much.

Bidding Work for Associations

Contractors also should be cautious and even more conscientious than usual when bidding work for homeowners' associations. A number of townhouse and condominium complexes were built in the 1980s which contained design or construction defects. Many of these defects recently started surfacing, leading homeowners associations' to seek out and make actionable claims against the original developers and contractors.

Unlike most individual homeowners, these associations often have the financial resources to carry through with investigation and litigation. Accordingly, if you work for homeowners' associations be aware that your work must be of the highest standards and your comprehensive general liability premiums should be paid up. You may also want to verify with your insurance carrier that your liability policy definitely covers work performed for condominium associations.

Subcontractor and Supplier Bidding Checklist

The Subcontractor and Supplier Bidding Checklist provides a simple way to organize and compare prices from subs and suppliers. It also serves as a phone log indicating when various subs were called and who has quoted you prices. Most categories on the form also contain notes that remind you about some common ambiguities with specific types of subcontractors and suppliers. These are important to address and clarify before you rely on any prices quoted. Again, fine-tune this form to suit *your* needs and experiences.

Insert the names and phone numbers of the subs you normally work with under the alphabetically listed categories of subcontractors and materials suppliers. Be sure to add any other categories that you frequently use.

After you have determined which subcontractor you want to work with, you can fill in the scope of work section and send him an unsigned subcontract for the work. This way, if you are reasonably sure of getting the job, and uncertain whether a sub will hold his price, you can get a written commitment from the sub prior to signing a prime contract with the owner. (Remember to make sure your subcontractors understand your insurance requirements for the job prior to bidding the job for you. See Chapter 5, Section 5.3 on the importance of verifying subcontractor insurance.)

8.1 ▪ Subcontractor and Supplier Bidding Checklist

JOB NAME: _____

ADDRESS: _____

_____

PHONE: _____

ARCHITECT'S NAME: _____

PHONE: _____

BID DEADLINE: _____

SUBCONTRACTOR/ SUPPLIER	PHONE NUMBER	DATES CALLED	AMOUNT

* Appliance Suppliers:

| _____ | _____ | _____ | _____ |
| _____ | _____ | _____ | _____ |

Notes: _____

_____

_____

(Note whether installation is included in price above. Have all finishes been specified and properly bid? Has the correct electrical, gas, and mechanical work been specified for the new appliances? Does the quoted price include tax and delivery?)

* Asbestos Abatement Contractors:

| _____ | _____ | _____ | _____ |
| _____ | _____ | _____ | _____ |

Notes: _____

_____

_____

* Asphalt/Paving Contractors:

| _____ | _____ | _____ | _____ |
| _____ | _____ | _____ | _____ |

Notes: _____

_____

_____

(Verify whether price includes header boards, weed and seed eradication, base rock, excavation, soil export, base compaction, and seal coat. Are striping and curbs required?)

SUBCONTRACTOR/ SUPPLIER	PHONE NUMBER	DATES CALLED	AMOUNT

* Building Supplies (including framing lumber package):

Notes: _____

_____

(Verify how long price will be held by supplier. Has sales tax been included?)

* Cabinets:

Notes: _____

_____

(Note whether installation, cabinet finishing, and hardware have been included in price above. Does installer carry worker's compensation insurance covering field installations?)

* Carpets/Vinyl/Hardwood Floors/Wallpaper/Window Coverings:

Notes: _____

_____

(Note whether demolition and removal of existing floor coverings, floor patching, preparatory work, pad, under-layment, and installation is included in price above. Will doors have to be cut down? Touch-up painting required after new floor covering installed? Does the quoted price include tax and any applicable freight charges?)

* Concrete Contractors:

Notes: _____

_____

(Verify whether backfill and finish grading have been included.)

* Concrete Demolition/Concrete Cutting Contractors:

Notes: _____

_____

(Is removal of concrete included in price above?)

SUBCONTRACTOR/ SUPPLIER	PHONE NUMBER	DATES CALLED	AMOUNT

* Concrete Pumpers:

Notes: _____

* Concrete Suppliers:

Notes: _____

(Does the quoted price include all sales tax? Any "short-load" extra charges?)

* Doors and Millwork Suppliers:

Notes: _____

(Is all baseboard, door, and window casing, specialty trim, closet poles, shelves, door hardware, etc. accounted for? Does the quoted price include all sales tax? Identify stain-grade vs. paint-grade vs. pre-primed materials.)

* Garage Door Contractors:

Notes: _____

(Verify whether installation, number of door operators, and sales tax are included in quoted price. If paint-grade, are doors primed?)

* Shower Door/Mirror Contractors:

Notes: _____

(Verify finish of metal and thickness of glass. Sales tax included?)

SUBCONTRACTOR/ SUPPLIER	PHONE NUMBER	DATES CALLED	AMOUNT
* Drywall Contractors:			
_____	_____	_____	_____
_____	_____	_____	_____
Notes: _____			

(Verify texture. Will subcontractor remove overspray texture from areas not to be textured?)

* Electrical Contractors:

_____	_____	_____	_____
_____	_____	_____	_____

Notes: _____

(Have all fixtures been included? Cost of setting all finish fixtures included in price above? Verify that cost of hooking up any temporary power has been included. Does electrician (or HVAC sub) supply exhaust fans?)

* Excavation/Demolition Contractors:

_____	_____	_____	_____
_____	_____	_____	_____

Notes: _____

(Have hauling dirt spoils/debris and dump fees been included?)

* Fire Sprinkler Contractors:

_____	_____	_____	_____
_____	_____	_____	_____

Notes: _____

(Verify that work from riser to street has been bid. Verify the type of finish trim.)

* Formica/Corian Contractors:

_____	_____	_____	_____
_____	_____	_____	_____

Notes: _____

SUBCONTRACTOR/ SUPPLIER	PHONE NUMBER	DATES CALLED	AMOUNT

* Framing Contractors:

Notes: _____

(Verify that all backing/blocking/openings required for all other trades and for railings and accessories have been included.)

* Insulation Contractors:

Notes: _____

* Landscape Contractors:

Notes: _____

* Masonry and Glass Block Contractors:

Notes: _____

* Mechanical/Gutters/Sheetmetal (HVAC) Contractors:

Notes: _____

(Verify that all gas and electrical for new mechanical equipment has been bid. Who supplies cutting, blocking, and roof patching for mechanical ducting and roof penetrations? Does electrician (or HVAC sub?) supply exhaust fans? Who supplies condensate drains? Plumber or HVAC sub? Who supplies foundation and attic vents? Lumber supplier or HVAC sub?)

SUBCONTRACTOR/ SUPPLIER	PHONE NUMBER	DATES CALLED	AMOUNT

* Painting Contractors:

Notes: _____

(Is the work bid per a written specification? If not, have painter provide a written specification of the labor and materials he has bid. In order to not void warranty on exterior doors, verify whether exterior wood doors and wood windows need to be primed immediately upon installation and before the rest of the exterior paint work is performed.)

* Plumbing Contractors:

Notes: _____

(Verify whether all fixtures have been included. Cost of setting all finish fixtures included in price above?)

* Rental Equipment Suppliers:

Notes: _____

* Roofing Contractors:

Notes: _____

(Verify if roofer supplies and installs sheetmetal on roof, roof jacks, and counter flashing. If specialty type of roof, should exterior be painted prior to installation of roofing materials?)

* Sanitation:

Notes: _____

SUBCONTRACTOR/ SUPPLIER	PHONE NUMBER	DATES CALLED	AMOUNT

* Scaffolding:

Notes: _____

* Stairs & Handrails (specialty):

Notes: _____

(Note whether installation/finishing/priming is included in price above.)

* Structural Steel Supplier:

Notes: _____

(Note whether installation is included in price above. Is steel primed? Special equipment required to install steel or unload steel?)

* Stucco Contractors:

Notes: _____

(Note whether this is a 2-coat or 3-coat stucco job. Is scaffolding required? If so, who supplies it?)

* Tile/Granite Contractors:

Notes: _____

* Truss Suppliers:

Notes: _____

(Note whether crane is required for loading trusses. Included?)

SUBCONTRACTOR/ SUPPLIER	PHONE NUMBER	DATES CALLED	AMOUNT

* Wallpaper Contractors:

Notes: _____

_____

(Note whether all areas to receive wallpaper are properly prepared to receive it. If not, who does prep work?)

* Window and Skylight Suppliers:

Notes: _____

_____

(Note whether specialty interior and exterior trim for uniquely shaped windows is included. Note whether jamb extensions are included for wood windows. Note lead time for windows. Does the quoted price include all sales tax? Verify finish of windows.)

* Woodstove & Spa Suppliers:

Notes: _____

_____

(Note whether installation is included in price above. Has gas or electrical been bid for these items? Does the quoted price include all sales tax? Is all work required for outside combustion air included?)

* Miscellaneous:

Notes: _____

_____

_____

_____

(Note whether installation and sales tax is included in prices above.)

The Estimating Worksheet helps the contractor to remember all kinds of job costs when he runs down the list verifying that he has included everything in his bid (many jobs have some unique costs which should be written in the miscellaneous categories).

Most items on the form are listed in the general sequence in which they come up on the job (e.g., permits, foundation, framing, finish, etc.). Remember, this is only a *sample* form. Add any additional categories of labor, materials, equipment, and miscellaneous costs that are frequently used on your jobs.

When you are ready to fill in your Estimating Worksheet to arrive at the bid price, you can simply transfer the information from the Subcontractor and Supplier Bidding Checklist, add in your own labor, miscellaneous costs, profit and overhead, and your bid will be complete.

You can have subcontractors verify their prices on one of your Subcontract Agreements (which is signed by the subcontractor but remains *unsigned* by you until you have a signed contract from the owner). Most suppliers will provide a short statement verifying a material price or lumber package price for a certain period of time. Be sure that you can rely on the prices from your subs and suppliers by requesting this type of statement.

At the end of the Estimating Worksheet, you'll see the ever-important line items for overhead and profit. Entire books are written on estimating and how to properly determine profit and overhead rates. If you are reading this now, you probably already have a good idea of how to estimate projects and what profit and overhead rates you should charge. If you have trouble with estimating, talk to your friends in the business and find out how they estimate. Also, buy a good book on estimating and study it. You may even want to consult an accountant who is familiar with construction.

Quite often, profit and overhead rates will be roughly determined by the size of the job and what the market will bear in your local area. The important thing to realize is that every business has overhead (e.g., trucks, tool acquisition and maintenance, insurance, professional services, office expenses, and perhaps rent, utilities, advertising, etc.).

These expenses are *not* direct job costs and somehow must be factored into your bidding so that they are paid for out of funds other than your own wages.

In addition, every business that is to survive both needs and deserves to make a fair profit. By accurately determining your actual direct job costs in the estimating phase and then adding a percentage for profit and overhead, you will help to ensure that your anticipated profits aren't mysteriously absorbed by unaccounted-for overhead expenses. The combined markup of direct costs for profit and overhead is typically between 10% and 30%, depending upon the size of the job, your gross annual volume and annual expenses, and the average rate in your area.

Scope of Work

An important note about estimating is worth mentioning here. One of the most critical parts of estimating (that can be overlooked in the rush of putting a bid together) is clarifying the scope of work in the contract documents (typically the plans, written invitation to bid, and any related specification or engineering reports).

You don't want to bid items that aren't part of the scope of work or required to perform the work. However, if you don't clarify confusing or unknown details during the bidding phase, you may find yourself arguing later about not paying for extra work that the owner "naturally" assumed you had included in your bid.

If the scope of work is significantly altered by the owner or architect in a phone conversation, you must specify this change clearly in your contract with the owner. If the bid deadline is verbally changed by any of the parties, it's a good idea to specify the new deadline in a memo to the owner and architect.

On a typical remodel bid there are usually 10 to 20 questions about items such as: details that are confusing, details you think may have been omitted, clarifications about what the owner will be supplying, missing structural connection details, missing elevations for interior cabinetry, which items should be allowance items, acceptable alternate materials and methods, etc.

Therefore, the first page of the Estimating Worksheet is titled "Questions and Clarifications for Owner and Architect." This part of the Estimating Worksheet will go a long way toward helping nail down any questions you may have about the scope of work.

By clarifying the scope of work as much as possible during the bidding phase — i.e., asking the owner and the architect lots of questions and incorporating those clarifications into your contract — you'll have fewer disputes during the course of the project.

8.2 ■ Estimating Worksheet

JOB NAME: _____

ADDRESS: _____

_____

PHONE: _____

ARCHITECT'S NAME: _____

PHONE: _____

OWNER'S NAME: _____

PHONE: _____

BID DEADLINE: _____

NUMBER OF BIDDERS ON THIS JOB: _____

I. QUESTIONS AND CLARIFICATIONS FOR OWNER AND ARCHITECT

1) Question to: ____Owner ____Architect *Date:_____

_____

_____

Response: _____

_____

2) Question to: ____Owner ____Architect *Date:_____

_____

_____

Response: _____

_____

3) Question to: ____Owner ____Architect *Date:_____

_____

Response: _____

_____

4) Question to: ____Owner ____Architect *Date:_____

_____

Response: _____

_____

5) Question to: ____Owner ____Architect *Date:_____

_____

Response: _____

_____

II. ESTIMATING FORM

A. OUTSIDE & GOVERNMENT FEES, RENTAL EQUIPMENT, AND UTILITIES:

ITEM	SUPPLIER/SUB	ALLOWANCE	COST
1. SURVEYOR:	_____	$ _____	$ _____
2. ARCHITECT:	_____	_____	_____
3. ENGINEER:	_____	_____	_____
4. PERMIT FEES:	_____	_____	_____
5. HOOK-UP FEES: (Water, Sewer, Electric, Phone, Cable)	_____	_____	_____
6. TEMP. ELECTRIC/GAS:	_____	_____	_____
7. TEMPORARY WATER:	_____	_____	_____
8. PORTA-POTTY:	_____	_____	_____
9. JOB-SITE PHONE:	_____	_____	_____
10. TEMP. FENCING:	_____	_____	_____
11. SCAFFOLDING:	_____	_____	_____
12. MISC. RENTAL EQUIP.:	_____	_____	_____
13. MISCELLANEOUS:	_____	_____	_____
14. MISCELLANEOUS:	_____	_____	_____
15. MISCELLANEOUS:	_____	_____	_____

B. DEMOLITION, DEBRIS REMOVAL, CLEANUP, FINAL CLEANING:

ITEM	SUPPLIER/SUB	ALLOWANCE	COST
1. ASBESTOS ABATEMENT:	_____	$ _____	$ _____
2. DEMOLITION LABOR:	_____	_____	_____

3. REGULAR JOB-SITE
   CLEANUP: _____ _____ _____

4. FINAL CONSTRUCTION
   CLEANING: _____ _____ _____

5. DUMPSTERS & DEBRIS
   HAULING: _____ _____ _____

6. CONCRETE
   CUTTING/REMOVAL: _____ _____ _____

7. DIRT/FILL REMOVAL: _____ _____ _____

8. DUMP FEES: _____ _____ _____

9. MISCELLANEOUS: _____ _____ _____

10. MISCELLANEOUS: _____ _____ _____

C. SITE WORK, SEPTIC, DRAINAGE, TRENCHING, ASPHALT:

ITEM	*SUPPLIER/SUB*	*ALLOWANCE*	*COST*
1. GRUB/CLEAN LOT; TREE REMOVAL:	_____	$ _____	$ _____
2. ROUGH GRADING:	_____	_____	_____
3. FINISH GRADING:	_____	_____	_____
4. UTILITY TRENCHING:	_____	_____	_____
5. FOUNDATION EXCAVATION:	_____	_____	_____
6. SEPTIC/SEWER EXCAVATION:	_____	_____	_____
7. SEPTIC SYSTEM:	_____	_____	_____
8. WELL:	_____	_____	_____
9. FRENCH DRAINS, SPLASH BLOCKS:	_____	_____	_____

10. RETAINING WALL
     EXCAVATION: _____   _____   _____

11. BACKFILL: _____   _____   _____

12. ASPHALT PAVING: _____   _____   _____

13. SIGNAGE: _____   _____   _____

14. LANDSCAPING: _____   _____   _____

15. SPRINKLER/IRRIGATION: _____   _____   _____

16. MISCELLANEOUS: _____   _____   _____

D. FOUNDATION, CONCRETE, FLATWORK, STEEL & MASONRY:

ITEM	SUPPLIER/SUB	ALLOWANCE	COST
1. FOUNDATION LABOR:	_____	$ _____	$ _____
2. FOUNDATION MISC. MATERIALS:	_____	_____	_____
3. CONCRETE MATERIALS: _____ YDS @ $_____	_____	_____	_____
4. CONCRETE WATERPROOFING:	_____	_____	_____
5. CONCRETE SIDEWALKS/PATIOS:	_____	_____	_____
6. CONCRETE FLOORS:	_____	_____	_____
7. CONCRETE DRIVEWAYS:	_____	_____	_____
8. CONCRETE RETAINING WALLS:	_____	_____	_____
9. REBAR & STRUCTURAL STEEL:	_____	_____	_____
10. METAL RAILINGS:	_____	_____	_____
11. MASONRY FIREPLACES:	_____	_____	_____

12. MASONRY BLOCK WALLS: _____ _____ _____

13. ORNAMENTAL MASONRY: _____ _____ _____

14. CONCRETE PUMPING: _____ _____ _____

15. MISCELLANEOUS: _____ _____ _____

16. MISCELLANEOUS: _____ _____ _____

E. FRAMING AND RELATED PHASES OF WORK:

ITEM	SUPPLIER/SUB	ALLOWANCE	COST
1. FRAMING MATERIALS:	_____	$ _____	$ _____
2. FRAMING LABOR:	_____	_____	_____
3. WOOD SIDING & EXTERIOR TRIM LABOR:	_____	_____	_____
4. TRUSSES:	_____	_____	_____
5. WINDOWS, SKYLIGHTS, GLASS DOORS:	_____	_____	_____
6. GARAGE DOORS:	_____	_____	_____
7. WOODSTOVES:	_____	_____	_____
8. HVAC, GUTTERS, SHEETMETAL:	_____	_____	_____
9. FIRE SPRINKLERS:	_____	_____	_____
10. ROUGH PLUMBING:	_____	_____	_____
11. PLUMBING FIXTURES:	_____	_____	_____
12. ROUGH ELECTRICAL:	_____	_____	_____
13. ELECTRICAL FIXTURES:	_____	_____	_____
14. INSULATION:	_____	_____	_____

15. DRYWALL: _____ _____ _____

16. ROOFING: _____ _____ _____

14. STUCCO: _____ _____ _____

15. MISCELLANEOUS: _____ _____ _____

16. MISCELLANEOUS: _____ _____ _____

F. FINISH PHASE WORK:

ITEM	SUPPLIER/SUB	ALLOWANCE	COST
1. WOOD DOORS, BASE, INTERIOR TRIM:	_____	$ _____	$ _____
2. FINISH CARPENTRY LABOR:	_____	_____	_____
3. DOOR HARDWARE & STOPS:	_____	_____	_____
4. GLASS BLOCK:	_____	_____	_____
5. INTERIOR PAINT:	_____	_____	_____
6. EXTERIOR PAINT:	_____	_____	_____
7. CABINETS:	_____	_____	_____
8. TILE:	_____	_____	_____
9. FORMICA:	_____	_____	_____
10. GRANITE:	_____	_____	_____
11. CUSTOM MILLWORK MATERIALS:	_____	_____	_____
12. WALLPAPER:	_____	_____	_____
13. TOWEL BARS, T.P. HOLDER, MEDICINE CHEST:	_____	_____	_____
14. MIRRORS, SHOWER DOORS:	_____	_____	_____

15. VINYL & UNDERLAYMENT: _____ _____ _____

16. CARPET & PAD: _____ _____ _____

17. HARDWOOD FLOORING: _____ _____ _____

18. CUSTOM WOOD RAILINGS: _____ _____ _____

19. APPLIANCES: _____ _____ _____

20. MISCELLANEOUS: _____ _____ _____

21. MISCELLANEOUS: _____ _____ _____

G. PROFIT, OVERHEAD, SUPERVISORY LABOR, MISCELLANEOUS EXPENSES:

ITEM	SUPPLIER/SUB	ALLOWANCE	COST
1. SPECIALIZED INSURANCE/BOND FEES:	_____	$ _____	$ _____
2. SUPERVISORY LABOR:	_____	_____	_____
3. LEGAL FEES:	_____	_____	_____
4. MISCELLANEOUS:	_____	_____	_____
5. MISCELLANEOUS:	_____	_____	_____
6. TOTAL DIRECT COST:			$ _____
7. OVERHEAD @ _____%:			$ _____
8. SUBTOTAL:			$ _____
9. PROFIT @ _____%:			$ _____
10. LUMP SUM TOTAL:			$ _____

Preliminary Cost Estimate

The following Preliminary Cost Estimate form can be used with owners who want a written preliminary "ballpark" estimate, but who do not have a complete set of plans to bid. It also can be quickly filled out to satisfy owners whom you suspect may be just "tire kicking." I don't recommend wasting a lot of time bidding these prospects, but sometimes just keeping the door open with a "ballpark" estimate form like this will turn an unlikely prospect into a real job at a future date.

In addition, this Preliminary Cost Estimate can be a marketing tool for converting a "competitive bid" project into a "cost-plus" type of project where you are hired as a general contractor working on a fixed percentage profit and overhead fee. (See Chapter 2 for an in-depth explanation of cost-plus agreements.)

You may want to follow up this Preliminary Cost Estimate with a letter that highlights the benefits of the cost-plus approach to building and how it may serve the needs of the owner better than the traditional competitive bidding approach.

Don't forget, this Preliminary Cost Estimate form is not a contract or an offer to contract — it is a "ballpark" estimate that is not binding on the contractor or the owner. It should never be given to an owner for signature because it is not a contract and has no General Conditions section or signature line for the owner.

8.3 ■ PRELIMINARY COST ESTIMATE

CONTRACTOR'S NAME: _____

ADDRESS: _____

_____

PHONE: _____

FAX: _____

LIC#: _____

DATE: _____

OWNER'S NAME: _____

ADDRESS: _____

_____

PROJECT ADDRESS: _____

_____

Because final plans and specifications have not been provided to Contractor for this project, the following Preliminary Cost Estimate has been prepared by Contractor. This is *not* a contract or an offer to contract. Once a complete set of plans has been provided to Contractor, or even sooner in some cases, a complete and binding Construction Agreement (including a definite price and Scope of Work) can be sent to Owner for review.

As an alternative, our company can work with you during both the design and construction phases on a "cost-plus-a-fee basis," thereby reducing total project costs and the time required to complete the work. Feel free to contact our office for more information about this approach to your project.

I. GENERAL SCOPE OF WORK DESCRIPTION (BASED ON CONTRACTOR'S ASSUMPTIONS ABOUT TYPICAL CONSTRUCTION DETAILS)

_____

_____

_____

_____

* **Total Preliminary Estimate:** $ _____

II. EXCLUSIONS

This Preliminary Cost Estimate does *not* include *labor or materials* for the following work:

_____

_____

_____

_____

Please contact me if you have further questions about any aspect of your project, or when you have a more fully developed set of plans or specifications. I hope to have the chance to work with you in the near future.

Respectfully Submitted,

_____

CONTRACTOR'S NAME
COMPANY NAME

8.3 ■ PRELIMINARY COST ESTIMATE

ANNOTATED

Charlie Contractor Construction, Inc.
123 Hammer Lane
Anywhere, USA 33333
Phone: (123) 456-7890
Fax: (123) 456-7899
Lic#: 11111

DATE: **August 1, 2001**

OWNER'S NAME: **Mr. & Mrs. Harry Homeowner**
ADDRESS: **333 Swift St.**
Anywhere, USA 33333

PROJECT ADDRESS: **same**

Because final plans and specifications have not been provided to Contractor for this project, the following Preliminary Cost Estimate has been prepared by Contractor. This is *not* a contract or an offer to contract. Once a complete set of plans has been provided to Contractor, or even sooner in some cases, a complete and binding Construction Agreement (including a definite price and Scope of Work) can be sent to Owner for review.

As an alternative, our company can work with you during both the design and construction phases on a "cost-plus-a-fee basis," thereby reducing total project costs and the time required to complete the work. Feel free to contact our office for more information about this approach to your project.

I. GENERAL SCOPE OF WORK DESCRIPTION (BASED ON CONTRACTOR'S ASSUMPTIONS ABOUT TYPICAL CONSTRUCTION DETAILS)

A. **Furnish labor and materials to remodel kitchen. Price based on average-grade new cabinets with layout similar to existing layout, average-grade appliances and plumbing fixtures, and new 2x4-foot skylight over dining area; includes new paint, standard tile countertops, and average-grade vinyl flooring with new underlayment:**

> ■ Make sure your Scope of Work is generally complete and describes the quality level of the finishes, such as "average or mid-range cabinets." Everyone has a different idea of what they want and sometimes memories are short, so state the quality range of the finishes you are using in your ballpark estimate to avoid future surprises.

* Total Preliminary Estimate: $ __18,600__

> ■ You'll never get the ballpark estimate just right, except by accident. It's not supposed to be a final number, so just get it as close as you can and explain you've provided the most realistic "ballpark" estimate possible based on the information provided to you and your past experience with similar projects.
>
> The fine-tuning of the project specifications and the future contract will be on hold until the owner takes the next step. Sometimes you may not hear back from the owner for 6 months or a year, then suddenly he'll call and say he is now serious about proceeding with the project and would like you to confirm your bid and meet with him to review the final project specifications.
>
> If work is slow, it never hurts to keep a file of these preliminary estimates and to call these potential customers once every couple of months to check on the status of their project.

II. EXCLUSIONS

This Preliminary Cost Estimate does *not* include *labor or materials* for the following work:

__New appliances, any electrical or drywall work (beyond drywall work in skylight well).__

> ■ Be sure to state the general items you have excluded from your "ballpark" bid. This helps avoid future "sticker shock" when the owner discovers *excluded* items in the contract which he thought you had agreed in your first meeting to *include*.

Please contact me if you have further questions about any aspect of your project, or when you have a more fully developed set of plans or specifications. I hope to have the chance to work with you in the near future.

Respectfully Submitted,

CHARLIE CONTRACTOR, PRESIDENT

CHARLIE CONTRACTOR, PRESIDENT
CHARLIE CONTRACTOR CONSTRUCTION, INC.

> ■ If your business is a corporation, be sure to sign your name followed by your corporate title and place the word, "Inc." after your company name. If you fail to do this, you may have personal liability under the Agreement.

CHAPTER 9

MECHANIC'S LIENS

The mechanic's lien is a claim created by state statutes for the purpose of securing payment for the value of work performed. The basic premise underlying a mechanic's lien is that one who furnishes labor or materials to a work of improvement and is not paid, should have a right against the landowner so that the landowner can not be unjustly enriched by the improvements installed on the property.

As such, the mechanic's lien right may be claimed by materials suppliers, rental companies, surveyors, subcontractors, contractors, architects, or engineers on residential or commercial construction projects, and on new construction, repair, or renovation projects. By filing a mechanic's lien, the claimant has, in essence, reserved the right to later sue the owner of the property for payment and force the sale of the property, if necessary.

The mechanic's lien is created and controlled by state law and attaches to the land and to the work of improvements on the land. In this respect, it is different from a contract claim which generally runs between the contracting parties. The rules governing mechanic's liens vary from state to state but in all cases are exacting, technical, unforgiving, and vary depending upon what type of claimant you are. (If you miss a deadline related to the filing of a mechanic's lien, all is not lost. You may have lost the ability to sue the owner on the lien, but you may still have a valid breach of contract action against the party you contracted with — as long as the statute of limitations for suing on a contract has not expired.)

The mechanic's lien is coupled with the "power of sale" which means, in a lawsuit to foreclose on the lien, the claimant can often petition the court to order the sale of the property in order to satisfy the amount of the lien. Also, if the property is sold after the lien is recorded, the new purchaser must take the property "subject to" the mechanic's lien.

In short, the buck stops with the landowner. If the one who furnishes labor or material to improve private land is not paid, that person will usually have a claim of lien against the improvements and the underlying land itself. In the event of nonpayment, filing a mechanic's lien usually does a good job of getting the attention of the landowner.

A Cloud on the Title

While exacting, a mechanic's lien is very easy to fill out and record compared to the amount of trouble it can cause the landowner and the effect it can have on the marketability of the property. The mechanic's lien places a "cloud" on the title of the property and as such has been argued in some states to be an unjust "taking" of property insofar as it can diminish (at least temporarily) the value of the property once it is recorded.

A landowner who has a mechanic's lien recorded against his property may have trouble selling, borrowing on, or refinancing his property. When a construction loan has been placed against a property, the mechanic's lien presents a serious problem to the owner and the construction lender who both need the property generally free and clear of liens in order to roll the loan over from construction financing to permanent financing.

Despite the trouble a lien causes, the vast majority of states have held that the recording of a mechanic's lien does not constitute an "unjust taking" or "slander of title" and therefore, to date, the right to quickly and easily file a mechanic's lien has been repeatedly upheld as being protected from lawsuits by the owner for "malicious prosecution."

One exception is when a mechanic's lien is recorded too late or grossly inflates the amount actually owing. In that case, the recorder of the lien does risk exposure to a malicious prosecution suit by the party injured by the recording of the "unperfectable" lien.

Many Questions To Consider

Because the mechanic's lien is created by state statute, there are many technical statutory requirements to consider before filing a lien. Among them are the following:

- Who can file a mechanic's lien?
- Is the mechanic's lien claimant a subcontractor or a prime contractor?
- Must a "preliminary notice" be provided prior to filing the mechanic's lien?
- What information must the mechanic's lien form contain?
- Must the mechanic's lien be verified by the lien claimant?
- How must the mechanic's lien be recorded or filed? With which government agency? What form of notice must be given to the owner?
- When must the mechanic's lien be filed? Can it be filed too early? When is the last day to file a mechanic's lien on the project?
- When does the time period for recording the mechanic's lien begin to run?
- What conditions must precede the filing of a mechanic's lien?
- What items can the mechanic's lien include (e.g., extra work items, attorney's fees, interest, etc.)?
- When must a lawsuit be filed on the mechanic's lien?
- What is the priority of mechanic's liens vs. other deeds of trust and encumbrances on a property?
- Should a *lis pendens* (see glossary for definition) be filed with the mechanic's lien complaint?
- How can you file a lien and foreclosure action and still preserve the claimant's right to carry through with arbitration of the dispute?
- Can the owner "bond around" a mechanic's lien?
- How can a landowner raise a defense to the filing of a lien if the landowner is not the one who orders the work of improvement?
- How can conditional and unconditional lien release forms for general release, progress, and final payments affect the lien? (See sample waiver and release forms in this chapter.)
- What is the effect of the owner's bankruptcy on mechanic's lien rights in state and federal courts?

- Can a Notice of Completion or Notice of Cessation of Labor reduce the time in which the mechanic's lien must be filed? Can the owner file successive notices of completion for different phases of the same project, thereby shortening lien periods for distinct phases of the work?
- How does the contractor assert Stop Notice or bond rights against a project performed on land owned by local, state, or federal government since ordinarily mechanic's lien rights may not be asserted against government-owned projects?

These questions are raised to get you headed in the right direction when it comes to the confusing and technical area of mechanic's liens. This book cannot offer specific answers or specific forms for your business because the answers and forms will vary from state to state. Therefore the reader is well-advised to contact an attorney in his state to obtain forms and information before attempting to use mechanic's liens in his business.

Remember that lien laws and filing deadlines are very unforgiving. The time line in which to file liens and then file a lawsuit on the lien is usually very brief and can change depending on such factors as your position in the construction process, when the project was substantially complete, and certain forms the owner can file.

A sample mechanic's lien is included on the facing page. This sample form has been adapted for use only in California and must not be used by the reader without first verifying that it is suitable for use in your state.

If you think you may have to file a lien, contact your attorney and briefly describe the situation to him early on, and be sure you don't miss the filing deadlines. It's often a good idea for you or your attorney to send a warning or notice letter to the owner advising him of your need to file a lien in order to protect your lien rights if payment is not promptly made.

Although the owner may be upset upon receiving this letter, this is often a courteous and helpful way to let the owner know you are serious about keeping all your legal options available in the event of nonpayment. Certain states may even require advance notice to the owner prior to recording the lien.

RECORDING REQUESTED BY _____

AND WHEN RECORDED MAIL TO _____ (NAME)

_____ (STREET ADDRESS)

_____ (CITY, STATE AND ZIP)

SPACE ABOVE THIS LINE FOR RECORDER'S USE

■ MECHANIC'S LIEN ■

SAMPLE ONLY
CONSULT YOUR LOCAL
ATTORNEY FOR PROPER
FORM FOR YOUR STATE

The undersigned, _____,
(NAME OF PERSON OR FIRM CLAIMING MECHANIC'S LIEN. CONTRACTORS
TO USE NAME EXACTLY AS IT APPEARS ON CONTRACTOR'S LICENSE.)

claims a mechanic's lien upon the following described real property:

City of _____, County of _____, California,

_____
(GENERAL DESCRIPTION OF PROPERTY WHERE THE WORK OR MATERIALS WERE FURNISHED.
A STREET ADDRESS IS SUFFICIENT BUT IF POSSIBLE, USE BOTH STREET ADDRESS AND LEGAL DESCRIPTION.)

The sum of $_____ together with interest thereon at the rate of ___% per annum from _____, 19__,
(AMOUNT OF CLAIM DUE AND UNPAID)   (DATE WHEN BALANCE BECAME DUE)

is due claimant (after deducting all just credits and offsets) for the following work and material furnished by

claimant: _____

_____
(INSERT GENERAL DESCRIPTION OF THE WORK OR MATERIALS FURNISHED TO PROJECT)

Claimant furnished the work and materials at the request of, or under contract with:

_____
(NAME OF PERSON OR FIRM WHO ORDERED OR CONTRACTED FOR THE WORK OR MATERIALS)

The owners and reputed owners of the property are:

_____

Firm Name: _____

By: _____

VERIFICATION

I, the undersigned, say: I am the _____ , the claimant of the foregoing
(PRESIDENT OF, MANAGER OF, PARTNER OF, OWNER OF, ETC.)

mechanic's lien. I have read said claim of mechanic's lien and know the contents thereof: the same is true of

my own knowledge. I declare under penalty of perjury under the laws of the State of California that the

foregoing is true and correct.

Executed on _____, 19__, at _____, California.
(DATE OF SIGNATURE)   (CITY WHERE SIGNED)

_____
(PERSONAL SIGNATURE OF THE INDIVIDUAL WHO IS SWEARING THAT
THE CONTENTS OF THE CLAIM OF MECHANIC'S LIEN ARE TRUE)

Typical Uses of Liens

There are a couple of situations where you are likely to need to use a mechanic's lien. The first is when the owner doesn't pay you for work that has been properly performed and the owner has no reason to withhold payment. In this case, after consulting with your attorney, you may decide to file a mechanic's lien. You'll need to be sure the lien is properly filled out and recorded in the right location and at the right time.

Next, you'll need to make sure that the lawsuit to foreclose the lien is properly filed and contains the proper allegations. If you have an arbitration clause in your agreement and want to arbitrate the dispute while preserving your mechanic's lien rights, your attorney will ordinarily file the mechanic's lien complaint and breach of contract action in civil court along with a motion to stay the civil court action pending the outcome of the arbitration.

The procedure of filing, staying, and preserving the mechanic's lien action while you proceed toward arbitration must be done properly and in a specific order so that you don't lose out on either your right to arbitrate or your mechanic's lien rights (and the right to foreclose on the property).

Once you have properly preserved your mechanic's lien rights pending the outcome of the arbitration, you can arbitrate the dispute. If the arbitrator makes an award in your favor, this award can then be enforced in civil court under the mechanic's lien action which you filed earlier. At that point, if you are not paid, you can proceed with foreclosure proceedings on the property in civil court under your initial civil court case.

If this sounds time-consuming and costly, that's because it normally is just that. As with most legal proceedings, you're better off taking steps to avoid this process if at all possible.

The second situation when you may need to use a mechanic's lien is when a supplier of one of your subcontractors places a mechanic's lien against your project. In most states, the subcontractors and materials suppliers must file some sort of Preliminary Notice that informs the owner that he has mechanic's lien rights. Anyone who doesn't have a contract directly with the owner (materials suppliers and subcontractors) is required to file this Preliminary Notice in many states.

Because the owner has no contract with these subcontractors and suppliers, he is often unaware of which people have lien rights against his property. The Pre-

liminary Notice requirement tells the owner who has lien rights (and therefore whom he should have the contractor get lien releases from) and is strictly enforced in most states. Failure to comply with the requirements of the Preliminary Notice statute may void any attempt of the supplier, subcontractor, or lower-tier subcontractor to execute on a lien against the owner's property.

Assuming the supplier has properly filed or sent out any Preliminary Notice required and has not been paid by your subcontractor (who has been paid by you for all work), the materials supplier may record the lien and commence a mechanic's lien foreclosure action against the property.

Doesn't this subject the owner to paying twice for the same work? You bet it does. The owner paid you once and you paid your sub, but your sub didn't pay his supplier? This is the problem with construction from the owner's perspective. He can end up paying twice for the same work if money intended for subs and suppliers is not passed on as it should be. This is why it is not unreasonable for an owner to request lien releases from subcontractors and materials suppliers who have met the Preliminary Notice requirements and preserved their mechanic's lien rights.

If you pay a sub who fails to pay his supplier and the supplier places a valid lien on the property, you'll obviously want to put pressure on the subcontractor to pay the materials supplier immediately in order to have the supplier release the lien. If the reason the sub can't pay the supplier is because he diverted the funds from your project to another or to his own personal use, you'll be faced with deciding whether or not to pay off the supplier (if that's even a financial possibility) in order to make the owner's property free of the lien. Then you have the option of suing the subcontractor for the amount you had to pay to remove the supplier's lien.

Depending on what state you're in, you may also have an action against the subcontractor's bond and want to make a complaint against the subcontractor's license with whatever agency governs contractor licensing in your state. You can't afford to pay a subcontractor twice for the same materials and you also share a legal responsibility with the subcontractor for keeping the owner's property free and clear of unwarranted liens.

It should be noted that the cost of carrying through with all the phases of a mechanic's lien foreclosure suit can be quite high and, depending upon the complexity of your dispute, the costs may exceed the amount of

PRELIMINARY 20-DAY NOTICE
(Private and/or Public Works)

SAMPLE ONLY
CONSULT YOUR LOCAL ATTORNEY FOR PROPER FORM FOR YOUR STATE

CALIFORNIA PRELIMINARY NOTICE
THIS IS NOT A LIEN. THIS NOTICE IS GIVEN PURSUANT TO SECTIONS 3097 AND 3098, CALIFORNIA CIVIL CODE.

NOTICE TO PROPERTY OWNER
IF BILLS ARE NOT PAID IN FULL FOR THE LABOR, SERVICES, EQUIPMENT, OR MATERIALS FURNISHED OR TO BE FURNISHED, A MECHANICS' LIEN LEADING TO THE LOSS, THROUGH COURT FORECLOSURE PROCEEDINGS, OF ALL OR PART OF YOUR PROPERTY BEING SO IMPROVED MAY BE PLACED AGAINST THE PROPERTY EVEN THOUGH YOU HAVE PAID YOUR CONTRACTOR IN FULL. YOU MAY WISH TO PROTECT YOURSELF AGAINST THIS CONSEQUENCE BY (1) REQUIRING YOUR CONTRACTOR TO FURNISH A SIGNED RELEASE BY THE PERSON OR FIRM GIVING YOU THIS NOTICE BEFORE MAKING PAYMENT TO YOUR CONTRACTOR OR (2) ANY OTHER METHOD OR DEVICE THAT IS APPROPRIATE UNDER THE CIRCUMSTANCES.
(THIS STATEMENT IS APPLICABLE TO PRIVATE WORKS ONLY)

YOU ARE HEREBY NOTIFIED:
THE NAME AND ADDRESS OF THE PERSON OR FIRM WHO HAS FURNISHED OR WILL FURNISH LABOR, SERVICES, EQUIPMENT OR MATERIAL OF THE FOLLOWING DESCRIPTION IS: DATE: _____

INDIVIDUAL OR FIRM
(NAME)
(ADDRESS)
(CITY) (STATE/ZIP)
BY:
(SIGNATURE)
(TITLE) (DATE)

(DESCRIPTION OF MATERIALS, LABOR, SERVICES OR EQUIPMENT FURNISHED OR TO BE FURNISHED)
(ADDRESS OF BUILDING, STRUCTURE, WORK OF IMPROVEMENT)
(CITY) (STATE/ZIP)

THE NAME AND ADDRESS OF THE PERSON WHO CONTRACTED FOR THE PURCHASE OF (PRIVATE WORKS) OR WHO WILL BE FURNISHED (PUBLIC WORKS) SUCH LABOR, SERVICE, EQUIPMENT OR MATERIAL IS _____

TO: OWNER OR REPUTED OWNER OR PUBLIC ENTITY

TO: SUB-CONTRACTOR

TO: ORIGINAL CONTRACTOR OR REPUTED CONTRACTOR

TO: LEASE HOLDER OR TRUST FUND

TO: CONSTRUCTION LENDER OR REPUTED CONSTRUCTION LENDER

ESTIMATED PRICE OF THE LABOR, SERVICES, EQUIPMENT OR MATERIALS DESCRIBED HEREON:
$_____

PROOF OF SERVICE AFFIDAVIT
(SECTION 3097.1 CALIFORNIA CIVIL CODE)
I, _____ declare that I served copies of the above PRELIMINARY NOTICE (check appropriate box)
a. ☐ By personally delivering copies to _____ (name(s) and title(s) of person served) at _____ (address)
on _____ 19___, at _____ a.m. / p.m.
(date) (time)
b. ☐ By First Class Certified or Registered Mail service, postage prepaid, addressed to each of the parties at the addresses shown above on _____ 19___.
I declare under penalty of perjury that the foregoing is true and correct.
Signed at _____, California, on _____ 19___.
ATTACH RECEIPTS OF CERTIFIED OR REGISTERED MAIL WHEN RETURNED _____
Signature of person making service

money you are owed. Nevertheless, all of these suits must start out with the timely recording of the mechanic's lien.

Probably 95% or more of the liens recorded against residential property by contractors end up being settled without going to trial, many without even having a complaint filed on the lien. Even if the cost of carrying through with the lawsuit is greater than the amount of money you are owed, it costs very little to file the lien and indicate your intentions to recover what you are owed through every legal means available to you. Your simple actions to protect your lien rights and your poker face just may get you paid without the need to carry through with a lawsuit.

Beware Unscrupulous Owners

You should be aware that one tactic pursued by a very small percentage of unreasonable owners is to either file or threaten to file a counterclaim lawsuit against the contractor for five or ten times the amount stated in the contractor's mechanic's lien. In some cases, the owner will allege outrageous and unfounded defects in the contractor's work just to attempt to deter the contractor from filing a legitimate lien claim. If this happens to you and the amount of money you are owed is between $5,000 and $15,000, you will be between the proverbial "rock and a hard place."

Only a crystal ball will tell you whether the owner really will carry through with his unfounded defensive lawsuit against you. Is the owner merely bluffing to get you to drop your lien or is he serious about carrying through with the lawsuit against you?

And only a crystal ball will tell you whether the owner will be able to come up with a convincing construction expert that will be able to document the thousands of dollars of *allegedly* defective work you performed. If you think you will soon be facing this situation, document and photograph as much of the project as possible to show the condition of the project when you were last working.

Depending upon how long and hard the owner fights you in the case, your legal bills may exceed the amount of your claim. Depending upon who the prevailing party is and whether you have an attorney's fees clause in your contract, you may have part or all of your attorney's fees reimbursed, or you may end up paying the owner's attorney's fees and part of his claim.

If the owner files this type of trumped-up counter-claim against you and you decide to hang in there and battle it out in court, the first question you should ask is, Do I have comprehensive general liability insurance that will pay my defense costs against the owner's counterclaim? If you do, your insurance company may end up paying your legal bills for all claims except your claims against the owner. There are potential conflicts that can arise if you pursue this option because your insurance company will normally want to settle the claim, and this can mean dismissing all or part of your affirmative claim against the owner. Discuss with your attorney the pros and cons of having your insurance company pay for your defense costs if and when the owner files a counterclaim against you.

If an owner wants to manipulate you out of the final 10% of the project funds and has the experience, funds, and intention to use the legal system to keep you from collecting your final few thousand dollars, tread lightly and call your attorney so that together you can assess the potential risks and benefits, as well as the costs, associated with your various options.

In my experience, well-organized and thorough contracts will often deter this type of owner from signing a contract with you. The type of owner who makes a practice of routinely manipulating contractors out of the final 10% to 15% of the project funds usually knows his chances of success are higher with a contractor who operates off a one-page contract and appears unsophisticated about business and legal issues. For this reason, whenever I encounter or hear about an owner who is hypersensitive to having a reasonably detailed contract, my first thought is that this may be the type of owner you are better off not working with.

It should be noted that even the contractor who files a small claims court action against an owner like the one described above can be unpleasantly shocked when the owner responds to the contractor's $4,000 small claims court action with a $35,000 superior court claim. Unfortunately, this can happen in many states.

When this happens, the contractor will suddenly be jerked out of small claims court where he expected to have a quick and inexpensive day in court without the presence of attorneys, and thrust into an expensive municipal or superior court action where he'll need to hire an attorney to defend himself against the owner's $35,000 claim. In this case, the thought of the original $4,000 claim against the owner becomes secondary in the mind of the contractor.

Waiving Lien Rights

Many contractors ask if they should waive their lien rights *prior* to receiving their final payment. The answer is *no,* ordinarily you should not unconditionally waive lien rights prior to receiving any payment — whether it's a progress payment or a final payment. Some states have adopted *conditional* lien release forms which make the lien waiver or release conditioned upon the owner's check being made good by the bank upon which it is drawn.

Title companies and lending institutions are notoriously insistent that contractors waive their lien rights prior to being issued the final payment. Again, resist waiving your lien rights unconditionally unless you

---

■ CONDITIONAL WAIVER AND RELEASE UPON PROGRESS PAYMENT ■

SAMPLE ONLY
CONSULT YOUR LOCAL ATTORNEY FOR PROPER FORM FOR YOUR STATE

Upon receipt by the undersigned of a check from _____ (Maker of Check) in the sum of $_____ (Amount of Check) payable to _____ (Payee or Payees of Check) and when the check has been properly endorsed and has been paid by the bank upon which it is drawn, this document shall become effective to release *** any mechanic's lien, stop notice, or bond right the undersigned has on the job of _____ (Owner) located at _____ (Job Description) to the following extent. This release covers a progress payment for labor, services, equipment, or material furnished to _____ (Your Customer) through _____ (Date) only and does not cover any retentions retained before or after the release date; extras furnished before the release date for which payment has not been received; extras or items furnished after the release date. Rights based upon work performed or items furnished under a written change order which has been fully executed by the parties prior to the release date are covered by this release unless specifically reserved by the claimant in this release. This release of any mechanic's lien, stop notice, or bond right shall not otherwise affect the contract rights, including rights between parties to the contract based upon a rescission, abandonment, or breach of the contract, or the right of the undersigned to recover compensation for furnished labor, services, equipment, or material covered by this release if that furnished labor, services, equipment, or material was not compensated by the progress payment. Before any recipient of this document relies on it, said party should verify evidence of payment to the undersigned.

Dated: _____ _____ (Company Name)

By: _____ (Title)

have been given a certified bank check or cashier's check in an amount equal to that of the lien release.

I strongly recommend that you find out if your state allows or has adopted specific conditional lien release forms so that you are not faced with the prospect of having unconditionally waived your lien rights only to find out two days later that the owner's check bounced and you are stuck with a worthless piece of paper and no more lien rights.

Some states have adopted forms that specifically address conditional and unconditional lien releases upon both progress payments and final payments. These dif-

---

■ Unconditional Waiver and Release Upon Progress Payment ■

SAMPLE ONLY
CONSULT YOUR LOCAL
ATTORNEY FOR PROPER
FORM FOR YOUR STATE

The undersigned has been paid and has received a progress payment in the sum of $_____ for labor, services, equipment, or material furnished to _____ (Your Customer) on the job of _____ (Owner) located at _____ (Job Description) and does hereby release ***any mechanic's lien, stop notice, or bond right that the undersigned has on the above referenced job to the following extent. This release covers a progress payment for labor, services, equipment, or materials furnished to _____ (Your Customer) through _____ (Date) only and does not cover any retentions retained before or after the release date; extras furnished before the release date for which payment has not been received; extras or items furnished after the release date. Rights based upon work performed or items furnished under a written change order which has been fully executed by the parties prior to the release date are covered by this release unless specifically reserved by the claimant in this release. This release of any mechanic's lien, stop notice, or bond right shall not otherwise affect the contract rights, including rights between parties to the contract based upon a rescission, abandonment, or breach of the contract, or the right of the undersigned to recover compensation for furnished labor, services, equipment, or material covered by this release if that furnished labor, services, equipment, or material was not compensated by the progress payment.

Dated: _____ _____ (Company Name)

By: _____ (Title)

***NOTICE: THIS DOCUMENT WAIVES RIGHTS UNCONDITIONALLY AND STATES THAT YOU HAVE BEEN PAID FOR GIVING UP THOSE RIGHTS. THIS DOCUMENT IS ENFORCEABLE AGAINST YOU IF YOU SIGN IT, EVEN IF YOU HAVE NOT BEEN PAID. IF YOU HAVE NOT BEEN PAID, USE A CONDITIONAL RELEASE FORM.

ferent lien release formats make sense and should be used whenever possible. If your state does not have such conditional lien release forms, ask your attorney how to obtain your progress payments and final payments without first waiving your lien rights.

Sample forms for a conditional and unconditional waiver and release of mechanic's lien rights are shown below. These lien releases come in two varieties: one for progress payments and one for the final payment.

The unconditional lien releases should only be used after the owner's check has cleared the bank. Before the owner's check clears, use only the conditional release. It

---

■ CONDITIONAL WAIVER AND RELEASE UPON FINAL PAYMENT ■

SAMPLE ONLY
CONSULT YOUR LOCAL ATTORNEY FOR PROPER FORM FOR YOUR STATE

Upon receipt by the undersigned of a check from _____ (Maker of Check) in the sum of $_____ (Amount of Check) payable to _____ (Payee or Payees) and when the check has been properly endorsed and has been paid by the bank upon which it is drawn, this document shall become effective to release any mechanic's lien, stop notice, or bond right the undersigned has on the job of _____ (Owner) located at _____ (Job Description). This release covers the final payment to the undersigned for all labor, services, equipment, or material furnished on the job, except for disputed claims for additional work in the amount of $_____. Before any recipient of this document relies on it, the party should verify evidence of payment to the undersigned.

Dated: _____ _____ (Company Name)

By: _____ (Title)

is every bit as good a release (once the owner's check clears), but if the owner's check bounces, you will not have waived your lien rights.

These sample forms have been adapted for use only in California and must not be used by the reader without first verifying that they are suitable for use in your state.

■ UNCONDITIONAL WAIVER AND RELEASE UPON FINAL PAYMENT ■

SAMPLE ONLY
CONSULT YOUR LOCAL
ATTORNEY FOR PROPER
FORM FOR YOUR STATE

The undersigned has been paid in full for all labor, services, equipment, or material furnished to _____ (Your Customer) on the job of _____ (Owner) located at _____ (Job Description) and does hereby waive and release ***any rights to a mechanic's lien, stop notice, or any right against a labor and material bond on the job, except for disputed claims for extra work in the amount of $_____.

Dated: _____ _____ (Company Name)

By: _____ (Title)

***NOTICE: THIS DOCUMENT WAIVES RIGHTS UNCONDITIONALLY AND STATES THAT YOU HAVE BEEN PAID FOR GIVING UP THOSE RIGHTS. THIS DOCUMENT IS ENFORCEABLE AGAINST YOU IF YOU SIGN IT, EVEN IF YOU HAVE NOT BEEN PAID. IF YOU HAVE NOT BEEN PAID, USE A CONDITIONAL RELEASE FORM.

CHAPTER 10

WARRANTIES

A written warranty is a very important part of every contract. The warranty clause attempts to allocate responsibility for the cost of repairing defects that arise in the project after it is completed. It is important to define and limit what the contractor is responsible for repairing and for how long the contractor remains liable to the owner for correcting such defects.

Warranties have been briefly discussed in the annotated versions of the agreements in this book. The warranty clauses in the agreements in this book should be carefully reviewed by you and your attorney to determine whether these clauses meet your specific needs and comply with your state law.

Express and Implied Warranties

When the project is completed and the contractor has been paid in full, he still has a legal obligation to promptly investigate any work that fails, or any defects that arise during the express warranty period (often the one-year period specified in the warranty section of the contract with the owner) and also during any warranty period that is required in your state.

If you receive a call from the owner about an alleged construction defect or failure, promptly respond to the call and investigate to determine whether you are responsible. You may need the assistance of an experienced construction attorney to help you make this decision.

In many states, depending upon numerous factors discussed below, the contractor's responsibility for the project may well extend past the one-year express warranty period specified in the contract.

Warranties generally can be divided into two main categories, express warranties and implied warranties.

An *express warranty* is a written or oral promise made in conjunction with a contract under which the contractor assures the quality, utility, or performance of the completed project for a stated period of time. The contractor essentially guarantees that the quality of the materials and workmanship will be of a specified or reasonable standard and will remain at that level for a stated period of time.

An *implied warranty* is a warranty that is imposed by state law upon the construction transaction. Different states treat implied warranties very differently in terms of what type of warranty is imposed and how long the owner (or even a subsequent owner in some cases) may invoke the implied warranty.

Many states place longer implied warranties upon a contractor who builds and sells new homes (a developer) than upon a contractor who enters into a remodeling or new home construction contract with an owner. Courts in many states have also determined that contractors who build and sell new homes must provide to the original purchaser of the home (and sometimes even subsequent purchasers) implied warranties of merchantability, habitability, and fitness for a particular purpose under the principles established by the Uniform Commercial Code (U.C.C.).

While the U.C.C. was written specifically to govern the sale of "goods" by a "merchant," many courts have referred to the language of the U.C.C. in determining the outcome of warranty disputes related to the construction business. The Implied Warranty of Merchantability and the Implied Warranty of Fitness for a Particular Purpose are defined in the Uniform Commercial Code in Sections 2-314 and 2-315 and can be summarized as follows:

- *Implied Warranty of Merchantability:* The goods must pass without objection in the trade and be generally fit for the purpose for which they are intended.

- *Implied Warranty of Fitness for a Particular Purpose:* When the seller has reason to know the purpose for which the goods are required and when the buyer is relying on the seller's skill or judgment to select or furnish suitable goods, there is (unless excluded or modified in a manner consistent with state or federal law) an implied warranty that the goods shall be fit for such purpose.

• *Implied Warranty of Habitability:* This implied warranty is routinely applied to new housing, where the builder/vendor warrants that he has complied with the local building codes in constructing the residence and that the residence was built in a "workmanlike manner" and is suitable for habitation. (For a discussion of "workmanlike manner," see Form 2.3, III.A.)

Contractors should take note that some states which are more consumer-protection oriented are beginning to apply similar types of implied merchantability, fitness, and habitability warranties to builders and remodelers as well as developers. These states require the remodeler and new home builder to offer implied warranties that their work is generally free of defects, has been completed in a workmanlike manner, and is generally suitable for the intended purpose for which it was built. The trend of many courts has been to expand the implied warranties available to the owner.

It is also important to recognize that, in many states, even when a one-year *express* warranty is agreed to by the parties in the contract, if a defect arises after one year and the applicable statute of limitations for the type of claim (e.g., breach of contract, negligence, misrepresentation, etc.) has not expired, the owner still may bring a lawsuit against the contractor. In some states, the contractor may be sued for construction defects for up to *ten* years from the date of substantial completion of the project. While the contractor's exposure under the statute of limitations is not endless, ten years is a long time!

Beware of Unfair Warranty Clauses

For obvious reasons, I would resist signing any warranty clause which states that the work will be performed to the "satisfaction of the owner." While many courts have held that this language may not be literally interpreted, why submit yourself to the vagueness and potential liability associated with this type of warranty clause?

I would also hesitate to sign on to "end result" type warranty clauses. This type of clause (normally drafted by the owner or his representative) cleverly attempts to make the contractor responsible, in part, for what is more properly within the realm of the design professional.

For example, when the owner furnishes the plans and specifications to the contractor and then wants the contractor to guaranty that "the basement will not leak," the owner has asked the contractor to guaranty an end result that may flow from the plans, not the workmanship. Other examples are "the exposed decks over the living space will not leak for 10 years" or "the project meets the requirements of all applicable building codes."

The prime obligation of the contractor is to follow the plans furnished by the owner and to complete the project in a workmanlike manner, not to be the guarantor of the owner's design. The contractor should not guaranty the performance or suitability of the owner's plans and specifications by promising an end result that may be beyond his control. The problem is that a failure to meet the desired end result may be due to a design defect, not a construction defect. Why assume liability for defects that may be caused by design errors or omissions?

As I've stressed earlier in the book (see Chapter 3) the contractor should be careful to strictly adhere to the plans unless he strongly believes that the plans don't meet code or won't achieve the intended result. Then the builder must be sure to document in writing any deviation from the plans and have the architect or designer (and owner) approve of the deviation from the plans before performing the work.

Hype Versus Warranties in Advertising

The contractor must always be careful about making promises — either orally or in writing — that he cannot keep. Many states have held that a contractor's comment to the owner that "this roof won't leak for the next 50 years" or "this concrete will never crack" have been later construed by the courts as additional express warranties provided by the contractor.

On the other hand, the law recognizes that not all sales claims are express warranties. Claims or opinions made by the builder that the buyer could not reasonably rely on when signing the contract are considered by the law to be mere sales hype or "puffing." For example, advertising that you are a "quality contractor" or "one of the best in the field" is more likely to viewed by the courts as sales hype — not an express warranty.

The kinds of activities that can create a warranty include any statement of fact or promise made by a seller to the buyer, any description of the goods, or any sample or model used by the contractor to make the sale. This might include printed advertisements and brochures, radio or television spots, contracts/proposals, business cards, job-site signs, vehicle signs, photos, models, plans, or specifications. The seller does not need to use words like guarantee or warranty to create an express warranty.

So, for example, if you advertise that you are a specialist in a certain field or that your roofs "never leak" and the owner relies on that representation in awarding you the contract, then you may be held to a higher standard by the owner, the court, or the arbitrator if a future depute arises over the quality of your work.

In addition to increasing your exposure to civil suits over warranty issues, providing false, deceptive, or misleading advertising is also against state and federal criminal law. For example, advertising or stating that you are licensed to perform a type of work that you are not licensed for would clearly be illegal false advertising and misrepresentation. Or claiming that your new technology for installing a product is "time-tested" when in fact the product and installation methods are brand new would likely be crossing the line between aggressive advertising and misrepresentation.

Exactly what is puffing and what is a warranty or misrepresentation is not always clear. Accordingly, it is wise to be conservative about the promises or representations you make in your advertising or in remarks casually spoken to the owner. In short, if you're not sure that your company can live up to a statement, don't make it — either verbally or in writing.

Patent and Latent Defects

Construction defects are generally divided into two categories, *patent* and *latent* defects. Patent defects are those that are generally apparent or can be discovered in the exercise of ordinary care and prudence by the owner. Latent defects are defects that are less apparent, perhaps concealed, and would not be discoverable by the ordinary care and prudence of the owner.

Many states indicate the time period in which an owner can file suit against a contractor depending upon whether the defect is considered a patent or a latent defect. While many states limit suits on patent defects to a shorter time period — perhaps two to four years from completion — some allow an owner to bring suit for a latent defect for up to ten years in some situations.

In addition to whether a defect is patent or latent,

the exact amount of time you could be liable to the owner will depend on a number of factors. These include what state you are in, how the warranty clause in your contract was drafted, whether you were the developer of the home or only the contractor, whether the alleged defect is being claimed by the person you contracted with or a subsequent purchaser, how long the owner waits to file the suit after discovering the defect, and how many years have elapsed since the completion of the project.

While the owner may not succeed with a lawsuit based on breach of warranty due to the limitations of the warranty clause, he may well succeed based on negligence or some other "theory of recovery" against the contractor. For this reason, the contractor must be concerned with the applicable statutes of limitations or repose related to construction in his state.

In many states the contractor must also accept that it is not likely that he will be able to fully limit his liability based on express warranty clauses and disclaimers. These liability limiting devices may help, but they will not bar the owner from filing suit, nor will they ordinarily bar the owner from recovering damages in a situation where the defect in the work appeared much earlier than typically expected and was due to the negligence of the contractor or one of his subcontractors.

Latent Defect Claims: Factors To Consider

If you are ever faced with an alleged warranty defect or the threat of a lawsuit based on a latent defect, don't assume that you can ignore it because the project was completed a long time ago. Also, if the house was sold to a subsequent owner, don't assume that you no longer have any exposure because you did not contract with the current complainant.

Instead, promptly investigate the owner's complaint, document your investigation in writing, and contact a construction attorney who is familiar with the statutes relating to construction defects in your state. Factors to consider include the following:

- Is the person claiming the defect the party you contracted with to perform the work or is he a subsequent purchaser of the property?
- Did you build *and sell* the house to the person alleging the defect or just build or remodel the house for the person alleging the defect?
- Were you the contractor and the designer, or did the owner furnish you with plans to follow in construct-

ing the project? Is the alleged defect due to design errors?

- Does your comprehensive general liability insurance cover a portion of the alleged damages? Does the owner have homeowner's insurance that would correct the damage or alleged defect? Has the claim been tendered to the owner's insurance company?
- Did your contract expressly limit your liability for the type of defect being claimed? Does your state law allow you to contractually limit your liability in this situation?
- Do state or federal warranty laws (e.g., Magnuson-Moss Warranty Act) govern the alleged defect?
- Is the defect patent or latent?
- How many years have elapsed since substantial completion of the project?
- How many years have elapsed since the claimant's discovery of the alleged patent or latent defect?
- Are other parties potentially liable for the defect (e.g., the architect, subcontractor, material supplier)? Should these other parties be called in to investigate at this time?
- Has the alleged defect arisen through inadequate or improper building maintenance by the owner?
- Did the alleged defect arise due to an unauthorized deviation in the plans and specifications by the contractor or was the project built according to approved plans and specifications?
- Does the alleged defect involve a violation of applicable building codes?
- Has the owner waived all or part of his rights by failing to inspect and claim a patent defect within the express warranty period?
- Has the owner waived his rights by accepting the alleged defect, through a failure to mitigate, reduce, or limit damages once he realized there was a defect?

If, after carefully investigating the alleged defect and potential liability, it's still difficult to accurately assess the contractor's liability, you should consider fixing the alleged defect as a "gesture of good faith." Consider whether you could make the repair for significantly less money and time than it would take to go through a legal proceeding.

As long as the cost is not too high, just fixing the problem sometimes makes more sense than battling over liability and suffering the loss of reputation that usually accompanies warranty disputes.

If this is the course of action you choose, explore whether other potentially responsible parties will contribute to a repair effort. Let them know that if they don't contribute they may be sued by the owner. Prior to making such a repair, you should discuss with your attorney whether a Settlement and Release Agreement is needed between you and the owner to release you from any further liability related to the defect.

Manufacturers' Warranties

Contractors should pay attention to which manufacturers' product warranties are shorter in duration than the express or implied warranties that may apply to their contract with the owner. If some manufacturers' warranties are shorter than the contractor's warranty, the contractor may want to factor into the cost of the project any additional risk or "insurance" he has to provide on these items.

It is possible to draft a warranty clause that "passes through" manufacturers' warranties from the contractor (the original purchaser of the products) to the owner upon either final completion or the expiration of the contractor's express warranty period. Whether such a clause would stand up in court would depend on the specific product warranties (are they transferable?) and on state consumer-protection laws.

Regardless of the legality of such a clause, it certainly would not help foster good customer relations. For example, most customers would not be happy if, three months after their addition was completed, they called the contractor with a legitimate warranty claim and were told to "call the manufacturer and have him make it right."

Even if the manufacturer's warranty clearly covers the alleged defect, I think the contractor should still quickly investigate the owner's complaint and assist the owner with correcting the problem if at all possible. This is simply good customer service, which will help build your reputation and win referrals.

After the expiration of the express warranty period provided by the contractor and manufacturer, you'll be in a more complicated area that is governed more by state, and in certain cases, federal law. Such disputes where the liability may be shared by some combination of the owner, contractor, and product manufacturer should be investigated by the contractor and discussed with an attorney familiar with warranty law in your state.

Magnuson-Moss Warranty Act

The Magnuson-Moss Warranty Act, 15 USC Section 2301, was enacted in July 1975. This is a complicated piece of federal legislation that attempts to regulate written warranties and service contract programs related to the sale of consumer goods. The Act also imposes liability on the sellers of any consumer product who breach either an express or implied warranty for transactions covered by the law.

Since the courts have not consistently agreed on how this law applies to the construction of a new home or a "substantial addition," the application of the Act has confused construction experts and legal experts, alike.

The warranty clauses in the sample agreements in this book *do not* attempt to conform to the requirements of the Act. This is because it is the author's opinion that the vast majority of work done under most agreements for a new house or substantial addition will not be covered by the Act. Furthermore, it is unusual for a homeowner to sue a contractor based on this law (although it is not impossible depending on the type of transaction).

If an owner wants to make a federal case (literally) out of an alleged warranty defect, however, he may be able to use this Act as the basis for his claim. One attraction for the complainant is that the Act allows for the recovery of attorney's fees.

What exactly is considered a "consumer product" under the Act? The law states that consumer products are personal property that is sold for "personal, family or household purposes." The warranties applicable to such products must conform to the Act unless the contractor entirely excludes consumer products from his warranty. The Act does not cover products that are purely in "commercial use" or intended for resale when purchased, nor does the Act cover the sale of real property (land or buildings). But, the Act does cover consumer products that are intended to be attached to or installed in real property.

Therefore property covered by the Act may start out as personal property (water heaters, furnaces, roofing materials, etc.), but later be transformed into a "fixture" when they are installed in a home. After this purely legal transformation from personal property into real property, these installed products are still covered by the Act.

In general, the Federal Trade Commission has attempted to clarify the application of the Act to construction by suggesting that if the consumer product has a function separate from the real property — like a water heater, furnace, or air conditioner — it is covered by the Act. But, if the consumer product has no function separate from the real property, such as 2x4 studs, wiring, etc., it is not covered by the Act.

The law clearly applies to certain construction-related products and services listed below. Contractors who install these types of materials or equipment as part of a "substantial addition" or new home are likely to find that *only the listed items* are covered by the Act. However, contractors who perform this kind of work in connection with an isolated home improvement or repair job, will likely find that their *entire transaction* is covered by the Act. Covered items include:

- Service contracts to maintain or repair "consumer products"
- Installation of smoke alarms and fire protection equipment, burglar alarms, intercoms, central vacuum, HVAC equipment, water heaters, and water treatment systems
- Retrofit windows
- Re-roofing
- Standard household appliances such as ranges, range hoods, refrigerators, dishwashers, and disposals

What should a contractor do when faced with how to provide warranties on "consumer products" covered by the Act? Because conformance with the written warranty guidelines of the Act are optional, he has the option of doing nothing. In this case, the terms of the Act relating to "consumer products" will automatically apply, obligating him to honor any written express warranties and implied warranties under state law or the federal Uniform Commercial Code. As a practical matter, this is the most common course of action.

Technically, the contractor also has the option of specifically excluding all covered "consumer products" from his warranty (not a great strategy from a public relations standpoint). A third option is to attempt to write an additional warranty clause covering only consumer products in a manner that conforms to the Act.

To make your warranty conform to the Act it must include the following information:

- the name of the person covered by the warranty
- what the warranty covers and what it excludes
- the length of the warranty period

- whether it is transferable to subsequent purchasers, what will be done and charged if the product becomes defective during the warranty period
- the procedure for making a warranty claim
- consumer protection information informing the homeowner about restrictions the law imposes on the contractor's attempts to limit implied warranties.

Because this is such a complex piece of legislation, you should consult an experienced construction attorney if you intend to draft a warranty clause covering "consumer products" in a manner that conforms to the Magnuson-Moss Act.

Additional Express Warranty Exclusions

Contractors may want to consider developing a list of items that are excluded from the contractor's warranty. These items could be placed directly in the warranty clause or made an addendum to use with large projects. A few of these excluded items are suggested in certain warranty clauses in this book. They include items that fall within the scope of normal owner maintenance, such as caulking that shrinks or minor stress fractures in materials as a result of the curing of lumber, concrete, or other materials.

Additional areas to be excluded from the warranty clause in your contract could include such items as: wood that checks or cracks, drywall or stucco that develops minor cracks, fading of paint or stains due to sunlight exposure, floor squeaks, damage caused to structure due to owner's activities or failure to maintain, and any items furnished or installed by owner.

Pay Attention to the Punch List

The contractor can prevent many warranty problems from developing by scheduling numerous thorough inspections during the construction and punch list phases. It is a good idea to make frequent quality inspections prior to covering up distinct phases of the project. Quickly correct items that aren't in conformance with the plans, codes, or workmanlike practices.

After the project is substantially complete, don't take forever to complete the punch list. Once again, delaying the punch list work will only leave a bad taste in the owner's mouth at the time both of you may have had just about enough of each other. In addition, this is the time that the owner is showing off his new project to friends and is being asked, "How was the contractor"?

Even though it can be difficult, the final punch list phase is a critical time for customer relations. Putting in some extra effort will pay off in terms of future referrals, getting your final check, and avoiding significant warranty claims down the road!

Customer Service Follow-Up

Some contractors make a note to call the owner both at six months and twelve months after the project is complete to see if any warranty work has arisen and to answer any questions the owner may have. While some contractors would view this as willfully and knowingly sticking their head back in the lion's mouth, taking the time to do this leaves a long-lasting positive impression in the mind of the owner that pays off in the long run with increased referrals.

CHAPTER 11

DISPUTE RESOLUTION

*A*lthough watching L.A. Law or Perry Mason reruns may be entertaining for many people, these shows do a poor job of showing contractors what to expect if they get involved with the legal system. What is it really like? In a word, expensive! It is not uncommon for a contractor to spend $10,000 to $25,000 to take an average small case ($30,000 or so) through the civil court system. For a $200,000 suit, the average cost of litigation rises to $25,000 to $50,000 or more.

The best legal advice you'll ever be given is to implement a few basic procedures in your business that will help keep you out of disputes and out of court. In fact, your goal should be to *avoid* getting into any legal disputes at all — large or small.

There are many different approaches to home building and remodeling. Many different contingencies can arise. Murphy's Law can visit the project in ways you've never imagined.

The technical complexities of construction coupled with the uncertainties of the personal and business relationships between the owner and the builder throughout a lengthy residential project provide fertile ground for many kinds of disputes.

It's no secret that disputes drain valuable time and money from your company and even from your personal life. Disputes keep you from enjoying your work. Disputes take away from the time you spend with your family or on your favorite pastime. In fact, a large dispute may temporarily make you hate your work and question your sanity for ever having picked up a hammer and a circular saw in the first place.

Without a doubt, the smart money is spent on implementing reasonable and practical procedures to prevent disputes.

With residential construction, the basis for many disputes lies in the unknown. The contractor doesn't know how the owner will react to certain events should they arise on the job, and the same is often even more true for the owner.

As the owners see their house torn apart —particularly on a large remodel — they experience inconvenience and stress. At the point when their "castle" seems to be in a state of complete chaos, they are apt to have questions, concerns, and expectations about your performance that they could never have anticipated at the outset.

Both the contractor and the owner typically have such unexpressed expectations, most of which they aren't even aware of until a particular situation or contingency arises. Then, once such a situation arises, the owner and contractor may suddenly find they have completely opposing ideas about the "correct" way to settle the matter. Right at this point, you have what I think is the basis for at least 75% of the disputes in this business — *disagreements based on unexpressed expectations due to a lack of written communication.*

Forget about who's at fault. Instead, ask yourself Who's in a better position to inform the other about what to expect? The contractor or the owner? Obviously, it's the contractor.

What can you do to keep these unexpressed expectations from turning a good job into one you'll regret ever having bid? The key is to do a better job of telling the owner what to expect! Prior to signing a contract, draw from your experiences (both good and bad) and let your expectations be known about how to handle the basic aspects of the business relationship.

Develop routine systems for tracking and managing the basics of the business (e.g., contracts, change orders, insurance, etc.) so you don't have to constantly reinvent solutions to the same old recurring problems.

An owner once called me with complaints about a contractor who misled her about the cost of work that was performed on a verbal T&M basis. The contractor couldn't seem to complete any one part of the project and promised completion dates that were consistently not met. Later on in the conversation she told me that the contractor was being sued by four other owners in the same area.

I asked her why. She said it had to do with complaints and misunderstandings over verbal T&M contracts. Apparently, this contractor had a hard time figuring out the source of his problems. I'm sure he blamed everyone else for disputes that he could have easily avoided by simply taking the time to obtain a signed contract before starting a job.

Much of this book has been an effort to provide ideas and procedures for organizing your company's basic business operations in order to avoid disputes and the erosion of job profits. When taken as a whole, these ideas and procedures become more than just a few separate techniques.

Together, they form a systematic approach to running a more successful company which allows you to spend more time enjoying your work while making a reasonable profit, and less time fighting with your customers.

It's tempting to say that two or three of the items below are the most important, but the fact is that all of them are critical to your long-term survival and success in this business.

Much of the focus of this book has been on the importance of using written agreements and communications simply because these are the first and most effective ways to define and document your expectations (many of which are contractually binding) with all the parties involved in the construction process.

Implementing certain routine business procedures to reduce small disputes is similar to providing basic safety training to your employees to reduce job-site injuries. You won't eliminate all disputes any more than you will eliminate all job-site injuries, but you will prevent many of them from ever occurring.

Regarding exposure to large lawsuits, implementing good fundamental business and legal practices is like putting fire sprinklers in a building. You can't guarantee a fire or major dispute will never start, but the likelihood of losing the whole structure or business is greatly reduced.

What routine business practices can you put in place to help avoid small disputes and reduce the po-tentially crippling exposure to large lawsuits? As discussed in prior chapters, a summary of these routine business practices should include the following:

(1) Careful Bidding and Precontract Work: Organize your preconstruction bidding process. Carefully review the plans and bidding documents prior to estimating every project. Don't guess at numbers because you've run out of time and the bid deadline is on top of you. Price the job high enough to make a fair profit — don't "buy" the job hoping you'll be able to make money on the extras or figure out ways to reduce your costs once you're into the job.

Don't even bid on jobs you aren't qualified and staffed to perform and don't bid on jobs if the completion date required by the owner is unrealistic. You'll make more money in any given year by passing up the jobs that aren't right for you. Learn to recognize the type of job that won't be profitable for your company and then let that type go.

(2) Good Contracts: First, *never* work off verbal contracts! Provide a contract that is fair, but detailed and complete — one that takes into account the risks and contingencies unique to the project you are bidding.

On every job, think about, address, and in your contract assign who is responsible for the unique risks of the job. For example, if you suspect that hidden structural damage may increase the scope of a remodeling job, specify in the contract that the owner will pay for any such extra work.

This is one of the unique and powerful advantages most residential contractors overlook. By having the opportunity to draft and furnish the contract, you also have the opportunity to establish many of the rules that will legally govern your business and legal relationship with the owner. Don't miss out on this opportunity — it's worth more than you might think.

If the owner insists on furnishing the contract, carefully review it. If you are uncertain about the meaning of clauses, have your attorney review it. Don't be afraid to draft alternate clauses, strike clauses you can't live with, and generally negotiate the owner's agreement.

Become proficient at drafting your own agreements developed for *your business* so that you can quickly customize them for each job.

(3) Payment Schedule (and Retention, if required): Include a payment schedule in your contract that keeps frequent, smaller payments coming in. More frequent, smaller payments are better than lumping them into two or three very large payments. Try to keep your final payment at an amount that is near or below the dollar limit of the local small claims court.

Because many disputes seem to mysteriously appear around the time of the final payment, you want the amount owing to you at that time to be as small as possible. By structuring the payment schedule so less money is owing to you at any one time, you reduce your exposure, and you reduce the chances of having larger monetary disputes turn into full-scale lawsuits.

Retention: If at all possible, avoid a payment retention altogether. If you must agree to one, try to have it not exceed 5%, and try to have 50% of the total amount withheld paid to you upon the "weathering in" of the structure. In general, never agree to a retention that exceeds 10%.

(4) Good Subcontracts: Most projects are the result of coordinated teamwork between the general and the subcontractors. Work off adequately detailed subcontracts when you subcontract work. Work with good subs who carry the proper insurance. Allow only experienced subcontractors who do quality work to bid on your projects.

(5) Proper Insurance: Carry the proper insurances to protect against both common and catastrophic risks (e.g., worker's compensation and comprehensive general liability insurance). Require the owner to carry builders risk or course-of-construction insurance on most projects.

(6) Communicate and Document: Communicate well and often with the owner and all others involved in the construction process. Return phone calls promptly. Document all important communications in writing in a brief, but professional, manner. If you later have to go to court or arbitration, your written documentation can be very important to the success of your case.

(7) Good Workmanship, Timely Performance, and Customer Service: Provide high quality construction and a high level of customer service. Complete the job on time. Complete punch list work right away. Don't work

like crazy to reach substantial completion in six weeks only to take another six weeks (and ten phone calls from the owner) to reach final completion. If part of your work clearly falls below an acceptable standard, replace it.

(8) Change Orders: Obtain written change orders prior to performing extra work. This is one of the most frequent areas of dispute, but the one most easily avoided. If you absolutely can't get it signed prior to starting the work, get the owner's verbal approval and make a written note of the conversation.

Then, if he drags his feet signing the change order after you deliver it, give him a letter telling him the work was done based on his verbal approval and that work may need to be suspended or slowed down pending his signing of the outstanding change order. Build a written record like this and most potential disputes over extra work will never appear.

(9) Depersonalize Minor Disputes: Having laid a good business foundation through a detailed agreement and written change orders, you'll find it much easier to depersonalize disputes. This is important because some owners will take personal offense at differences of opinion over, for example, what is a legitimate change order, whether the work is of adequate quality, or what is the correct interpretation of the contract documents.

Referring back to the contract and reminding the owner that a particular area was already addressed in the contract will often help quickly resolve such disputes. People usually go along with what's in writing as long as the dispute is not a large one.

(10) Preserve Your Right To Make Claims: Become familiar with mechanic's lien laws in your state and be sure to preserve your lien rights and not miss your filing deadlines. If disputes flare up, be sure you understand and strictly adhere to the dispute resolution and notice procedures in your contracts. Also, contact your attorney early on before committing to a legal course of action on your own, such as stopping the work or terminating the contract.

(11) Compromise and Settlement: As much as I like to see contractors get what they are due, some degree of compromise is appropriate to many disputes. An inflexible, "all or nothing" approach on larger disputes is

a good way to provide lots of extra income to the lawyers at the expense of the contractor.

No textbook answer can tell you when to fight and when to compromise in every situation. What's critical is the ability to analyze the factual and legal elements of the dispute and, as objectively as possible, determine the correctness of your legal position and your odds of success. You will typically benefit from some good legal help in this area.

Fill in your attorney honestly on all the good and bad aspects of the job. Don't be embarrassed to tell your attorney about work that wasn't performed well or the potential defects you see in the work. He is bound by law to maintain attorney/client confidentiality. The more facts he has, regardless of whether you think them favorable or unfavorable to your side, the better he can advise you.

No matter how wrapped up you are in the correctness of your position or your commitment to justice being done (which probably means full payment to you or not having to perform corrective work), don't forget that there is a real and significant cost to settling disputes. Forget about revenge. Now's the time to be practical and to evaluate the cost of pursuing what you think you're entitled to and your odds of success. You must weigh the potential benefits against the costs. Remember, every lawsuit is a costly gamble.

While not always possible, the cheapest way to settle disputes is to never have them arise in the first place. The second cheapest way to settle them is to deal with them quickly and fairly without the involvement of attorneys. Some degree of compromise generally plays a positive role. There's no sense winning the battle and losing the war.

However, when you are convinced that no quick and easy settlement is possible, you can often get good results from an attorney experienced with construction who is more interested in your long-term interests than in obtaining a litigation retainer. Find a good lawyer so he's available if and when you need one.

(12) Check Your Insurance When A Claim Arises: If you are sued in court, have an arbitration action brought against you, or are brought into any kind of legal action (even by way of cross or counterclaim), the first question you should ask yourself is, Do I have insurance that may cover some of the costs and potential losses? Ask yourself the following questions:

- Will my insurance carrier provide money for a covered loss in the event of a judgment against me?
- Will my insurance carrier provide money toward a settlement offer in order to have the action against me dismissed?
- Will my insurance carrier provide money for all or part of my legal defense?
- Will my insurance carrier pay my regular construction attorney to handle the case? (This is not typical, but may be possible in some cases, especially where your insurance company has an actual or potential conflict of interest, such as holding a policy for the owner. Discuss the pros and cons of this with your attorney if the situation arises.)

If faced with a large dispute, legal defense costs and settlement money can be critical to the survival of your business. Don't overlook the possibility that you may have already paid for all or part of these costs through your insurance premiums. Some attorneys not familiar with the construction business may forget to fully explore this area with you.

Strategizing a Dispute

Having implemented good basic business procedures, some disputes are still inevitable. Disputes come in all different sizes, shapes, and levels of complexity. They may be with the owner, architect, subcontractor, supplier, your insurance company, or even with the Contractor's State License Board.

When a dispute arises, I recommend an early consultation with your construction attorney so that you can get a fresh perspective on options for resolving the dispute. This consultation might take only a few minutes for a small dispute. With a good attorney, two heads are almost always better than one at this crucial turning point prior to the time you've committed to a definite course of action.

Suspending the Work or Terminating the Contract

Normally, when a contractor considers shutting down a job, it's because the owner has failed to perform according to the terms of the contract. This normally means he has missed a payment, but it can also mean he has blocked access to the site or withheld information the contractor needed to advance the work. The ongoing cash flow of the payment schedule is critical to the success of most projects and the contractor can't be expected to "finance" the job for any extended period of time.

However, a contractor should consult his attorney and consider a number of factors very carefully before suspending the work or terminating the contract. You will need a valid legal justification to do either — or you risk exposing yourself to greater liability. Without a valid legal right or "excuse" for suspending the work, you will likely be considered under the law to be in breach of contract and therefore subject to a lawsuit from the owner.

You also need to quickly and honestly assess if the owner's refusal to pay is based on an alleged defect in your performance. In other words, does the owner have reasonable legal grounds for withholding payment? If so, consider with your attorney the likely consequences of stopping work at this point.

It is also critical that you follow the procedures for work stoppage set forth in the dispute-resolution clause of your contract. Whatever the particulars, it's important that you give the owner written notice of the relevant facts, your legal basis for walking off the job, and what you expect the owner to do to "cure" his breach of contract.

Owner breaches (and contractor's as well) can be put into at least two basic categories, *minor* and *material* breaches. How the law defines minor and material will be endlessly argued by the lawyers and the parties.

The key difference is that a material breach would give one party the right to suspend the work or consider the contract terminated. A minor breach would not give rise to suspending or terminating performance under the contract. Determining whether a breach is material, however, is not easy. The same action could be a material breach on one job and a minor breach on another job.

For example, is a late progress payment grounds to stop work? It all depends. If the owner is three days late with a 15% draw payment on a job that is 60% complete, this is probably *not* a material breach of contract at this time and you would not have clear legal justification to stop the work.

If the owner is ten days late, however, you're more likely justified in stopping the work — as long as your contract gives you that authority and you provide proper notice of default to the owner.

You almost certainly need to return to finish the work if the owner makes his payment in a reasonable amount of time. However, after a month or more goes by without payment, without a valid legal justification, the owner's minor breach may rapidly turn into a material breach of contract which justifies your terminating the contract for cause — due to the lack of payment by the owner.

The scenario above illustrates how both the contractor and the courts may view certain acts or omissions of the owner differently depending upon how long the act or omission continues, whether the owner presents a reasonable legal justification, the severity of the hardship caused by the owner's act or omission, and other pertinent factors.

One factor the courts have considered is how much of the contract has been performed by the parties at the time of the breach. If the breach occurs very early on, it may be viewed as more of a material breach than if it occurs after the parties have performed most of their obligations under the contract.

Approach very carefully the area of suspension of work or termination of the contract due to the owner's default. Err on the side of reasonableness and always give the owner the opportunity to cure his default. Always give the owner written notice of his default and make sure you follow the notice and dispute resolution procedures set forth in your contract.

If you are threatened with or served with notice by the owner that you have breached the contract and should not return to the job, the same factors discussed above will need to be considered (i.e., Did your act or omission constitute a material or minor breach of contract? Did you have an opportunity to correct the alleged breach? Has the owner followed any notice provisions in the contract? etc.).

I strongly recommend that you consult your attorney to discuss these factors if you need to suspend the work or the owner kicks you off the job (or threatens to). Your response at this point in time is critical to your future success in resolving the dispute. Creating a paper trail now is also more critical than ever.

Stopping the work may keep you from financing the job, but it may also wind up bringing on a lawsuit. At times stopping work is the necessary course of action. At other times it should be delayed or the contractor should simply finish the work under written protest and then perhaps seek damages later from the owner.

Keep Your Lawyer Backstage

I recommend keeping letters from lawyers out of the process for as long as possible. Most lawyers love to write letters. It gives them a good feeling for the dispute and runs up the billable hours. However, once both sides marshall their attorneys and begin the letter writing process, the tension between the contractor and owner escalate and the stakes get raised.

It's often helpful to keep your lawyer behind the scenes until your personal negotiation efforts with the owner appear to be entirely unsuccessful. This may happen very quickly, or it may not. It just depends on your relationship with the owner and your negotiating skills. While a strongly worded, but reasonable, letter from an attorney may be just what you need to move some smaller disputes toward settlement, with other cases this may just be the spark that incites the owner to get his lawyer and construction expert involved to see just how many real or imagined construction defects they can find in your work.

With a large dispute involving the collection of a large payment from the owner, you'll need to not only evaluate the potential costs of collection, but also the likelihood that the defendant will file for bankruptcy or have so little equity in the property that even if you foreclose on the property, you may not get enough money to justify the cost of an expensive legal action.

If you have followed the advice earlier in the book to draft payment schedules that call for numerous smaller payments, you should rarely find yourself in a situation where the owner owes you a great deal of money. This is a perfect example of the way in which preventative legal actions can help insulate you from serious legal problems.

Along the same lines, avoid having a payment retention altogether, if possible. If you must agree to one, try to have 50% of the amount withheld paid to you upon the weathering in of the structure. Never agree to a retention that exceeds 10% of the cost of the project.

A Test of Endurance — Not Justice

About 90% of the lawsuits filed never go to trial. And the vast majority of arbitration "demands" never reach the final arbitration hearing. Many lawsuits are primarily a test of endurance. The party who can hang in there the longest and continue to pay their attorneys often wins almost by default. This, of course, has nothing to do with concepts of justice, right and wrong, or any other notion that should ideally govern the outcome of a dispute. It's just the practical reality of the legal system.

The legal system is imperfect. It usually takes a lot of money and endurance to achieve a favorable judgment, and even then, the result is not always a fair one — and it is never guaranteed.

Obviously, the longer that attorneys are involved, the more expensive things become for both sides. Since you really can't count on getting attorney's fees and costs reimbursed in most lawsuits, before you start a legal battle, it's wise to consider whether you'll be able to afford to finish it. A $30,000 dispute can easily cost you $10,000 to $25,000 to see through to completion. A large suit — $50,000 to $200,000 — can run you $25,000 to $50,000. If you don't have the stomach or the wallet to support that kind of figure, once again, some type of compromise needs to be considered early on.

Look Carefully Before You Leap

Making the decision to have your attorney write a letter to a customer about a dispute represents a milestone in the breakdown of the owner/contractor relationship. The decision to file a lawsuit or a formal demand for arbitration is a much more serious milestone yet and should not be approached without a serious exploration of the potential costs involved and the likelihood of recovery or possible loss, including an inquiry into the defendant's ability to pay the judgment you'd like to get.

In short, you need to approach and evaluate every legal dispute as if it was a potential business investment. While this may involve some crystal ball gazing, so do most investments. Consider whether the investment of time and money is worth the likely return. If not, don't "buy" the lawsuit. Carefully explore this area with your attorney.

Negotiate, settle, and cut your losses! Most people agree with this philosophy after they've spent $20,000 on a legal dispute, gotten a judgment, and failed to collect a penny. Next time they would do things differently and would place a much lower value on "getting even."

No matter how much you're told by your attorney that you have a good case, something can always go wrong. New facts can come to light, the judge, jury, or arbitrator can just fail to see it your way. An arbitrator can fail to apply the law. A $2,500 small claims court case can be removed to superior court and combined with the owner's $25,000 lawsuit against you. Once you commit to this path, the outcome is unpredictable, but nearly always very expensive.

Collateral Damage to the Contractor

Sometimes the collateral damage of a lawsuit can even outweigh the benefits of whatever you eventually collect. The case may drain so much of your time, attention, and capital that your business suffers while you fuel the litigation machine. With large disputes, you may find it difficult at times to concentrate on anything besides the litigation and the possible outcomes.

In other cases you may have made some technical violation of the law or your Contractor's State License Law which costs you civil penalty fines with the state, and a disciplinary contractor's bond, or provides a possible technical defense to your otherwise rock solid claim in court.

Suffice it to say that no matter what approach you take, the waters of dispute resolution are treacherous and the night can be very long. Take all the preventative steps that you can to avoid these troubled waters. Increase your company's resistance to disputes and lawsuits by following the fundamental procedures outlined at the beginning of this chapter — the costs will be insignificant compared to the alternative.

Places To Settle a Dispute

Every once in a while I see attorneys and clients on both sides make a huge, drawn-out, and costly federal case out of a very minor dispute (which either side can easily afford to absorb financially). The whole mess is nothing more than an expensive spitting contest.

Having seen enough of these cases, I've been struck with the idea that justice might be better served if the court simply arranged for a speedy boxing match between the disputing parties (or their attorneys or appointed agents) and whoever came out on top would be granted an immediate fair and favorable decision in the lawsuit with no option to appeal.

While this approach would probably be frowned upon nowadays, it would be far less costly for both sides (even the losing side) and would be over in just a few minutes compared to the modern lawsuit that can drag on for years and then take yet a few more years on appeal. The saying "fight for your rights" would take on new meaning in this context.

At the present time, however, contractors have only the following venues to select from for dispute resolution. Any alternative dispute resolution process you select should be clearly specified in your contract in the dispute resolution clause. Discuss this area with your attorney to decide which of the following venues you think will most advantageous for your business:

- Settle prior to filing a civil complaint or a demand for arbitration.
- Have the dispute heard either in small claims court, a higher level civil court, or federal court in some cases.
- By contractual agreement, submit the dispute to alternative dispute resolution (ADR). The most popular forms of ADR for residential construction are non-binding mediation and binding arbitration.

Small Claims Court

If you don't agree to binding arbitration in your contract, and any efforts to negotiate or mediate the dispute fail, you'll probably wind up in the civil court system. Small claims court is the lowest court in the civil court system and is typically a fast and easy place to settle disputes.

Luckily for you, lawyers *can't* represent you in small claims court in many states. In states where lawyers *can* represent you, the cost of paying them to come to court may or may not be worth it (depending upon the maximum dollar limit of the court).

Taking a lawyer into a small claims hearing (if allowed in your state) may even create the perception that you are "over-staffed" given the smaller dollar limit of the disputes. People who represent themselves in small claims court usually do pretty well. Judges aren't looking for lawyerly presentations. They're looking for facts on which to base a decision.

The policy underlying small claims court is that people should have a fast and economical place to resolve minor disputes. The maximum amount you can sue for varies from state to state. In many states it is $2,500. In some states it is as high as $5,000 (or even higher in a couple of states) and it may be as low as $1,000. Some states require the parties to make an effort at mediation prior to showing up in court for the small claims hearing.

The small claims court typically has no jurisdiction over mechanic's liens. Don't plan on bringing a mechanic's lien foreclosure action in small claims court unless this is specifically permitted in your state — see your local attorney or small claims court adviser for confirmation.

Filing the paperwork to initiate the case is incredibly simple. The court can often serve the lawsuit by certified mail. The filing fee is usually around $10. The combined cost of filing the lawsuit and serving the complaint can run you around $25 or $30 in many areas. That's pretty reasonable for your "day in court."

One problem with small claims court is that you don't always get your day in court with a real "judge." In some areas you don't get a judge at all, but a lawyer who is sitting temporarily to hear cases.

In some areas, the court calendar is booked solid and you'll literally have only a few minutes to present your case, even though you may want a full hour to present your evidence, testimony, and perhaps even a witness. Sometimes the judge doesn't even appear to spend any time reading your case statement or reviewing the evidence you submit.

The results in small claims court are unpredictable. But what do you want for $25? Results in any area of dispute resolution are somewhat unpredictable. You can usually get a case heard within a few weeks or a couple of months and there is generally no appeal for the plaintiff who loses the case. The defendant may have an appeal if he loses the case. Check with your attorney or small claims court adviser for the rules in your local area.

I favor small claims court for small disputes simply because it is fast and economical. However, two words of caution — *be prepared!* Many plaintiffs sue for the maximum dollar limit of the court which happens to be $5,000 in my state. The defendant on the other side often thinks he too has a legitimate $5,000 claim against the party bringing the suit.

This means that each side has $10,000 at risk (the $5,000 they are seeking and the $5,000 they may be forced to pay out if they lose). With $10,000 at stake, I recommend you consult with an attorney experienced in the area of the dispute to get some advice on preparing a short statement of claim and perhaps a brief oral outline. You don't want to go into the courtroom talking about issues that may seem important to you, but may be totally irrelevant to the judge. This is a common mistake people make all the time and it often decreases their odds of receiving a favorable judgment.

The basic approach to presenting a case in small claims court is to take evidence and an argument to court that supports your case and also presents a defense against the other side's claims. You may want to prepare a brief written statement summarizing your claim and a defense which incorporates any relevant documents. You hand this material to the judge and the opposing side when the hearing commences (or otherwise in accordance with your local rules).

Be sure to *keep it simple!* State your main reason for being in court right away upon being given an opportunity to speak. For instance, if you are owed $4,500 by an owner and all work has been completed, except for one item which the owner told you he would excuse in exchange for some extra work, you might tell the judge the following:

Introductory Statement of Claim: "My name is Charlie Contractor. Despite my repeated requests for payment, Harry Homeowner has refused to pay me the balance of $4,500 which has been due and payable on our construction contract for the last 60 days."

Expanded Statement of Claim: "On November 1, 1999, I entered into a contract with Harry Homeowner to build an addition onto his home. Here is a copy of the contract and a copy of a letter I sent to the owner on August 1, 1999, showing all payments made toward this contract and a balance due of $4,500.

I have received no payments from the owner since the date of this letter, and the $4,500 is still due and payable. The owner has breached the contract by refusing to pay this amount. I completed all work under this contract in a timely and professional manner. The building department signed off and approved all my work on this project on July 15, 1999. Here is a copy of the signed-off building permit. Here is a photograph of the completed addition.

There was a small punch list on the job which is typical for all construction projects. Enclosed is a copy of the corrective punch list dated July 15, 1999, which has been initialed by the owner indicating that all corrective work has been completed by my company and approved by the owner."

Defense to Owner's Claim: "One item that I did not complete is the painting of the kitchen and bathroom. My performance of this work was excused because the owner asked if he could do this work and, in exchange, receive a new fence at the rear of his property. I agreed to his request. Here is a letter wherein I agreed to that request and here is a photograph of the new fence."

Conclusion: "I respectfully ask the court to order the defendant to pay me the $4,500 he still owes me based on the evidence submitted to the court which shows the defendant breached the contract by failing to pay me the balance of money owing under the contract."

A statement similar to the one above (*and, perhaps even more importantly, copies of the documents referred to above*) is about all you should need for a case like this. If the judge wants more information, he'll ask for it. Listen carefully to what the judge asks you. Don't be so lost in your own thoughts that you can't respond carefully and concisely to what the judge is asking you. He may be looking for critical information that you have available but failed to mention or to furnish him.

You should bring a brief outline to help keep you focused and to help you get back on track if you stray off the topic. But don't read your statement word for word. It's better to prepare a brief outline and then tell the story briefly, in your own words. Before going to court, practice your presentation on a friend or spouse.

While you generally don't want to have a lawyer represent you in small claims court, it is usually worthwhile to get some legal help in analyzing your claim(s), identifying evidence that will support your claim(s), and anticipating counterclaims and evidence to defeat the other side's counterclaims.

The law is an old-fashioned and imperfect machine that loves conformity and expects standardized approaches to case presentation. The courts normally shun most "creative approaches" to case presentations. Your attorney may not be as creative in presenting the case as you would be, but he'll probably do a better job of identifying and prioritizing the relevant legal issues.

Also, it is likely that you will need to obtain statements or cost estimates prior to the hearing. Or, you may want to bring a witness with you to testify in court. Bringing evidence into court such as contracts, change orders, accounting summaries, corrective estimates, copies of canceled checks, invoices, letters, photographs, or even a witness, in some cases, is one of the most important steps you can take to obtain a favorable decision — assuming you have a good case.

The judge will have a difficult time determining credibility by just listening to you, so he'll want to see bits of evidence in order to piece the facts together prior to making a legal ruling. If you don't bring in these documents which corroborate your testimony, the judge will have a much harder time ruling in your favor.

It's important that any money you spend on an attorney to prepare for small claims court be reasonable relative to the amount of money you are seeking or defending against. If you have $10,000 at risk on a construction defects case with numerous issues, spending several hundred dollars obtaining good legal advice can be a good investment. However, if you are seeking only $1,000 and there is no counterclaim against you, you'd obviously want to keep any legal expenses at a bare minimum.

As previously mentioned, one risk with small claims court is that the defendant will file a counter claim in a higher court and ask that the two cases be joined. This maneuver has the effect of moving your case to a higher court with all its associated costs. Consider this risk prior to filing in small claims court. This doesn't happen often, but when it does it's a very shocking and costly surprise.

Arbitration Vs. the Civil Court System

If the amount in controversy is above the maximum dollar limit of the local small claims court, you may be headed for a higher state court or even a federal court in some cases. Your attorney will help you with the question of which court to file in. Almost always, he'll file the complaint for you (if you are the plaintiff) or file the answer if you are the defendant.

If your contract calls for arbitration, you may be headed toward some form of binding arbitration. The American Arbitration Association has become a standard administrator of arbitration services for many participants in the construction industry. Because negotiating your way through the arbitration process and the civil court system are similar in many ways, they are treated together in this section with their key differences noted.

Arbitration is a process of dispute resolution in which a neutral third party (or panel) renders a decision after a private hearing at which both parties have an opportunity to be heard and present evidence. Agreements to arbitrate have been declared valid by nearly all courts, and in most cases the decision is binding and enforceable. Arbitration clauses must be carefully worded, however, or they may not cover all types of disputes that can arise out of a contract.

Many contractors favor arbitration because it is usually faster and somewhat cheaper than the civil court system. The arbitration process can usually be fully completed in three to five months for most residential construction disputes, compared to one to two years or more for a case to make it through the civil court system (more if the decision is appealed). The rules of evidence are far more relaxed with arbitration and a party's right to appeal is severely limited. You can have a lawyer represent you in an arbitration hearing just as in a civil court hearing.

Arbitration seeks to avoid the formality and some of the expenses associated with the civil court system. The greater informality of arbitration is generally more conducive to examining residential construction disputes. Another advantage of arbitration is that you can usually select an arbitrator with some experience in the construction business — compared to a judge or jury who may not know the difference between a joist and a jamb, or a flashing and a footing.

If you want to arbitrate future disputes rather than litigate them, you need to make sure your contracts call for arbitration — this is the standard way to be assured of the right to arbitrate. However, this right is easily waived if you file a claim in court — even small claims court — or if you fail to raise the arbitration clause in your contract as a defense immediately upon being served with a civil complaint. This is a technical area of the law that must be carefully handled to make sure you don't lose your right to have a dispute arbitrated. Discuss this process in detail with your attorney before you are in the midst of a dispute that needs to be resolved.

While arbitration has some distinct benefits for construction disputes, it is far from perfect. One criticism of arbitration is that arbitrators often split the difference between the monetary award being sought by the parties.

Another criticism is that the law is not necessarily strictly applied by the arbitrator in arriving at a "fair" decision. In addition to considering the law, the arbitrator may also take into account the testimony of experts, the credibility of the parties, and darn near anything else he may think relevant in arriving at his decision. The arbitrator also brings with him to the hearing his own biases regarding certain issues which can't possibly be fully disclosed or known about prior to

selecting the arbitrator. Of course, a judge or jury may also have similar hidden biases.

Despite these potential shortcomings, I still think arbitration is preferable to litigation in the court system for most construction disputes.

Initiating the Lawsuit or Arbitration

If you work near the state or county line, check with your attorney to determine the rules regarding where suits can be brought. As a matter of convenience, you'll ordinarily want to be able to have a case brought before a court that is near to your place of business.

The civil lawsuit officially begins with the filing of a complaint. The complaint can be fairly brief and ordinarily lists which court has jurisdiction, the basic facts of the dispute, and the causes of action against the opposing side. The complaint must be properly served on the defendant. This is ordinarily done by personal service using a process server.

The arbitration process, on the other hand, is initiated by the filing of an arbitration *demand* with the agency administering arbitration.

Once you are served, *immediately* contact your attorney and give him a copy of the complaint or demand for arbitration. The legal response to the complaint is called the "answer" or "response." Every allegation in the complaint must be responded to, at least briefly, in the answer. You may well want to file a counterclaim (claim against the party bringing the action against you) or a crossclaim (claim asserting someone else is responsible for the claim being brought against you) in conjunction with your answer or response.

In many states the time allowed to file the answer to the complaint or arbitration demand is very short and if you fail to file a timely answer, you may suffer a judgment against you by default.

Your Insurance Company's Potential Duty To Defend You

If you are served a complaint or arbitration demand unexpectedly, you should immediately contact your attorney to discuss all the issues around dispute and settlement strategies covered earlier in this chapter. In addition, you should immediately determine whether you have, or had in effect at the time of the alleged damages, an insurance policy that will pay for the defense of this lawsuit. If you don't ask for this, you won't get it.

Under most comprehensive general liability policies, the insurance company may have clear legal duty to defend the contractor. This duty to defend may even be triggered by small events such as very minor property damage at the job site, which may account for only a small portion of the entire claim. Generally, the contractor has to prove merely the *potential for coverage under his policy.*

In order to *not* provide a legal defense for the contractor, the burden of proof is on the insurance company which must show the *absence of such potential* for coverage under the policy. When there is an ambiguity in this area, the ambiguity is often construed in favor of providing a defense to the contractor, with a "reservation of rights" by the insurance company.

If you are entitled to a defense by your insurance company, they will defend you against covered claims, but they will usually not go to court on behalf of your affirmative claims or your counterclaims. For example, they will not ordinarily help you try to get money from the owner for a breach of contract claim.

If your insurance carrier agrees to defend you in a suit, expect a reservation-of-rights letter from the insurer. This letter will reserve the carrier's right to tell you at a future date (depending upon new facts that emerge in discovery) that they have reconsidered and decided that no coverage is afforded you under the policy. In that case, the insurer would still pay for the defense, but might come after the contractor for money for the settlement or damages assessed in court.

The insurance company's duty to defend in a lawsuit or arbitration may be easily triggered, while its duty to provide coverage is harder to trigger and may be hotly debated. As a practical matter, with smaller disputes, the carriers often want to get in and out quickly, settle the case, and avoid the potential exposure and costs of a trial. With larger cases, however, the carrier may look very closely at ways (such as legal technicalities) they can find to deny policy coverage.

Due to the high costs of attorneys and construction experts, don't wait to notify your insurance carrier about a lawsuit or arbitration or you may be waiving your rights under the policy. Once again, the potential duty of your insurance company to provide your legal defense discussed above applies not only to situations where you are sued in civil court. It also normally applies when someone brings an arbitration action against you either directly or by way of cross or counterclaim.

Discovery

Once the complaint is filed and answered, or the arbitration demand has been filed, the *discovery* process begins. The discovery process allows the two parties in a dispute to learn about the facts supporting each other's case, following set procedures. In the court system, discovery rules are established by the court.

With arbitration, discovery may be done on a more limited basis with the opposing attorneys agreeing to the terms of the process under the watchful and encouraging eye of the arbitrator. The discovery process may be less thorough in arbitration than in the civil court system. The arbitrator has less authority to compel discovery than does a judge in the civil court system.

The purpose of discovery is to give the opposing parties the opportunity to confirm suspected facts or learn new facts that will help them predict the likely outcome of the case. Based on facts that come out through discovery, many disputes are settled.

Discovery can be very short and simple with cases that are factually very clear. Or, discovery can be long and costly if the facts of the case are very complicated and many different parties and witnesses are involved. With large construction cases, hundreds or thousands of pages of documents, and numerous parties or witnesses can make the discovery process complicated and costly.

Several common discovery devices are used in both civil litigation and arbitration. They are *interrogatories*, *depositions*, and *requests to produce*.

Interrogatories are a set or series of written questions about the facts of the case, normally prepared by an attorney, and answered in writing, normally under oath by an opposing party, a witness, or some other person having information about the case.

If you are given written interrogatories to answer, you will have time to review the questions and answer them with the assistance of your attorney prior to sending them back to the opposing side. You will normally have a couple of weeks or longer to do this, so the response is not given under pressure, like a deposition.

A deposition is an oral interview where the testimony of a witness, party, or other person having information about the case is taken by an attorney. Questions relating to the case are asked of the witness by the attorney. The witness may also have an attorney present.

All parties in the case often have their attorneys present at the deposition. A court reporter normally transcribes every word spoken at the official deposition into a record which is then referred to by the attorneys in their preparation of the case and in their future examination of witnesses.

You might be asked to give your testimony at a deposition and be asked all kinds of questions ranging from your educational experience and work history to a full probing of the facts of the case. You need to spend some time with your attorney preparing for the deposition if you will be the *deponent*, or the one answering the questions.

If your side calls the deposition, the combined costs of your attorney and the court reporter will easily run in the $225 to $300 *per hour* range for an average deposition. Then after the deposition has been taken and the transcript produced by the court reporter, your attorney will spend more time (and more of your money) reviewing the transcript of the deposition.

Requests to produce are normally prepared by an attorney, and request the other side to produce certain documents related to the case. If you are given a request to produce documents, you will normally have a couple of weeks or longer to review your files and either send copies of the documents or send the originals for the other side to review and copy.

The basic discovery devices described above are used both in the civil court system and the arbitration system. Interrogatories and requests to produce are ordinarily less expensive than depositions due to the fact that depositions can go on for hours and you will be paying not only your lawyer, but often the court reporter and transcript fee as well.

Expert Witnesses

A critical part of preparing for any construction dispute is obtaining an expert witness with a strong background in the area of the alleged damages. The expert witness will evaluate the construction, prepare a report that summarizes his findings, and be available to back up that report with credible testimony at the trial or arbitration hearing if required. An expert's testimony may be used to aid in negotiating a settlement. Often an expert witness will testify about whether the contractor's work complied with building code and met the applicable standard of care.

An expert can be anyone with a good reputation who possesses the education, experience, and training to make him a credible witness regarding the specific technical issues contested in your dispute. Your attorney should have a list of experts who can testify in the construction area. Some construction experts are retired contractors or engineers; others are actively working and combine expert testimony work with their regular career. With very technical construction/engineering issues, an engineer may be a more effective expert witness than a general contractor.

If you are sued by the owner over alleged defective work, you will want an expert to review the owner's claims and hopefully demonstrate and conclude that the owner's allegations are unfounded. If you are being represented by an attorney through your comprehensive general liability insurance carrier, the insurance carrier should pay the cost of your expert witness.

Be sure that your expert has experience testifying at hearings. An experienced attorney can sometimes confuse or discredit an expert witness who is not accustomed to testifying at hearings. If this occurs, the expert that probably cost you $50 to $125 per hour can end up doing you more harm than good.

The importance of this part of case preparation can't be overestimated. While the parties always color the facts in their favor, detailed testimony from a credible expert witness can often tip the balance and help convince a judge or arbitrator to rule in your favor.

The Trial or the Arbitration Hearing

The vast majority of disputes settle out before reaching the trial or arbitration hearing. This occurs for many different reasons. Discovery can change a party's perception of their odds of success at trial and encourage settlement.

The pretrial expense of attorneys, motions, experts, and discovery can drain the parties of the money they need to advance a trial or hearing. An insurance carrier can agree to pay a settlement amount that is accepted by the other side. And, often, the anticipated high costs of a trial or hearing (typically one day to a week with many residential disputes) places both parties in a position where they are finally more willing to compromise and reach a final settlement.

If no settlement takes place, you'll get to spend a day or more in court or at an arbitration hearing listening to your attorney make an opening statement, present witnesses in your favor (including your expert witness), introduce exhibits or documents you want admitted as evidence to support your case, cross examine the other side's witnesses, and deliver a closing statement or brief.

Whew! Lots of work for everyone involved. Lots of time you'll not be able to go to work to earn money to pay for this thing. Lots of money spent not knowing if you'll have to pay or be paid. From a business standpoint, this is rarely a reasonable course of action.

Post Judgment Concerns

When the judgment comes in the civil court system, you may be able to appeal if the finding is against you. With arbitration, in the absence of extreme misconduct on the part of the arbitrator (which is very rare and usually has to do with improper conflicts of interest that weren't disclosed by the arbitrator), you ordinarily will have extremely limited grounds to appeal a decision you don't like. This can be good or bad depending upon the decision and depending upon how much of a litigation war chest you still have left.

If a large judgment is either awarded in your favor or against you, the issue of bankruptcy will nearly always be raised. If the judgment is against you and you can't pay it, you will inevitably discuss bankruptcy with your attorney. If the decision is in your favor and the other side can't pay it, they will be discussing bankruptcy with their attorney. Either way, filing bankruptcy (or threatening to) often prevents many large judgments from being paid.

In the face of threatened bankruptcy, sometimes the prevailing party will agree to accept as little as 25% of the amount of the judgment or award in order to encourage the other side to not file bankruptcy. As hard as it is to get a judgment in your favor, you can't spend the piece of paper that has the word "JUDGMENT" on it. The really hard part of many lawsuits and arbitrations is collecting on the judgment. Sometimes the money is just not there.

If you wind up having to arbitrate or litigate a dispute, take an active role in reviewing the work of your attorney. He may know the law better than you, but you know the facts of the dispute and the construction business better than he does. The facts and how they are argued wins many cases. If your attorney doesn't understand the facts or is incapable of understanding the technical aspects of the dispute, you've got a serious problem.

You'll need to constantly review the attorney's work and fill him in on any facts you think he may be missing or may have misunderstood. Disclose everything to your attorney. Keep no secrets about the case or you may be sabotaging your own case without even knowing it. Your attorney has a legal duty to keep privileged communications confidential and argue the merits of your case. However, he also must be prepared to counter weaknesses in your case as the need arises.

Finally, don't hesitate to ask your attorney throughout the process what it may cost to proceed to the next level of the arbitration or civil court proceeding. And ask about the financial risks and potential rewards of proceeding. Study the numbers and evaluate your options as a business decision. If it doesn't make business sense, go home, lick your wounds, and get on with your business and your life.

Mediation

Mediation has proven very successful in resolving construction disputes at a cost that is often far lower than arbitration or litigation. The American Arbitration Association reports that as many of 90% of the cases submitted to their mediation program are successfully resolved.

Mediation is a non-binding, voluntary dispute-resolution process where a trained neutral third party reviews the dispute between the opposing parties and attempts to help them reach their own mutually agreeable resolution or settlement without the need for a costly lawsuit or an arbitration hearing.

Compromise is usually encouraged as a means of resolving the dispute. The neutral mediator does not have the authority to make a decision or award that is binding on the parties. This is not the function of the mediator.

A skilled mediator and willing participants are essential to the success of the mediation process. Because mediation is voluntary and legally non-binding, if the parties aren't sincere about resolving the dispute, the mediation can end up being an additional expense and seem like a waste of time. Statistics show, however, that mediation is often successful and greatly reduces the time and expense of resolving many disputes.

If you want to mediate disputes, it's a good idea to put a mediation clause (in addition to your other dispute resolution clauses) in your agreement. It might read something like this:

"In the event of a dispute that arises out of this agreement, the parties agree that either side may submit the dispute to a private, non-binding mediation before a neutral, independent mediator and that the parties to the agreement agree to participate in the mediation and split the costs of the mediator."

You can then designate the American Arbitration Association as the administrator of the mediation, or you can appoint some other local mediation group or an experienced local attorney, architect, contractor, engineer, etc., who agrees in advance to mediate any disputes, if and when they arise.

I favor small claims court for small disputes because, in general, it allows the parties a fast and inexpensive day in court. However, where mediation is successful, it can serve a similar purpose (quickly and inexpensively resolving the dispute) and leave many owners in a position where they feel much less hostile toward the contractor — he might even still get a referral.

If you need to mediate a dispute that involves technical construction issues, use a mediator with experience in the construction business.

American Arbitration Association

The American Arbitration Association (AAA) is a public service, non-profit association offering a broad range of dispute resolution services for all types of business transactions. It has become the most popular administrator of Alternative Dispute Resolution, as these processes are called, on a national and even international level.

The AAA has a large pool of well-trained arbitrators and mediators from every type of profession who sit as arbitrators and mediators, usually on a part-time basis in conjunction with their normal career work. You may have a panel of three arbitrators hear a dispute, but to keep costs down, only very complicated and high-dollar commercial cases typically have more than one arbitrator.

By agreeing to arbitrate a construction dispute through AAA arbitration, you are ordinarily agreeing to have the dispute arbitrated under a standardized set of Construction Industry Arbitration Rules developed by the American Arbitration Association.

These rules are in the process of being revised and an amended set of rules may be issued by the AAA by the time you are reading this. Call your local American Arbitration Association office (they have offices in most major cities in the United States) and obtain a copy of the rules.

The AAA charges an administrative fee for placing the parties together with the arbitrator or mediator. This fee can range from several percent of the amount in controversy on a small dispute (under $25,000) to one percent on claims up to about $250,000. The percentage is smaller for larger claims.

In addition to the AAA administration fee, the parties will likely also need to pay the arbitrator. The arbitrator's fees are administered through the AAA. These fees can be anywhere from $100 to $250 an hour, depending upon the arbitrator. Find out if the arbitrator will serve the first day of the hearing without compensation.

A standard AAA arbitration clause in your contract may look like this:

"Any controversy or claim arising out of or relating to this contract, or the breach thereof, shall be settled by arbitration in accordance with the Construction Industry Arbitration Rules of the American Arbitration Association, and judgment upon the award rendered by the arbitrator may be entered in any court having jurisdiction thereof."

Conclusion

Honest people sometimes just disagree. And, then again, there are those few people out there you should never try to work for.

By implementing some good fundamental preventative legal procedures into your business, you'll have fewer disagreements and retain more of your anticipated job profits. If and when disputes do arise, you'll be better prepared to deal with them. *The Contractor's Legal Kit* is meant to be used as a tool to help achieve these goals, with the review and guidance of your local attorney.

The book's purpose has not been to tell you specifically how to structure your business and legal affairs, but rather to get you thinking about the legal and business issues that have a constant and profound effect on your business. Hopefully, reading this book will help you preserve and enhance your ability to make and enjoy a living as an independent craftsman, builder, and building contractor.

CHAPTER 12

SAFETY CONCERNS

It goes without saying that the most important reason to be concerned about job-site safety is to prevent the inevitable pain and suffering of the injured worker. This injured worker could be your best friend, a long-time co-worker, or even yourself.

What does job-site safety have to do with legal concerns? Plenty. Focusing on job-site safety cannot only reduce injuries to employees, it can also avert the associated financial loss that accompanies every job-site injury. Someone always has to pay either directly or indirectly when a job-site injury occurs. If the job-site injury is a major one, you may end up losing your business, home, and livelihood.

Unfortunately, many contractors seem unable or unwilling to pay attention to this aspect of their business until they've been faced with either a visit from their OSHA inspector or a trip to the hospital to visit a seriously injured employee. This attitude is like playing Russian roulette.

Even when taking precautions, sooner or later an accident will happen and, for both legal and ethical reasons, you need to be able to look back and know you did all you reasonably could have done to prevent the accident. If you take steps to avert job-site accidents, you'll encounter far fewer of them and the ones you do encounter will tend to be of a less serious nature. Also, the fewer accidents you have, the lower your worker's compensation rate will be.

On the other hand, carelessness in this area can end your career in the construction business if you are the victim, or if one of your employees is seriously injured through a job-site injury that you could have avoided.

If a worker is hurt on the job in the course of his work, the employer's worker's compensation insurance is triggered as long as the employer was not extremely negligent and derelict in his duty to provide a safe working environment.

If the contractor was extremely negligent in meeting OSHA standards (or other applicable safety standards set by the state), he may suffer all or part of the financial responsibility for the injured employee's loss. Certainly he will open himself up to the liability of a civil lawsuit by the injured worker.

Remember, certain very basic duties to provide a safe work site are non-delegable. If one of your employees or a subcontractor's employee is injured as a result of your non-compliance with basic safety standards, despite the existence of a worker's compensation policy that covers the injured worker, you might still be sued for the resulting damages.

Damages for a permanently disabled worker could range from $250,000 to well over $1 million depending upon the seriousness of the injury and the cost of past and future medical treatments. Facing this type of lawsuit will be a time-consuming nightmare for the average residential contractor.

Inherently Dangerous Work

What do the statistics generally indicate about job-site accidents in the construction industry?

The bad news first. As many as one in ten construction workers in residential construction will be injured in a given year — some with major injuries. Many will not be able to return to work the next day due to their injury. Some will never return to work again or ever go home. Fatalities occur most commonly from falling, being struck by an object with great force, cave-ins, and being electrocuted (often falling after being shocked).

The good news is that residential construction appears to have less inherent dangers than commercial or industrial construction. Consistently implementing a good safety program can prevent many accidents, lower insurance costs, and reduce exposure to lawsuits.

At one time or another, we've all smashed our fingers, nicked our hands, shut down an eye with flying debris, or wound up with sore backs at the end of the day. However, the trauma of a severe job-site accident will leave a sick feeling in your stomach if you ever encounter one. After the nausea starts to fade, the first question you ask yourself as an employer and friend of the injured is, "How could I have prevented that accident from occurring"?

Why is this the first question that is inevitably asked? The reason must be because we all know that many job-site accidents can be prevented.

Construction always has and always will involve many inherent dangers. These dangers are easy for the customer to overlook and unfortunately, very easy for the busy contractor to overlook. The odds are that most residential contractors will never see an OSHA inspector on their job sites. However, this does not reduce the inherent dangers of the job site, nor does the absence of the OSHA inspector reduce the duty of the contractor to provide safe working conditions.

Worker's Compensation Is Essential

What can you do to reduce the financial risks associated with job-site injuries? First, make sure that you always have a worker's compensation policy in effect. The basic purpose of the worker's compensation policy is to provide medical treatment to your employees who are injured while in the course and scope of their employment.

Every employer is required by law to provide this insurance for all his employees. You also need to make sure that all your subcontractors with employees maintain worker's compensation policies. If your subcontractor does not maintain a worker's compensation policy and one of his workers is injured on your job, your policy may be relied on to cover the loss. You can guess what this will do to your rates.

If you fail to provide this insurance for your employees, you will be breaking a very important law and will subject yourself to guaranteed legal liability in the event one of your employees is injured. If you want to see the "long arm of the law" reach out and seize all types of personal assets, watch what happens when the government attaches a contractor's property if they think they may need it to pay off an injured worker's claim because the contractor had no worker's compensation policy. It's not a pretty sight!

Make Safety a Priority

The second thing you can do is to let your subs know through your contract and through your regular dealings with them that you place an extremely high priority on job-site safety. OSHA regulations state, in essence, that the general contractor is primarily responsible for maintaining job-site safety. However, regarding the subcontracted portion of the work, the general contractor and the subcontractor can agree on how to comply with safety obligations as long as these obligations are met.

While the general contractor may attempt to make the subcontractor responsible for his own (the subcontractor's) safe operations, the general contractor also has a duty to regularly monitor and supervise the job and to immediately correct or have corrected *any* safety violations which come to his attention (or should have come to his attention if he had been paying attention), regardless of who caused the safety hazard.

The third and perhaps most important thing a general contractor can do to reduce the financial risks associated with job-site injuries is to implement a safety program which includes:

• routine safety inspections
• teaching employees how to recognize and avoid unsafe job conditions
• providing job-site first-aid kits
• keeping a clean job site
• providing basic personal protection equipment to employees.

In fact, OSHA requires that you take all of these preventative steps!

General contractors must not delegate the responsibility for providing a safe work environment to their low-level employees. Even regarding subcontracted portions of the work, they can't count on entirely delegating this duty to their subs. Rather, contractors must monitor job-site conditions themselves, and promptly correct safety violations.

You may have a reasonable excuse for certain accidents occurring on your job, but you will have no excuse (and a much poorer defense against potential lawsuits) if you fail to take reasonable steps on a regular basis to *prevent* job-site injuries.

Safety Checklist

Having used guilt, gloom, and doom in the introduction above to persuade you to take this issue seriously, here is a list of common concerns and steps you can take to reduce the odds of making OSHA's most-wanted list:

- Set an example yourself by working safely and by putting safety first in your job-site activities. Don't allow unsafe conditions to continue for a minute longer once you notice them. Pull the plug on the tool without a guard, cut the cord on the tool without a grounding prong, throw out that ladder that looks unsafe, take that questionable nail gun in for service if you can't keep it from misfiring or jamming, dismantle the patched-together scaffolding, etc.

- Let your position on safety be known and suddenly your employees will have a standard to meet which they know you expect of them. This will go a long way toward increasing safety and decreasing the possibility of catastrophic job-site injuries.

- Make a checklist of safety concerns and make it mandatory reading for every new employee. Also, keep copies of it at job sites and have employees periodically review it. A periodic reminder like this may discourage a reckless act by one of your employees which he knows he really shouldn't attempt.

- Devote a few minutes of time on a regular basis to job-site tailgate safety meetings. Keep a brief written log of these meetings.

- Keep the names and phone numbers of the nearest hospital and emergency services in a location at the job site that everyone is aware of.

- Let it be known that you will enforce safety rules and won't tolerate anyone working in a manner that is obviously dangerous to himself and others. Have a few harsh words with the "macho" young worker who thinks himself invincible or the seasoned journeyman who's always done it his way — the dangerous way.

- Contract in your field of expertise. Working with tools and types of construction you are not experienced with increases the risk of job-site injury.

- Don't put dangerous power tools in inexperienced hands. Provide carefully supervised training for entry-level workers. Don't loan dangerous power tools to people who are not 100% qualified to use them.

- Keep a clean job site.

- Provide personal protection equipment such as safety glasses, ear plugs, hard hats, and suitable dust masks to all employees. Insist they be used!

- Insist that employees wear proper clothing, gloves, and footwear.

- Pay particular attention to the safety habits of your new employees and subs to assure they are meeting your job-site safety expectations.

- Train employees in proper lifting methods. Encourage the use of common sense in lifting and moving materials. Rent lifting equipment when necessary — it's cheaper in the long run than crushing your employees under the weight of that steel I-beam.

- Don't use scaffolding unless you're sure it meets OSHA and state safety standards. Make sure guard rails on scaffolding are always in place. Because falls are one of the most common and most serious job-site injuries, doing otherwise is a very bad gamble. If you aren't sure whether your equipment meets state and OSHA standards, find out before using it again.

- Don't allow ladders to be used unsafely. Again, falls from ladders that perch on top of unstable surfaces, lean at odd angles, or fail to rise high enough above the roof line are common causes of serious injuries.

- Don't allow tools to be perched on top of areas where they can fall, slide, or become dislodged and fall on top of others.

- This is not a joke: Consider contacting OSHA and having them send you one of their information officers to provide information and inspect a job site. The officers shouldn't issue citations since their job is to provide information that will keep the real OSHA guys from coming out to your job. Check with OSHA to find out if they have this program in your area.

- Remodeling is often done with the owner and his family around the job site. Establish safety rules that make it clear that the owner and his family (especially children) are to stay out of the work area at all times and should move with increased awareness of the safety hazards when they must enter the work area.

- When it comes to speed vs. safety, always put safety first!

An accident is just that. Not all of them are avoidable, but experience has proven without a doubt that many accidents *are* avoidable. By actively focusing on job-site safety, many potential job-site injuries will be prevented on your projects. By preventing accidents, you will prevent human suffering, reduce your insurance costs, and greatly limit your exposure to catastrophic lawsuits. You can't prevent all accidents, but you have an ethical and legal obligation to implement a safety program that will reduce them.

GLOSSARY OF COMMON TERMS
RELATED TO CONSTRUCTION LAW

*T*he following glossary of terms is meant to give the reader a basic understanding of commonly used words and phrases in the realm of construction law. These definitions are by no means comprehensive and the facts of every dispute can influence which legal term(s) apply.

These definitions have been excerpted with permission from "Black's Law Dictionary" (sixth edition), published by West Publishing Company. The author is grateful for their cooperation in allowing these excerpted definitions to be reprinted.

American Arbitration Association: National organization of arbitrators from whose panel arbitrators are selected for labor and commercial disputes. The Association has produced a Code of Ethics and Procedural Standards for use and guidance of arbitrators.

Arbitration: A process of dispute resolution in which a neutral third party (arbitrator) renders a decision after a hearing at which both parties have an opportunity to be heard.

An arrangement for taking and abiding by the judgment of selected persons in some disputed matter, instead of carrying it to established tribunals of justice, and is intended to avoid the formalities, the delay, the expense and vexation of ordinary litigation.

The majority of the states have adopted the Uniform Arbitration Act. A major body offering arbitration services is the American Arbitration Association.

Breach of Contract: Failure, without legal excuse, to perform any promise which forms the whole or part of a contract. Prevention or hindrance by party to contract of any occurrence or performance requisite under the contract for the creation or continuance of a right in favor of the other party or the discharge of a duty by him. Unequivocal, distinct and absolute refusal to perform agreement.

Brief: A written statement prepared by the counsel arguing a case in court. It contains a summary of the facts of the case, the pertinent laws, and an argument of how the law applies to the facts supporting counsel's position.

Cause of Action: The fact or facts which give a person a right to judicial redress or relief against another. The legal effect of an occurrence in terms of redress to a party to the occurrence. A situation or state of facts which would entitle party to sustain action and give him right to seek a judicial remedy in his behalf.

Completion Bond: A form of surety or guaranty agreement which contains the promise of a third party, usually a bonding company, to complete or pay for the cost of completion of a construction contract if the construction contractor defaults.

Contract: An agreement between two or more persons which creates an obligation to do or not to do a particular thing. Its essentials are competent parties, subject matter, a legal consideration, mutuality of agreement, and mutuality of obligation.

Damages: A pecuniary compensation or indemnity, which may be recovered in the courts by any person who has suffered loss, detriment, or injury, whether to his person, property, or rights, through the unlawful act of omission or negligence of another. A sum of money awarded to a person injured by the tort of another.

Damages may be compensatory or punitive according to whether they are awarded as the measure of actual loss suffered or as punishment for outrageous conduct and to deter future transgressions. Nominal damages are awarded for the vindication of a right where no real loss or injury can be proved. Generally, punitive or exemplary damages are awarded only if compensatory or actual damages have been sustained.

Compensatory or actual damages consist of both general and special damages. General damages are the natural, necessary, and usual result of the wrongful act or occurrence in question. Special damages are those "which are the natural, but not the necessary and inevitable result of the wrongful act."

Default: By its derivation, a failure. An omission of that which ought to be done. Specifically, the omission or failure to perform a legal or contractual duty; to observe a promise or discharge an obligation (e.g., to pay interest or principal on a debt when due); or to perform an agreement.

Defendant: The person defending or denying; the party against whom relief or recovery is sought in an action or suit or the accused in a criminal case.

Defense: That which is offered and alleged by the party proceeded against in an action or suit, as a reason in law or fact why the plaintiff should not recover or establish what he seeks. That which is put forward to diminish plaintiff's cause of action or defeat recovery. Evidence offered by accused to defeat criminal charge.

A response to the claims of the other party, setting forth reasons why the claims should not be granted. The defense may be as simple as a flat denial of the other party's factual allegations or may involve entirely new factual allegations. In the latter situation, the defense is an affirmative defense.

Deposition: The testimony of a witness taken upon oral question or written interrogatories, not in open court, but in pursuance of a commission to take testimony issued by a court, or under a general law or court rule on the subject, and reduced to writing and duly authenticated, and intended to be used in preparation and upon the trial of a civil action or criminal prosecution.

A pretrial discovery device by which one party (through his or her attorney) asks oral questions of the other party or of a witness for the other party. The person who is deposed is called the deponent. The deposition is conducted under oath outside of the courtroom, usually in one of the lawyer's offices. A transcript (word-for-word account) is made of the deposition.

Discovery: Trial practice. The pretrial devices that can be used by one party to obtain facts and information about the case from the other party in order to assist the party's preparation for trial. Under Federal Rules of Civil Procedure (and in states which have adopted rules patterned on such), tools of discovery include: depositions upon oral and written questions, written interrogatories, production of documents or things, permission to enter upon land or other property, physical and mental examinations and requests for admission. Rules 26-37.

Term generally refers to disclosure by defendant of facts, deeds, documents or other things which are in his exclusive knowledge or possession and which are necessary to party seeking discovery as a part of a cause of action pending, or to be brought in another court, or as evidence of his rights or title in such proceeding.

Dispute: A conflict or controversy; a conflict of claims or rights; an assertion of a right, claim, or demand on one side, met by contrary claims or allegations on the other. The subject of litigation; the matter for which a suit is brought and upon which issue is joined, and in relation to which jurors are called and witnesses examined.

Estop: To stop, bar, or impede; to prevent; to preclude.

Estoppel: "Estoppel" means that a party is prevented by his own acts from claiming a right to detriment of other party who was entitled to rely on such conduct and has acted accordingly.

Estoppel is a bar or impediment which precludes allegation or denial of a certain fact or state of facts, in consequence of previous allegation or denial or conduct or admission, or in consequence of a final adjudication of the matter in a court of law.

It operates to put party entitled to its benefits in same position as if thing represented were true. Under law of "estoppel" where one of two innocent persons must suffer, he whose act occasioned loss must bear it. Elements or essentials of estoppel include change of position of parties so that party against whom estoppel is invoked has received a profit or benefit or party invoking estoppel has changed his position to his detriment.

Fraud: An intentional perversion of truth for the purpose of inducing another in reliance upon it to part with some valuable thing belonging to him or to surrender a legal right. A false representation of a matter of fact, whether by words or by conduct, by false or misleading allegations, or by concealment of that which should have been disclosed, which deceives and is intended to deceive another so that he shall act upon it to his legal injury. Any kind of artifice employed by one person to deceive another.

Elements of a cause of action for fraud include false representation of a present or past fact made by defendant, action in reliance thereupon by plaintiff, and damage resulting to plaintiff from such misrepresentation.

Indemnify: To restore the victim of a loss, in whole or in part, by payment, repair, or replacement. To save harmless; to secure against loss or damage; to give security for the reimbursement of a person in case of an anticipated loss falling upon him. To make good; to compensate; to make reimbursement to one of a loss already incurred by him. Several states by statute have provided special funds for compensating crime victims.

Indemnity: Reimbursement. An undertaking whereby one agrees to indemnify another upon the occurrence of an anticipated loss. A contractual or equitable right under which the entire loss is shifted from a tortfeasor who is only technically or passively at fault to another who is primarily or actively responsible.

Interrogatories: A set or series of written questions drawn up for the purpose of being propounded to a party, witness, or other person having information of interest in the case.

A pretrial discovery device consisting of written questions about the case submitted by one party to the other party or witness. The answers to the interrogatories are usually given under oath, i.e., the person answering the questions signs a sworn statement that the answers are true.

Latent Defect: A hidden or concealed defect. One which could not be discovered by reasonable and customary inspection; one not apparent on face of goods, product, document, etc.

Defect of which owner has no knowledge, or which, in exercise of reasonable care, he should have had no knowledge.

Lis Pendens: A pending suit. Jurisdiction, power or control which courts acquire over property in litigation pending action and until final judgment.

Liquidated Damages: The term is applicable when the amount of the damages has been ascertained by the judgment in the action, or when a specific sum of money has been expressly stipulated by the parties to a bond or other contract as the amount of damages to be recovered by either party for a breach of the agreement by the other. The purpose of a penalty is to secure performance, while the purpose of stipulating damages is to fix the amount to be paid in lieu of performance. The essence of penalty is a stipulation as in terrorem while the essence of liquidated damages is a genuine covenanted preestimate of such damages.

Liquidated damages is the sum which party to contract agrees to pay if he breaks some promise and, which having been arrived at by good faith effort to estimate actual damage that will probably ensue from breach, is recoverable as agreed damages if breach occurs.

Damages for breach by either party may be liquidated in the agreement but only at an amount which is reasonable in the light of the anticipated or actual harm caused by the breach, the difficulties of proof of loss, and the inconvenience or nonfeasance of otherwise obtaining an adequate remedy. A term fixing unreasonably large liquidated damages is void as a penalty.

Magnuson-Moss Warranty Act: Federal statute (15 U.S.C.A. sec. 2301 et seq.) requiring that written warranties as to consumer products must fully and conspicuously disclose in simple and readily understood language the terms and conditions of such warranty, including whether the warranty is a full or limited warranty according to standards set forth in the Act.

Mechanics' Lien: A claim created by state statutes for the purpose of securing priority of payment of the price or value of work performed and materials furnished in erecting, improving, or repairing a building or other structure, and as such attaches to the land as well as buildings and improvements erected thereon. Such lien covers materialmen, tradesmen, suppliers, and the like who furnish services, labor or materials on construction or improvement of property.

AUTHOR'S NOTE: A mechanics' lien, among other things, typically includes a written statement signed and verified by the claimant or by the claimant's agent which states the following:

(a) the amount of claimant's demand (after deducting credits and offsets);

(b) the name of the owner or reputed owner, if known;

(c) the kind of labor, services, equipment or materials furnished by the claimant;

(d) the name of the person by whom the claimant was employed or to whom the claimant furnished the labor, services, equipment or materials (the contractor who hired you if you are a subcontractor or the owner who hired you if you are the prime contractor), and;

(e) a description of the site sufficient for identification.

IMPORTANT REMINDER: Immediately contact an attorney for mechanics' lien requirements in your state if you think you may need to file a lien. Requirements and recording deadlines of mechanics' liens are very strict and inflexible. You should become familiar with the requirements and time lines related to mechanics' liens in your state. Exact requirements vary from state to state.

Mediation: Private, informal dispute resolution process in which a neutral third person, the mediator, helps disputing parties to reach an agreement. The mediator has no power to impose a decision on the parties.

Miller Act: Federal statute which requires the posting of performance and payment bonds before an award may be made for a contract beyond a certain amount for construction, alteration or repair of a public building or public work of the U.S. government.

Misrepresentation: Any manifestation by words or other conduct by one person to another that, under the circumstances, amounts to an assertion not in accordance with the facts. An untrue statement of fact. An incorrect or false representation. That which, if accepted, leads the mind to an apprehension of a condition other and different from that which exists. Colloquially it is understood to mean a statement made to deceive or mislead.

In a limited sense, an intentional false statement respecting a matter of fact, made by one of the parties to a contract, which is material to the contract and influential in producing it. A "misrepresentation," which justifies the rescission of a contract, is a false statement of a substantive fact, or any conduct which leads to a belief of a substantive fact material to proper understanding of the matter in hand, made with intent to deceive or mislead.

Negligence: The omission to do something which a reasonable man, guided by those ordinary considerations which ordinarily regulate human affairs, would do, or the doing of something which a reasonable and prudent man would not do.

Negligence is the failure to use such care as a reasonably prudent and careful person would use under similar circumstances; it is the doing of some act which a person of ordinary prudence would not have done under similar circumstances or failure to do what a person of ordinary prudence would have done under similar circumstances. Conduct which falls below the standard established by law for the protection of others against unreasonable risk of harm; it is a departure from the conduct expectable of a reasonably prudent person under like circumstances.

The law of negligence is founded on reasonable conduct or reasonable care under all circumstances of particular case. Doctrine of negligence rests on duty of every person to exercise due care in his conduct toward others from which injury may result.

Patent Defect: In sales of personal property, one which is plainly visible or which can be discovered by such an inspection as would be made in the exercise of ordinary care and prudence.

Performance Bond: Surety bond which guarantees that contractor will fully perform contract and guarantees against breach of contract. Proceeds of bond are used to complete contract or compensate for loss in the event of nonperformance.

Prime Contractor: The party to a building contract who is charged with the total construction and who enters into subcontracts for such work as electrical, plumbing, and the like.

Remedy: The means by which a right is enforced or the violation of a right is prevented, redressed, or compensated.

Statute of Limitations: A statute prescribing limitations to the right of action on certain described causes of action or criminal prosecutions; that is, declaring that no suit shall be maintained on such causes of action, nor any criminal charge be made, unless brought within a specified period of time after the right accrued.

Stop Notice Statute: An alternative to the mechanics' lien remedy that allows contractors, suppliers, and workers to make and enforce a claim against the construction lender, and in some cases against the owner, for a portion of the undisturbed construction loan proceeds.

Substantial Performance Doctrine: A doctrine in commercial reasonableness which recognizes that the rendering of a performance which does not exactly meet the terms of the agreement (slight deviation) will be looked upon as the fulfillment of the obligation, less the damages which result from any deviation from the promised performance.

This doctrine which is widely applied in building contracts and which is intended to prevent unjust enrichment, provides that where contract is made for agreed exchange of two performances, one of which is to be rendered first, substantial performance, rather than strict, exact or literal performance, by first party of terms of contract is sufficient to entitle such party to recover on such performance.

Substantial performance of a contract is shown when party alleging substantial performance has made an honest endeavor in good faith to perform his part of the contract, when results of his endeavor are beneficial to other party, and when such benefits are retained by the other party; if any one of these circumstances is not established the performance is not substantial, and the party has no right of recovery.

Tolling: To suspend or defeat temporarily as the statute of limitations is tolled during the defendant's absence from the jurisdiction and during the plaintiff's minority.

Waiver: The renunciation, repudiation, abandonment, or surrender of some claim, right, privilege, or of the opportunity to take advantage of some defect, irregularity, or wrong.

Waiver is essentially unilateral, resulting as legal consequence from some act or conduct of party against whom it operates, and no act of party in whose favor it is made is necessary to complete it. And may be shown by acts and conduct and sometimes by nonaction.

Express Warranty: A promise, ancillary to an underlying sales agreement, which is included in the written or oral terms of the sales agreement under which the promisor assures the quality, description, or performance of the goods.

A written statement arising out of a sale to the consumer of a consumer good pursuant to which the manufacturer, distributor, or retailer undertakes to preserve or maintain the utility or performance of the consumer good or provide compensation if there is a failure in utility or performance; or in the event of any sample or model, that the whole of the goods conforms to such sample or model.

Implied Warranty of Merchantability or Fitness for Particular Purpose: A promise arising by operation of law, that something which is sold shall be merchantable and fit for the purpose for which the seller has reason to know that it is required.

Implied Warranty of Fitness: "Implied Warranty of Fitness" means that when the retailer, distributor, or manufacturer has reason to know any particular purpose for which the consumer goods are required, and further, that the buyer is relying on the skill and judgment of the seller to select and furnish suitable goods, then there is an implied warranty that the goods shall be fit for such purpose.

Warranty of Habitability: Under "Implied Warranty of Habitability," applicable to new housing, builder-vendor warrants that he has complied with the building code of the area in which the structure is located and that the residence was built in a workmanlike manner and is suitable for habitation.

INDEX

NOTE: The entire blank and annotated forms are indexed by name. Specific clauses, however, are indexed only in annotated forms and are labelled as "contract language."

INSTRUCTIONS FOR LEGAL FORMS ON DISK

*A*ll the forms and agreements that appear in this book with a number preceding the title, such as **1.2 Medium-Form Fixed Price Agreement**, are contained on the computer diskette found on the inside back cover. The few forms not found on the disk are ones that are governed by state law and, therefore, must be obtained by the reader locally. These include mechanic's liens, lien releases, and notices of cancellation and rescission.

The installation disk can be loaded onto any IBM-compatible computer equipped with a 3½-inch disk drive and running DOS or Windows. To obtain a 5¼-inch disk or a Macintosh-compatible disk, please use the coupon at the end of this section.

Once you install the forms on your computer's hard drive, you can edit them with your word processor to tailor them to your particular business and to a particular job. The forms are intended to provide the original purchaser of the book with a starting point for developing contracts and forms appropriate to his or her own business. The contracts should be reviewed by an experienced construction attorney in your state prior to use.

Getting Started

The files on the installation disk are compressed, so you must follow the installation procedure described below to copy them onto your hard drive. To start the installation procedure, insert the disk in the appropriate drive and do the following:

In Windows:
Under the **File** menu, select **Run** and type the command **a:setup**; then click on **OK**.

In Dos:
At the C prompt (C:\), type the command **a:install**; then press **ENTER**.

If the installation disk is in your B drive rather than your A drive, the correct command will be:
b:setup (Windows) or **b:install** (DOS).

Then follow the prompts on the screen to select the correct word processing format to install on your hard drive. The 21 files take up between 200,000 and 600,000 bytes (200 to 600 kilobytes) of space, depending on which format you choose. The exact size of the files appears in the Installation Program opposite each word processing format.

Selecting a Format

When selecting a word processing format, remember that newer versions of a program (typically identified by a higher "release number") will usually work with files from earlier versions. For example, if you use WordPerfect 6.0, you can select the format for Word-Perfect 5.0. Similarly, Windows versions of a program typically work with earlier DOS versions of the same program.

If the format(s) you select do not work with your word processor, simply re-install the program and select ASCII. Also select ASCII if your word processor does not appear as an option. Since there are over 50 word processing programs on the market (and many more if you count individual versions), we could only include the most common choices. By choosing the ASCII format, you will lose some of the nice touches, such as bold and italic lettering, but otherwise the forms should work with virtually any DOS-based or Windows-based word processor.

No matter what format you choose, you may have to make minor modifications to the files to accommodate the particular font and margins you select. The files are all formatted with a 60-character line in the common default fonts of most word processing programs.

Naming Conventions

Once you install the files, you will note that the file names all begin with the word "FORM" followed by two numbers separated by a hyphen. These numbers match the number of the corresponding form in the book. The file tags (the three characters following the period in the file name) correspond with the appropriate word processor. For example, the file named FORM2-1.MSW indicates that this is Form 2.1 Cost-Plus-Fixed-Fee Agreement, and it is in the Microsoft Word format.

The names and numbers of all the forms and contracts found on the disk are shown at the end of this section — along with their page numbers in the book. You can refer to this list when scanning through your file names.

After completing the installation, it's a good idea to make a backup disk storing all the forms in their original condition for your word processing format. That way, if your hard drive crashes or you accidentally modify the form, it's easy to go back and find the original on the backup disk.

Renaming Forms in Use

When using a form for a given job, you will want to first copy the form and save the copy under a new name. That way, you always have a copy of the original on your hard drive.

In general, you should give the new computer file a unique name that helps you identify it at a glance. One way is to change the file tag to match the name of the customer or job. So, for example, if you wanted to use Form 1.3 Long-Form Fixed Price Agreement on the Jones job, you might rename the file FRM1-3.JON. Or you can flip it around and name it JONES.1-3. It really doesn't matter as long as you pick a system and stick to it.

Once you've named the new file, you can modify and fill it in as you wish for that particular job without affecting the original copy.

Customizing Forms

You are now free to edit, modify, or fill in the forms as you wish. You can also move portions of one form into another. For example, you may want to take the Change Order clause from the Long-Form Fixed Price Agreement and insert it into the Short-Form version for a certain job.

In many cases, you will want to type much of the specific job information — such as the Scope of Work or Payment Schedule — right on your word processor. Then simply print out the form or contract and fill in any remaining information by hand. If you wish to make changes to the form or agreement after negotiating with your client, sub, etc., simply keystroke the changes and print out a new copy.

The Importance of Version Control

If you create more than one version of the same form for the same job, you are well advised to assign each version an identifying number. You should do this both in the computer file name and within the form itself at the top of the document. This way, you won't lose track of which version of the form you are looking at, whether you are looking at the computer screen, hard copy, or the file name. This is sometimes called "version control."

So, for example, you might name the first version of the Jones contract FRM2-1V1.JON (you can use up to eight characters before the period, up to three after). The "V1" stands for Version 1, so the file name tells you that this is Version 1 of Form 2.1 for the Jones job. The next version would be FRM2-1V2.JON, and so on.

At the top of the form itself, you should use an identifying code that matches the file name. You may also want to include a date as further identification. For example, your first version might be tagged "FRM2-1V1.JON/7/7/96" at the top of the form, and your second version created a week later would be named "FRM2-1V2.JON/7/14/96." This way you have a clear record of which computer file is the latest version and which hard copy version corresponds to which computer file.

Using a computer can greatly speed the process of generating, modifying, and updating forms, contracts, and other business documents. But it can also speed the process of creating a confusing mess if you don't systematically label multiple versions of the same contract. So make sure you have a good naming system in place before you go to work with these forms.

Tech Support

If, after following the procedures and prompts in the installation program, you are unable to install the legal forms, call 802/434-4747 and ask for Legal Forms Tech Support.

If you require a 5¼-inch DOS/Windows disk, or a Macintosh disk, please use the coupon below to order the correct disk.

Forms on Disk Listing

The following listing includes all the forms appearing on the computer disk, and shows you how to find the corresponding form in the book. The three-letter file tags are shown as ***, since they vary depending upon which word processor you choose when installing the program.

✂ -

The Contractor's Legal Kit ■ Forms on Disk for Macintosh and Other Disk Formats

The original purchaser of this book can order another disk format using this form (or a photocopy). The order **must include the original disk** along with a check, money order, or a credit-card number to cover the $5.00 shipping and handling fee.

Check One   **Shipping & Handling**

❑ Macintosh Disk 3½″ (high-density) $5.00

❑ IBM-compatible 5¼″ (high-density) $5.00

Send order and payment to:
Builderburg Group, Inc.
P.O. Box 435
Richmond, VT 05477

Form of Payment

❑ MasterCard   ❑ Visa   ❑ Check/Money Order enclosed

Card number _____ Exp. date _____

Name _____

Street Address _____

City _____ State _____ Zip _____

Phone Number (if any questions) _____